Equine Reproductive Physiology, Breeding and Stud Management

Mina C G Davies Morel

Farming Press

First published 1993

Copyright © Mina C G Davies Morel 1993

ISBN 0 85236 255 2

A catalogue record for this book is available
from the British Library

Published by Farming Press Books
Wharfedale Road, Ipswich IP1 4LG, United Kingdom

Distributed in North America
by Diamond Farm Enterprises,
Box 537, Alexandria Bay, NY 13607, USA

Cover design by Liz Whatling
Front cover photograph FLPA/Silvestris

Typeset by Galleon Photosetting
Printed and bound in Great Britain
by Butler and Tanner Ltd, Frome and London

Equine Reproductive Physiology, Breeding and Stud Management

Contents

Colour sections appear between pages 4 and 5, 100 and 101

Acknowledgements

I would like to thank all those people who have helped in many ways with the production of this book. I would like to thank individually Mr Guy Lewis for the production of the graphs and his advice on the diagrams; Alison Bramwell, Riceal Tully and Grainne Ni Chaba for drawing the diagrams; Mr Alan Leather for his help with the photographs; and Steve Rufus (Technikon Pretoria, South Africa), the Derwen International Welsh Cob Centre, the National Stud, the Thoroughbred Breeders Equine Fertility Unit, the Dyfnog Stud and the many others who either allowed me to take photographs or kindly provided me with them. Thanks must also go to my father Mr Jack Davies for his expert advice and proofreading. Lastly, but by no means least, I must thank my husband Roger and the rest of my family for their support and encouragement throughout the production of this book.

To my sons Christopher and Andrew,
whose births coincided with that of this book

Introduction

The horse, as we know it today, is the result of evolutionary change over millions of years. The first evidence for the existence of *Equus*, the ancestor of the present-day horse, is apparent in North America from where it migrated to the rest of the world before the land masses separated. The *Equus* left in North America subsequently became extinct but those that had migrated to other areas of the world became isolated in parts of Europe, Asia and Africa. The initial evolutionary adaptation of these animals was for survival within the different environments in which they found themselves, be it open plains, forest, wetland etc. As a result four distinct types of *Equus* became apparent: *Equus caballus*, *Equus hemionus* (the onager and kiang), *Equus asinus* plus the zebra and quagga. It is *Equus caballus* that is the direct ancestor of the horse we know today.

Further evolution of the *Equus caballus*, again driven by the environment in which it lived, resulted in the emergence of three types: the steppe type or wild Przewalskii horse, the oriental or plateau type animal and the occidental or forest type. Up until this stage the development of the horse had been driven entirely by the demands of survival of the fittest within their particular environment. However, in the last 20,000 to 30,000 years man has had an increasing association with the horse and hence affected its evolutionary development.

The first evidence of man's association with the horse dates from 25,000 years ago; 40,000 horse skeletons dating from this period have been discovered in Solutre, France. It is highly probable that this association was for food and it is not until 4500 B.C. that positive evidence suggests the beginnings of domestication in areas of China and Mesapotamia. From about 3000 B.C. onwards the horse really began to come into its own, replacing the onager and oxen in many areas as the major driving animal. Domestication of the horse not only involved the perfection of handling techniques, training, appropriate implements, harness and saddlery but also required significant changes in the characteristics of the horse to suit the various functions to which it was put.

Breeding and stud management began when man discovered that he could influence the characteristics of any offspring by appropriate selection of mare and stallion. The first evidence of selective breeding is seen in the Near East civilisations around 1400 B.C. where selective breeding by man led to the development of animals suitable to carry rather than just pull the weight of a man. Until this time the horse had been a very small animal only capable of being driven. By 750 B.C. Assyrian horses showed evidence of considerable selection with the emergence of a riding animal. Selective breeding of horses then became more widespread and was carried out in various parts of the world, often in isolation, and with different aims in mind.

This additional influence of selection for man's requirements, plus the continual natural selection for survival, resulted in the three main types of horse mentioned previously. The steppe or wild horse, which has largely escaped the effects of domestication and as such has really remained unchanged, is today virtually extinct. Oriental horses, developed in the Mediterranean/African areas, were fast, light and hardy animals to be used mainly for fort invasion and sortie type warfare typical of these civilisations. These animals are now termed hotblooded, for example the Arab. Conversely, the colder areas of the world, of north/mid Europe, saw the development of the third, heavier and slower type of horse, capable of surviving such harsh climates. These were the occidental or coldblooded horses. Again, largely as a result of man's selection, this group has further divided to form the Great horse of the Middle Ages, capable of carrying considerable weights, and the Celtic pony or horse, lighter in build and more agile, used for general transport and for working the land.

The last few hundred years have seen the most dramatic influence of man on the breeding of horses for type and function, which has led directly to the vast array of breeds and types registered today. The evolution of these types is a continual and on-going process and even in the last 50 to 75 years considerable changes in the type of horses commonly seen have been achieved. This is purely a result of the horse's change of function from use for power and transport at the beginning of this century, to a means of sport and leisure predominant today. This alteration in man's requirements of the horse has resulted in the development of and vastly increasing demand for such types as the warmblood horses, not previously very popular. The warmblood is the result of the initial crossing of hotblood and coldblood types, to produce a horse of power and strength, while retaining speed and agility.

Man's dictation of the type of horses required and, therefore, bred has not only resulted in a change in characteristics of the common horse but also in its management. The fashion for particular types of horses results in a demand for certain blood lines that have proved, by previous performance, to produce the desirable type of horse. The stallion, which can sire many foals in a single year, is especially susceptible to the whims of fashion, and those of the desirable type can, therefore, demand very high service fees, prices in the region of £50,000 not being uncommon in the Thoroughbred racing world. These animals thus become very valuable and their management reflects this, being increasingly geared towards mini-mising any risks to their wellbeing, especially any factors affecting reproductive potential. Financial considerations are, therefore, a major determining factor in the breeding management of horses today, and the highly intensive and controlled breeding schemes/management practices used have been developed accordingly.

As I hope will become apparent in this book, this is by no means always to the advantage of the horse and its breeding ability. To a large extent the greater man's influence, largely to protect his own financial investment, the greater the detrimental effect on reproductive performance. This then develops into a self-perpetuating situation where man feels the need to interfere/manipulate equine breeding further to compensate for the reduction in reproductive performance. Such a situation will continue as long as some horses and their potential for offspring command high prices. Such a situation is evident today within the Thoroughbred industry and has led to highly intensive and controlled breeding techniques. The converse is evident in smaller pony breeding studs, running animals of much lower value. In such enterprises the financial risks are, therefore, lower and much less intensive methods of breeding are employed. Systems nearer to the natural state are practised and as such, reproductive success is improved, but the number of coverings per season is limited along with the potential financial return.

Today, the breeding of horses is largely driven by man. The management techniques employed within the industry, therefore, depend largely upon financial considerations and fashion. The discerning individual breeder needs to be aware of the principles behind the numerous management practices and their advantages and disadvantages, in order to draw his own informed conclusions on their applicability to his/her own circumstances. In order to gain a full understanding of the practices used the breeder should have a knowledge of the reproductive anatomy and physiology of the equine. Such understanding will also allow new developments in stud farm management, especially in the areas of embryo transfer, artificial insemination and hormonal manipulation, so that informed decisions on their applicability to various systems can be made. This book aims to provide such information in a format applicable to the practising equine breeder but in adequate scientific detail to also provide a text for individuals wishing to study the equine at a higher academic level.

SECTION ONE

Reproductive Anatomy and Physiology

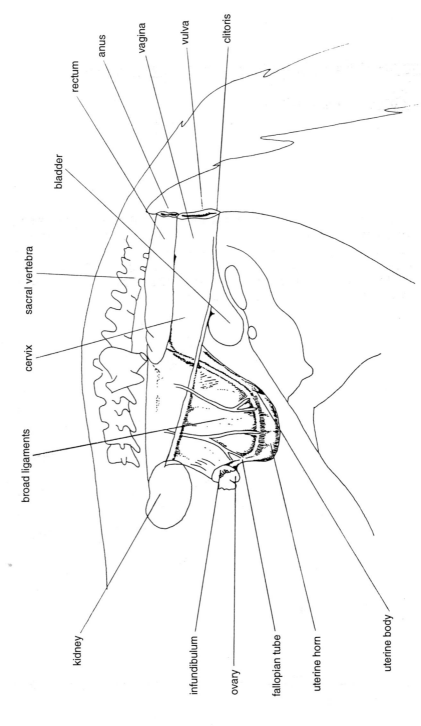

anus

rectum

vagina

vulva

clitoris

bladder

sacral vertebra

cervix

broad ligaments

kidney

infundibulum

ovary

fallopian tube

uterine horn

uterine body

Figure 1.1   The mare's reproductive system, as viewed from the side.

CHAPTER 1

Reproductive Anatomy of the Mare

THIS chapter will deal with the reproductive anatomy of the mare detailing the individual structures along with their function. Colour plate 1 (facing page 4), taken after slaughter, shows the major reproductive organs of the mare, and Figures 1.1 and 1.2 illustrate these diagrammatically. Each of these structures will be dealt with in turn in the following sections.

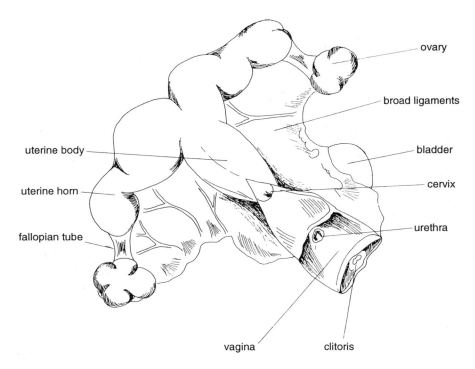

ovary

broad ligaments

bladder

uterine body

cervix

uterine horn

urethra

fallopian tube

vagina clitoris

Figure 1.2   The mare's reproductive tract, a diagrammatic representation of Colour plate 1.

The Vulva

The vulva (Colour plate 2) is the external area of the mare's reproductive system, protecting the entrance to the vagina. The outer area is pigmented skin with the normal sebaceous and sweat glands along with the nerve and blood supply normally associated with the skin of the mare. The inner area is lined by mucous membrane and is continuous with the vagina. It is situated approximately 7 cm below the anus. Below the entrance to the vagina in the lower part of the vulva lie the clitoris and the three sinuses associated with it. These sinuses are of importance in the mare as they provide an ideal environment for the harbouring of many venereal disease (VD) bacteria. Hence this area is regularly swabbed in mares prior to covering by the stallion, and indeed in the Thoroughbred industry such swabbing is compulsory.

The Perineum

The perineum is a rather loosely defined area in the mare, but includes the outer vulva and adjacent skin along with the anus and its surrounding skin. In the mare the conformation of this area is of clinical importance as it plays an important part in the protection of the genital tract from the entrance of air. Mal-conformation in this area predisposes the mare to a condition termed pneumo-vagina, or vaginal windsucking, in which air is sucked in and out of the vagina through the open vulva. Along with this passage of air also goes the passage of bacteria which bombard the cervix, exposing it to unacceptably high levels of contamination which it is often unable to cope with, especially during oestrus when it is less competent. Passage of bacteria into the higher, more susceptible parts of the mare's tract may result in bacterial infections such as contagious equine metritis (CEM) and other venereal diseases, further details on the causes of which are given in Chapter 19. All adversely affect fertilisation rates.

Protection of the Genital Tract

As mentioned above, adequate protection of the genital tract is essential to prevent the adverse affects of pneumovagina. There are three seals within the tract: the vulval seal, the vestibular or vaginal seal and the cervix. These are illustrated in Figure 1.3.

The vulval seal is formed by the perineum area and the sides of

Reproductive Anatomy of the Mare

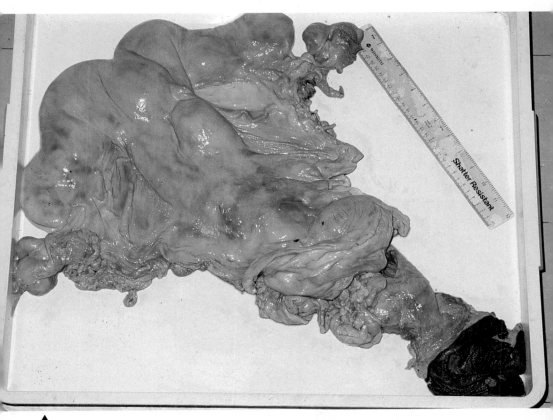

▲
1   The mare's reproductive tract.

2   The vulval area of the mare.   ▶

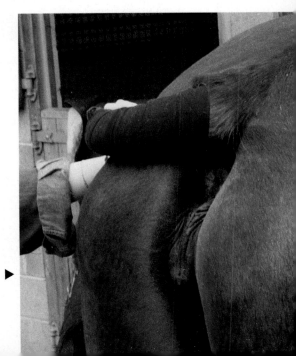

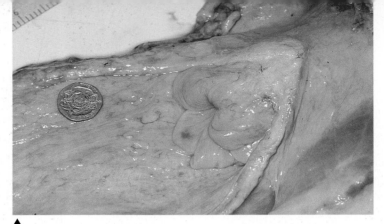

3 *The internal surface of the mare's vagina. (The coin in the colour plates measures 21 mm in diameter.)*

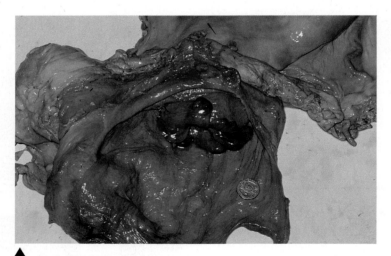

4 *The oestrus cervix protruding into the vagina of the mare.*

5 *The internal surface of the uterus of the mare.*

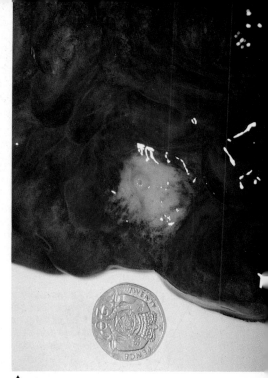

▲
6 The highly folded surface of the cervix in the mare.

▲
7 The utero-tubular junction in the mare, as seen from the uterine horn side.

8 The convoluted fallopian tube (oviduct) of the mare running through the mesentery from the uterine horn on the right to the ovary on the left.
▼

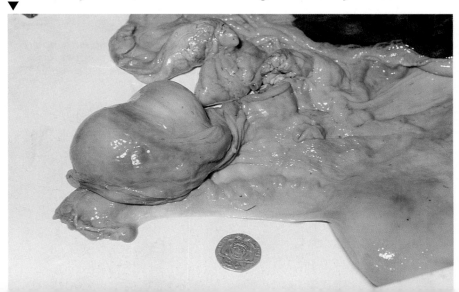

9 The ovaries of the mare. Note the difference in size between the ovary on the left, which is inactive, and the one on the right, which is active. ▲

10 The concave surface of a sectioned mare's ovary is shown in the centre of the photograph. This surface is free of attachment to the broad ligaments and is the location of the ova fossa and entry points for the ovarian nerve and blood supply. ▼

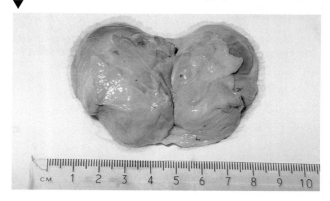

11 A cross-section through a mare's inactive ovary (on the left) and an active ovary (on the right). Note the dark mass of old corpora lutea in the centre of the active ovary and the large follicle to the top right of this ovary. ▼

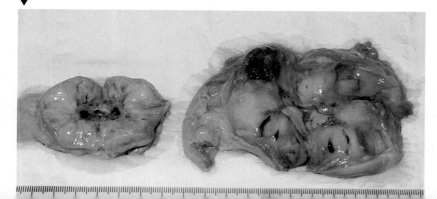

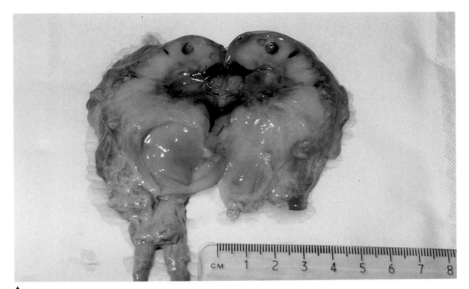

▲
12   A cross-section through a mare's active ovary illustrating the presence of a large follicle (3 cm in diameter) at the bottom of the ovary.

13   A cross-section through a mare's active ovary illustrating a large corpus luteum at the top of the ovary.
▼

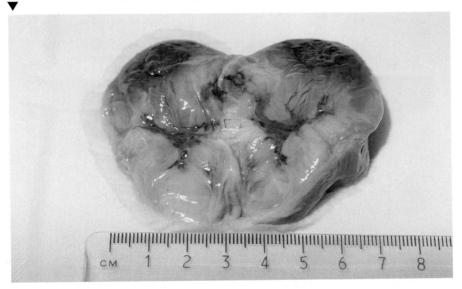

Reproductive Anatomy of the Stallion

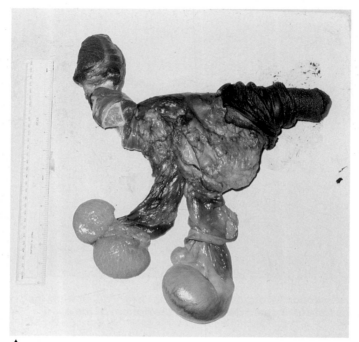

▲
14 *Cross-section through the main body of the penis (top left). The remainder of the penis and the accessory glands are not included.*

15 *A detail of the stallion testis illustrating the position of the epididymis lying across the main body of the testis.*
▼

Development of the Foetus

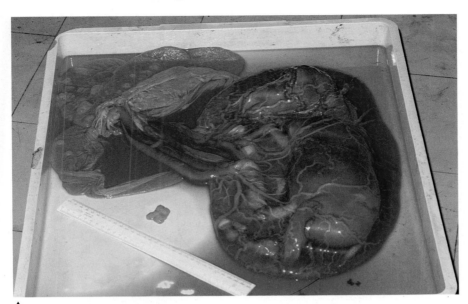

▲
16 Equine foetus at 200 days of gestation, illustrating the foetus within the amniotic sack, with the umbilical cord connection to the chorio-allantois which has been removed from around the foetus and is lying in the top left-hand corner. The as-yet very small hippomane can be seen just above the ruler.

17 An 80-day-old foetus.
▼

18 A 160-day-old foetus. The hippomane can be seen in front of the forelegs.
▼

19 *A 220-day-old foetus illustrating the sequential removal of the outer allanto-chorion layer, followed by the amnio-chorion leaving the foetus with umbilical cord attached. In the bottom plate the hippomane can be seen just in front of the hind fetlock.*

▼

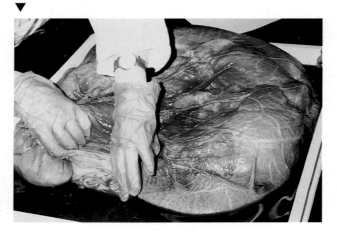

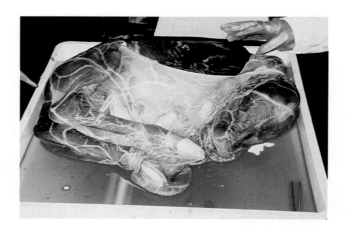

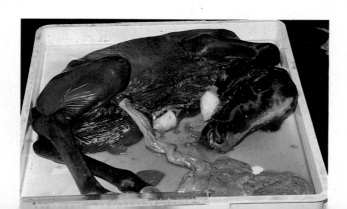

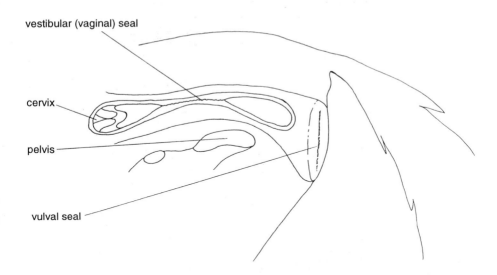

*Figure 1.3   The seals of the mare's reproductive tract during non-oestrus
(dioestrus).*

the vulva. The vestibular seal is formed by the roof of the posterior vagina, the floor of the pelvic girdle and the hymen, if still present. The cervical seal is made by the tight muscle ring at the entrance to the cervix. This series of seals is affected by the conformation of an individual and also by the stage of the oestrus cycle as illustrated in Figures 1.4 and 1.5.

The ideal conformation is achieved if 80% of the vulva lies below the pelvic floor. A simple test can be used to assess this. If a plastic tube is inserted through the vulva into the vagina and allowed to rest on the vaginal floor, the amount of vulva lying below this tube should be approximately 80% in well-conformed mares. This technique is illustrated diagrammatically in Figure 1.6.

If the pelvic floor is too low, then the vulva tends to fall towards the horizontal plane as seen in Figure 1.4C. This opens the vulva up to contamination by faeces, increasing the risk of uterine infection.

The effect of the oestrous cycle on the reproductive tract will be detailed in Chapter 3. In summary, oestrus results in the slackening of all three seals due to a relaxation of the muscles associated with the reproductive tract, especially the cervix. However, high oestradiol levels at the time of oestrus enhance the mare's immunological response, thus reducing the chance of uterine

Figure 1.4 *The effect of conformation on the competence of the vulval, vestibular (vaginal) and cervical seals in the mare.*

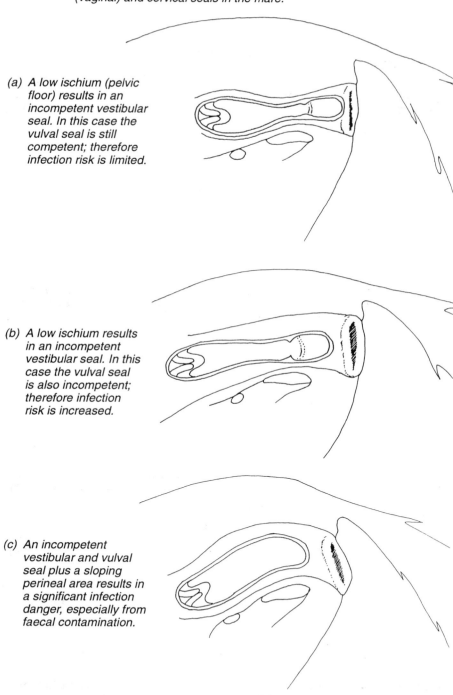

(a) *A low ischium (pelvic floor) results in an incompetent vestibular seal. In this case the vulval seal is still competent; therefore infection risk is limited.*

(b) *A low ischium results in an incompetent vestibular seal. In this case the vulval seal is also incompetent; therefore infection risk is increased.*

(c) *An incompetent vestibular and vulval seal plus a sloping perineal area results in a significant infection danger, especially from faecal contamination.*

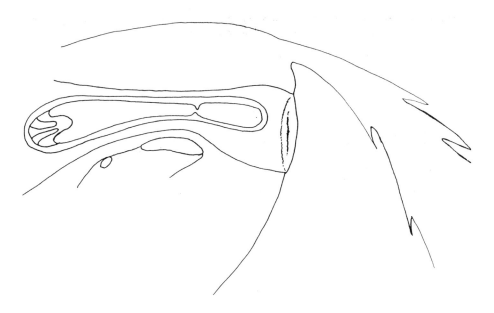

Figure 1.5   The effect of oestrus on the competence of the vulval, vaginal and
cervical seals in the mare, causing a relaxation of the vestibular seal
and, therefore, an increase in infection danger.

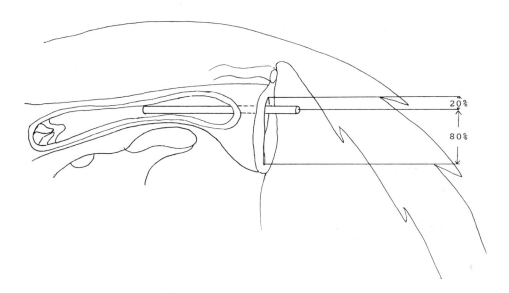

Figure 1.6   Diagram illustrating assessment of the mare's vulva and vaginal floor
positions in an attempt to ascertain the likelihood of pneumovagina.

infection. The effect of poor conformation of the perineal area may be alleviated by a Caslick's vulvoplasty operation, which was first developed by Dr Caslick in 1937 (Caslick, 1937). The lips on either side of the upper vulva are cut, and the two sides are then sutured together. The two raw edges heal together as in the healing of an open wound and hence seal the upper part of the vulva. The hole left is adequate for urination but small enough to prevent the passage of faeces into the vagina.

The chance that a mare requiring a Caslick's operation will pass on the trait to its offspring is reasonably high. This, coupled with the fact that the operation site has to be cut to allow mating and foaling, casts doubt on whether such mares should be bred.

Mares that have been repeatedly cut and sewn up become increasingly hard to perform a Caslick's operation on, as the lips of the vulva become progressively more fibrous and, therefore, difficult to sew up. In such cases a procedure termed a Pouret may be carried out (Pouret, 1982). This is a more major operation and involves the realignment of the anus as well as the vulva, the anus being pulled out to lie in line with the vulva.

This perineal malformation is particularly prevalent in Thoroughbred mares and tends to be exacerbated in mares that lack condition as well as in multiparous, aged mares. Its continued existence is largely due to man's selection of horses for performance and not reproductive competence.

The Vagina

The vagina of the horse is on average 18–23 cm long and 10–15 cm in diameter. The walls are normally apposed forming the vestibular seal. The hymen, also associated with this seal, divides the vagina into anterior and posterior sections, the urethra from the bladder opening just anterior to the hymen. The vagina is mainly covered by the peritoneum and is surrounded by loose connective tissue, fat and blood vessels. The walls of the vagina are muscular with a mucous coat lining, which is very elastic, allowing the major stretching required at parturition (Colour plate 3).

The vagina acts as the first protector and cleaner of the reproductive system, secreting acid which kills bacteria. The vaginal acid secretion, however, has the disadvantage of also attacking the epithelial cell lining of the vagina, necessitating a protective mucous secretion. A further disadvantage of the acid secretion is that sperm are also susceptible to a low pH. Thus, at ejaculation, sperm are deposited in through the cervix to minimise the spermicidal effect of the vagina. The exact composition of vaginal secretion is controlled by the cyclical hormonal changes of the mare's reproductive cycle.

The Cervix

The cervix lies at the entrance to the uterus. It is a tight, thick-walled sphincter, acting as the final protector of the system. In the sexually inactive, dioestrus state, it is tightly contracted, white in colour and measures 6–8 cm long by 4–5 cm in diameter; cervical secretion is also minimal. The muscle tone and, therefore, its size, along with its secretion, are again governed by cyclic hormonal changes. The muscle tends to relax during oestrus or heat, and there is an increase in secretion easing the passage of the penis into the entrance of the cervix and the passage of the sperm through the cervix at ejaculation. The oestrous cervix is pink in colour and may be seen protruding into the vagina (Colour plate 4).

The lining of the cervix consists of a series of folds or crypts, as shown in Colour plate 5. These crypts tend to slow the passage of the sperm, acting as a slow-release mechanism, allowing sperm to pass into the upper tract over a period of 24 hours, even though they are deposited originally in a matter of seconds. This mechanism increases the chances of the ova and sperm meeting at the opportune time for successful fertilisation.

The Uterus

The uterus of the mare is a hollow muscular organ joining the cervix and the fallopian tubes. It is attached to the lumbar region of the mare by two broad ligaments, out-foldings of the peritoneum either side of the vertebral column. This provides the major support of the reproductive tract (Figure 1.7) and can be divided into three areas: the mesometrium attached to the uterus; the mesovarium attached to the ovary and the mesosalpinx attaching to the fallopian tubes (Ginther, 1992).

The uterus is divided into two areas, the body of the uterus and the uterine horns. The body of the uterus normally measures in the region of 18–20 cm long and 8–12 cm in diameter and divides into two uterine horns which are approximately 25 cm long which reduce in diameter to between 1–2 cm as they approach the fallopian tubes. The size of the uterus is affected by age, and older mares, especially those that are multiparous, tend to have larger uteri. The uterus of the mare is termed a simplex uterus bipartitus due to the relatively large size of the uterine body compared to

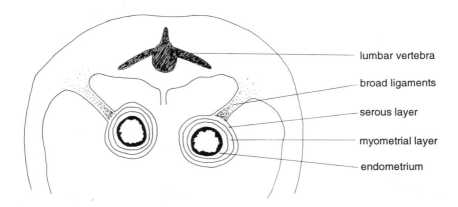

Figure 1.7   Cross-section through the abdomen of the mare illustrating the support of the reproductive tract provided by the broad ligaments.

other farm livestock, where, in size, the uterine horns are the predominant feature. The lack of a septum dividing the uterine body is also notable.

The uterine wall consists of three layers: an outer serous layer continuous with the broad ligaments, a central myometrial or muscular layer, and an inner endometrium or mucous membrane lining. The central myometrial layer consists of outer longitudinal muscle fibres and inner circular muscle fibres, and it is this layer that allows the considerable expansion of the uterus during pregnancy and provides the force for parturition. The inner endometrium is comprised of luminal epithelial cells, stroma of connective tissue and associated epithelial glands and ducts. The activity and, therefore, appearance of these gland ducts are dependent on the cyclical hormonal changes (Colour plate 6).

The Utero-tubular Junction

The utero-tubular junction is the constriction separating the end of the uterine horns from the beginning of the fallopian tubes. The site of fertilisation is immediately through this junction on the fallopian tube side. Fertilisation can only take place here, as only fertilised ova can pass through this junction and on to the uterus for implantation and further development. Fertilised ova actively control their own passage, possibly via a localised secretion of oestrogens, and overtake unfertilised ova as they pass. Unfertilised ova remain on the fallopian tube side of the utero-tubular junction and gradually degenerate over the next few months (Colour plate 7).

The Fallopian Tubes

The mare has two fallopian tubes or oviducts of 25–30 cm in length, which are continuous with the uterine horns. The diameter of these tubes varies slightly along their length, being 5–10 mm at the isthmus end, that nearest the uterine horn, and reducing to 2–5 mm at the ampulla, nearest the ovary (Colour plate 8). The fallopian tubes lie within peritoneal folds which make up part of the broad ligaments and have walls very similar in structure to the uterus but thinner. The three layers composing the wall are an outer fibrous serous layer, continuous with the peritoneal folds; a central myometrial layer of circular and longitudinal muscles fibres; and an inner mucous membrane. In addition, the ampulla end of the fallopian tubes has a fimbrae-lined inner mucous membrane, and the fimbrae waft directionally down the fallopian tube towards the utero-tubular junction, encouraging the ova to travel in a similar direction. The ampulla end of the fallopian tube ends in the infundibulum, a funnel-like opening of the fallopian tube close to the ovary.

The infundibulum is closely associated with the part of the ovary termed the ova fossa, which is unique to the mare and is the only site of ovulation; in other mammals ovulation may occur over the whole surface of the ovary. The infundibulum is, therefore, relatively hard to distinguish in the mare, not being so evident as a funnel structure surrounding the whole ovary. The infundibula are lined, like the ampulla, by fimbrae which attract and catch the ova, guiding them towards the entrance of the fallopian tubes. This area in the mare is prone to blockage by embryological remnants and developing cysts, thus reducing fertilisation rates.

The Ovaries

The two ovaries in the mare are situated ventrally to the fourth and fifth lumbar vertebrae. They make the total length of the reproductive tract in the region of 50–60 cm. In the sexually inactive stage, i.e. during the non-breeding season, the mare's ovaries measure 2–4 cm in length and 2–3 cm in width and are hard to the touch due to the absence of developing follicles. During the sexually active stage, when the mare is in season, they increase in size to 6–8 cm in length and 3–4 cm in width; they are also softer to the touch due to the development of fluid-filled follicles (Colour plate 9). Older, multiparous mares tend to show larger ovaries up to 10 cm in length.

As illustrated in Colour plate 10, the mare's ovaries are bean shaped. They have a convex surface or border, which is attached to the mesovarian section of the broad ligaments and is the entry point for the blood and nerve supply, and a concave inner surface, which is free from attachment.

The whole ovary is contained within a thick protective layer, the tunica albuginea, except for the centre of this concave surface at the point called the ova fossa. The stroma of the ovary in the mare is made up of the inner medulla and outer cortex. Ova release at ovulation occurs through this fossa; in addition, all follicular and corpora lutea development occurs internally, within the medulla of the ovary. The mare differs in these aspects from other mammals, in which the medulla and cortex are reversed and ovulation occurs over the surface of the ovary and all follicular and corpora lutea development occurs on the outer borders of the ovaries. Rectal palpation as a clinical aid to assess reproductive function in the mare is not, therefore, as easy as in other farm livestock, for example the cow. However, with the advent of ultrasound, reproductive assessment of ovarian characteristics in the mare is now quite accurate.

Follicular Development and Ovulation

The ovary is made of two basic cell types: interstitial cells (stroma) that provide support and germinal cells that produce the ova. The number of potential ova contained within the female ovary is dictated in utero prior to birth; subsequently no addition to the pool of ova can be made. These very immature ova are termed oogonia, and there are many more than an individual animal will use within her reproductive lifetime. These oogonia, with a full complement of chromosomes (64) and surrounded by a single layer of epithelial cells, are termed primordial follicles. At birth the ovary contains many thousands of these primordial follicles. After birth and prior to puberty some of the oogonia start development to primary oocytes, and these, surrounded by their follicular epithelial cells or granulosa cells, undergo the first stages of meiosis. They then await puberty, when hormone secretion from the anterior pituitary drives their further development.

From puberty onwards primary oocytes develop and complete the final stages of meiosis at varying rates designed to ensure that a regular supply of developed follicles is available for ovulation every

21 days during the breeding season. Not all primary follicles are destined to ovulate, as many are wasted along the way, degenerating and becoming atretic, and normally just one reaches the stage ready for ovulation.

As the primary follicle develops, its surrounding epithelial cells differentiate into follicular epithelial cells that secrete follicular fluid that fills the cavity surrounding the oocyte. The follicle increases in size as fluid accumulation increases. The primary oocyte now also increases in size and develops a thick outer jelly-like layer, the zona pellucida; it is then termed a secondary oocyte with a haploid number of chromosomes (32). The secondary oocyte now becomes associated with one inner edge of the follicle and lies on a mound of follicular cells called cumulus oophorus. The stroma immediately surrounding the follicle becomes organised into a double-lined membrane, the theca membrane. The inner layer is vascularised and the outer layer is not. The follicles then develop to a 3 cm stage, when they are termed graafian follicles (Figure 1.8).

This seems to be a critical size in equine follicular development, for now a decision is made as to which of the developing follicles is destined to develop further for ovulation and which ones will degenerate. In 75% of cases only one follicle is destined to ovulate

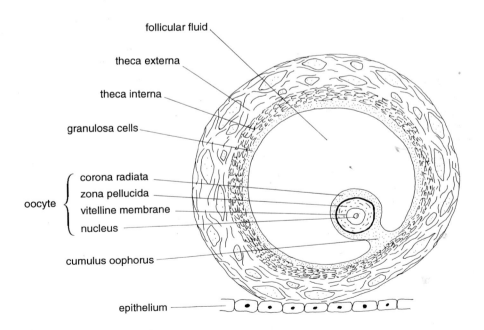

Figure 1.8 · The equine graafian follicle.

and hence develop further. In 25% of cases two follicles may develop further and ovulate. This decision seems to be governed by their response to high circulating oestrogen and androgen levels, though the exact mechanisms are unclear.

Follicular diameter in those follicles destined to ovulate increases to greater than 3 cm and at the same time they seem to move within the stroma of the ovary and orientate themselves to await ovulation through the ova fossa. The mature follicle then releases its ova and follicular fluid at ovulation through the ova fossa to be caught by the infundibulum and passed down the fallopian tube for potential fertilisation. Ovulation of follicles of less than 3 cm diameter does occur but is the exception.

After release of the secondary oocyte and follicular fluid the old follicle collapses and the theca membrane and remaining follicular epithelial cells become folded into the old follicular cavity. Bleeding occurs into the centre of this cavity, forming a clot. This clot, the theca cells and any follicular epithelial cells make up the corpus luteum or yellow body. The corpus luteum (CL) is then invaded by blood capillaries and fibroblasts. It is initially a reddish-purple colour. As the CL ages it becomes browner in colour and if the mare is not pregnant regresses to a yellow and then white colour (corpora albucans) as it becomes non-functional, and the luteal tissue is replaced with scar tissue (Figure 1.9). Colour plates 11–13 show sections through a mare's ovary illustrating the presence of developing follicles and corpora lutea.

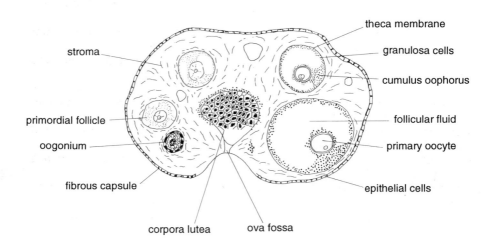

Figure 1.9   Follicular development and ovulation within the ovary.

Conclusion

It can be concluded, therefore, that the reproductive tract of the mare is specifically designed to maximise the chance of successful fertilisation. It is also ideally suited to provide the nourishment and support required by a developing embryo.

References and Further Reading*

Caslick, E.A. (1937) The vulva and vulvo-vaginal orifice and its relation to genital tracts of the Thoroughbred mare. Cornell Vet., 27, 178.

Ginther, O.J. (1992) Reproductive Biology of the Mare, Basic and Applied Aspects. 2nd Ed. Pub Equiservices, 4343, Garfoot Road, Cross Plains, Wisconsin 53528, USA.

Kenney, R.M., Condon W.A., Ganjam, J.K. and Channing, C. (1979) Morphological and biochemical correlates of equine ovarian follicles as a function of their state of viability or atresia. J.Reprod.Fert.Suppl., 27, 163.

Le Blanc, M.M. (1991) Diseases of the reproductive system: the mare. In: Equine Medicine and Surgery. 4th Ed. Volume II. Ed. Colahan, P.T., Mayhew, I.G., Merritt, A.M. and Moore, J.N. American Veterinary Publications Inc. Goleta, California.

Osbourne, V.E. (1966) An analysis of the pattern of ovulation as it occurs in the reproductive cycle of the mare in Australia. Aust.Vet.J., 42,149.

Pascoe, R. R. (1979) Observations on the length and angle of declination of the vulva and its relation to fertility of the mare. J.Reprod.Fert.Suppl., 27, 299.

Pascoe, R.R. Pascoe, D.R. and Wilson, M.C. (1987) Influence of follicular status on twinning rate in mares. J.Reprod.Fert.Suppl., 35, 183.

Pouret, E.J.M. (1982) Surgical technique for correction of pneumo- and urovagina. Equine Vet.J., 14, 249.

Powell, D.G., David, J.S.E. and Frank, C.J. (1978) Contagious equine metritis. The present situation reviewed and a revised code of practice for control. Vet. Rec., 103, 399.

Rossdale, P.D. and Ricketts, S.W. (1980) Equine Stud Farm Medicine. 2nd Ed. Baillère Tindall, London.

Simpson, D.J. and Eaton-Evans, W.E. (1978) Developments in contagious equine metritis. Vet. Rec., 102, 19.

Simpson, D.J. and Eaton-Evans, W.E. (1978) Sites of C.E.M. Vet. Rec., 102, 48.

Sisson, S. (1953) In: The anatomy of the Domestic Animal. Ed. J.D. Grossman. 4th Ed. Saunders, Philadelphia and London.

* See page 436 for full names of publications cited.

Sorensen, A.M. Jr. (1979) Animal Reproduction Principles and Practices. McGraw-Hill Book Company.

Vogelsang, M.M., Kraemer, D.C., Potter, G.D. and Scott, G.G. (1987) Fine structure of the follicular oocyte of the horse. J.Reprod.Fert.Suppl., 35, 157.

Witherspoon, (1975) The site of ovulation in the mare. J.Reprod.Fert. Suppl., 23, 329.

CHAPTER 2

Reproductive Anatomy of the Stallion

THIS chapter details the anatomy and function of the stallion reproductive system. Colour plate 14 shows the reproductive system of the stallion after slaughter and Figure 2.1 illustrates the system diagrammatically detailing all the major organs. These organs will be discussed individually below.

The Penis

In the resting position the penis of the stallion lies retracted and hence protected within its sheath or prepuce out of sight; it is held in this position by muscles. The prepuce is a double-folded covering to the penis that folds back on itself to give two folds of protection. Within the inner fold of the prepuce lies the end of the penis, the glans penis, giving this sensitive area additional protection. Protruding by 5 mm from the centre of the glans penis, or rose, is the exit of the urethra. Around this protrusion lies the urethral fossa, and below it a dorsal diverticulum, both of which are often filled with smegma, a red-brown secretion of the prepubital glands lining the prepuce, and epithelial debris. These areas also provide an ideal environment for bacteria, often harbouring VD bacteria such as *Klebsiella aerogenes*, *Haemophilus equigenitalis* and *Pseudomonas aeroginosa*.

The penis of the stallion is attached to the lower pelvic bone by two roots; these extensions of the penis join together to form the body. Here at its origin the penis is held in position by a ligament attachment to the pelvis. The urethra running from the bladder connects with the vas deferens and runs between the two roots before entering the body of the penis. The body of the penis contains a large percentage of erectile, as opposed to fibrous, tissue and as such it is termed heamodynamic. Figure 2.2 illustrates a cross-section view through the body area of the penis.

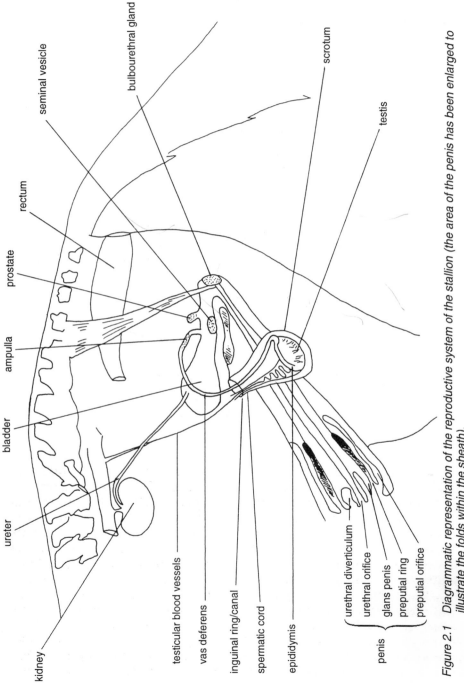

Figure 2.1   Diagrammatic representation of the reproductive system of the stallion (the area of the penis has been enlarged to illustrate the folds within the sheath).

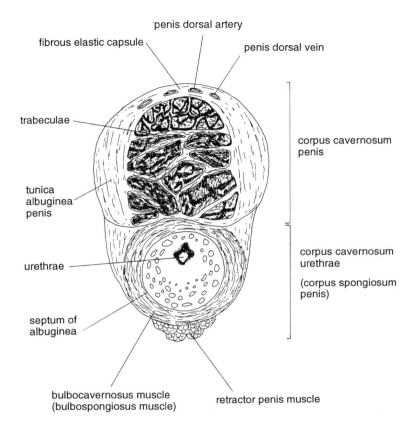

Figure 2.2   Cross-section through the penis of the stallion.

This diagram illustrates that the main body of the penis is divided into two sections, the lower corpus cavernosum urethrae and the upper corpus cavernosum penis. Through the corpus cavernosum urethrae runs the urethra surrounded by some trabeculae (sheets of connective tissue) enclosing areas of erectile tissue, all enclosed within the bulbocavernosus muscle. The corpus cavernosum penis, which is the largest part of the penis, contains a dense network of trabeculae and associated muscle tissue and scattered cavities, making up the major erectile tissue of the penis. The corpus cavernosum penis is contained within an elastic muscle sheet or capsule which allows the doubling in size seen at erection. Finally running along the bottom of the penis is a retractor muscle, used on leaving the mare to speed up the return of the penis within the prepuce. The glans penis is a continuation of the corpus cavernosum urethrae and as such contains some trabeculae. These are more elastic than in the rest of the penis, allowing the vast expansion (up to three times) of this area immediately post ejaculation.

During the process of erection the first reaction is a relaxation of the penile muscles allowing the penis to extrude from its sheath. Blood engorgement of the erectile tissue of the penis results in an initial turgid pressure followed by intromission pressure. It is essential that the penis is at intromission pressure before entering the mare; if not, it may cause permanent damage to the stallion or at least reduce his enthusiasm for future covering. The increase in blood pressure within the penis is a result of central nervous system stimulation, in response to the presence of a mare on heat or other sexual stimulation. As a result of central nervous system (CNS) stimulation, the muscles at the root of the penis contract, pulling it up against the pelvic bone. This cuts off the venous return, increasing the blood held within the penis, and at the same time CNS stimulation causes the arteries to dilate, hence engorging the erectile tissue.

Ejaculation is a result again of CNS response to the mare and also nerve impulses from the glans penis. As a result the muscle walls of the epididymis and the accessory glands contract, passing sperm plus seminal fluid up to the penis. Exit of semen from the penis is by contraction of the muscle fibres within the penis as well as contractions of the vas deferens. Ejaculation is in the form of 6 to 9 jets of semen, the initial 3 jets containing the majority of the biochemical components.

At full erection the penis doubles in size to 80 to 90 cm in length and 10 cm in width compared to 40 cm in length and 5 cm in width at rest. After ejaculation the glans penis triples in size, helping to prevent initial leakage of semen from the mare. The glans penis has to return to near normal size before the stallion is able to leave the mare. Failure to allow the glans penis to return to normal can cause damage to the mare and/or stallion. Problems may be encountered in overzealous stallions that show enlargement of the glans penis prior to entry into the mare; in such cases intromission is not safe until the glans penis has returned to its normal size. Such problems are especially evident in stallions of high libido, especially if they have only a limited workload.

The Accessory Glands

The accessory glands are a series of four glands situated between the end of the vas deferens and the roots of the penis. These glands are responsible for the secretion of seminal plasma, the major fluid fraction of semen. Seminal plasma provides the substrate for conveying the sperm to the mare and ensuring maturation. It provides energy in the form of lactate and protection from changes in osmotic pressure (citrate) and from oxidisation (ergothionine). It also contains a gel that forms a partial clot in semen, helping to minimise the escape of semen from the female tract after deposition.

All males have this series of accessory glands, and the relative importance of the secretions of each within the seminal plasma of different species is reflected in their relative size.

The Bulbourethral Glands

The bulbourethral or Cowpers glands are the accessory glands situated nearest the roots of the penis. They are paired and oval in structure, approximately 5 cm by 3 cm and lying either side of the urethra. Their clear fluid secretion, which can be seen during sexual stimulation of the stallion, helps to clean out the urine and bacteria collected in the urethra prior to ejaculation. Their secretion also acts as a lubricant easing the passage of sperm. Due to the potential contamination of the bulbourethral gland secretions with stale urine and bacteria, this initial fraction of semen is not collected for use in artificial insemination.

The Prostate

The prostate gland is a bilobed structure with a single exit to the urethra, situated between the bulbourethral glands and the ampulla. Prostate secretions in the stallion are alkaline and high in proteins, citric acid and zinc. The significance of high levels of zinc is not clear, but citric acid is known to act as an osmotic pressure buffer maintaining the same osmotic pressures in the seminal fluid as within the sperm heads, thus preventing the potential bursting of the sperm heads due to an influx of water from the surrounding solution. The protein component is thought to add to the characteristic odour of semen.

The Ampulla

The ampulla glands are paired outfoldings of the vas deferens situated where it meets the urethra from the bladder. The secretion of the ampulla glands contains a high concentration of ergothionine, an anti-oxidising agent, protecting susceptible chemicals in semen from oxidisation.

The Vesicular Glands

The vesicular glands or seminal vesicles are also paired in structure; they lie on either side of the bladder and are about 16 to 20 cm in length. They are lobed and can be compared with large walnuts in external appearance. They secrete a major amount of seminal plasma, with a high concentration of potassium, protein, citric acid and gel. The gel component is secreted in the later stages of ejaculation, being deposited in the mare after most of the sperm.

The Vas Deferens

The vas deferens connects the epididymis of the testis to the urethra before passing the accessory glands and on into the penis. It has a diameter of approximately 1 cm with a thick muscular wall made up of three layers of muscle: an inner circular, middle oblique and outer longitudinal (Figure 2.3). These muscle layers actively propel the sperm plus surrounding fluid from the testis to the penis. The lumen of the duct is small and lined with folds especially at the epididymis end, maximising the surface area and so aiding sperm storage and the reabsorption of testicular fluids.

The vas deferens, the testicular nerve and blood supply and the cremaster muscle all pass out of the body cavity through the

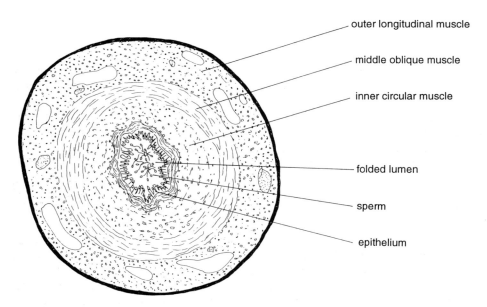

Figure 2.3   Cross-section through the vas deferens of the stallion.

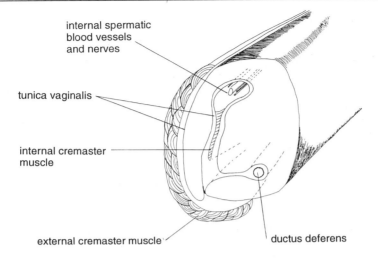

internal spermatic
blood vessels
and nerves

tunica vaginalis

internal cremaster
muscle

external cremaster muscle

ductus deferens

Figure 2.4   Cross-section through the spermatic chord of the stallion.

inguinal canal and together make up the spermatic cord (Figure 2.4). The cremaster muscle, which is divided into internal and external sections, is responsible for drawing the testis up towards the body cavity in response to cold, fear etc.

The Epididymis

The epididymis in the stallion lies over the top of the testis as illustrated in Colour plate 15 and Figure 2.5. It consists of long convoluted tubules and is subdivided into three sections, the head, the body and the tail. The head of the epididymis is connected by several ducts to the rete testis; as it continues towards the tail end these ducts merge and form a single duct, the vas deferens. The lining of these tubules is very similar to that of the vas deferens, with many folds. These folds have additional microvilli, increasing the surface area still further and so facilitating the reabsorbtion of testicular secretions in order to concentrate the sperm and therefore increase its storage capacity.

It is essential that all sperm spend a period of time, approximately 48 hours, within the epididymis in order to mature. Such maturation is essential so that released sperm are capable of further development (capacitation) within the female tract, enabling them to fertilise the awaiting ova. If they are not passed up to the vas deferens as a result of ejaculation they degenerate and are

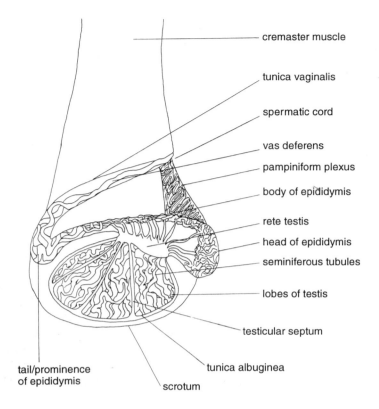

Figure 2.5 *Diagrammatic illustration of a cross-section through the testis of the stallion.*

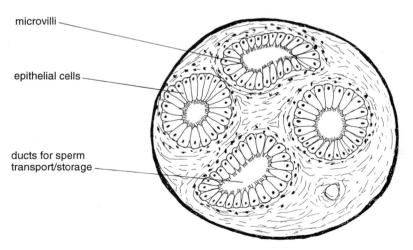

Figure 2.6 *A cross-section through the epididymis of the testis of the stallion.*

reabsorbed, allowing a continual supply of fresh sperm to maintain fertilisation rates (Figure 2.6).

The Testis

The testes are the site of sperm production and the major site of testosterone production in the stallion. Colour plate 15 shows in detail the structure of the testis.

The testes hang outside the body of the stallion in order to maintain a temperature of approximately 3°C below body temperature (ie 35–36°C rather than 39°C). Sperm production is maximised at this lower temperature. Increases in testicular temperature due to disease or inflammation of the scrotum, testis or epididymis result in a significant decrease in spermatogenesis. This decrease in sperm production is transitory but the duration of testicular dysfunction is related to the duration of temperature elevation. The testis temperature is normally controlled by means of the cremaster muscle, which can retract the testis in towards the body cavity, along with the abundance of scrotal sweat glands and the arterio-venous countercurrent heat exchange mechanism provided by the pampiniform plexus (Roberts, 1986; Friedman, Scott, Heath, Hughes, Daels and Tran, 1991).

The pampiniform plexus allows the testicular artery to divide into a dense capillary network before it enters the testis. As such it comes into close contact with the returning venous supply, also divided up into a capillary network. Blood entering the testis, therefore, loses its heat to the venous supply. Such an arrangement ensures that the testicular artery cools down prior to entry into the testis and the testicular vein is warmed up prior to its re-entry into the main body.

The testes lie within a skin covering, termed the scrotum, under which lies the tunica vaginalis, which is continuous up into the body cavity around the spermatic cord.

In the foetus, the testes descend from a position near the kidney, through the inguinal ring and into the scrotum at, or soon after, birth. The failure of the testes to drop fully or the descent of only one may not mean a stallion is infertile as sperm may still be produced by the testes within the body cavity; sperm count, however, is normally much reduced. A stallion with only one descended testicle is termed a rig (Figure 2.7).

The testes of the stallion normally lie with the long axis horizontal unless drawn up towards the body, when they may turn slightly. The long axis is normally in the region of 6–12 cm, the height and width being 4–7 cm and 5 cm respectively, with a weight of 300–350 g a pair. Their size increases allometrically with general body growth

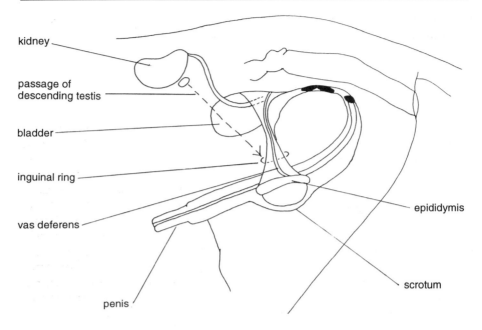

Figure 2.7 *The anatomy of a cryptorchid or rig with undescended testis. The dotted line shows the normal passage of the descending testis.*

until final body size has been reached at approximately five years of age.

Under the scrotal skin lies the tunica vaginalis and under this each separate testis is surrounded by a fibrous capsule, the tunica albuginea. Sheets of this fibrous tissue invade into the body of the testis and divide it up into lobes. Each lobe is a mass of convoluted seminiferous tubules with intertubular areas, which are shown in Figure 2.5. Each area is responsible for one of two functions. The Sertoli cells lining the seminiferous tubules act as nurse cells nourishing and aiding the developing spermatozoa in the lumen of the tubules. In addition, they are also phagocytic, digesting degenerating germ cells and residual bodies: they secrete luminal fluid and proteins, and also form the blood–testis barrier and provide cell to cell communication (Dym and Madhwa Raj, 1977). The Leidig cells found in the intertubular spaces around the seminiferous tubules secrete the hormone testosterone, which in turn drives sperm production and the development of general male bodily characteristics and behaviour.

The number of Sertoli cells varies with season, being significantly greater during the breeding season, when they are actively involved in sperm production. This increase in Sertoli cell number is accompanied by a corresponding increase in seminiferous tubule length (Johnson and Nguyen, 1986).

Sperm

As discussed, sperm are produced within the seminiferous tubules and are nursed by Sertoli cells. They start as very underdeveloped germ cells or spermatogonia attached to the wall of the semi-niferous tubules. These then progressively develop into primary spermatocytes, secondary spermatocytes, spermatids and finally mature sperm (Figure 2.8). As this development occurs they migrate, along with their Sertoli cells, away from the wall of the seminiferous tubules towards the open lumen. At the mature sperm stage they are freed of their Sertoli cells and are pushed along the seminiferous tubules by their rhythmic contractions and secretory fluid. By this stage they have lost a considerable amount of cytoplasm but have developed tails. The tails are not functional until maturation of the sperm in the epididymis has occurred.

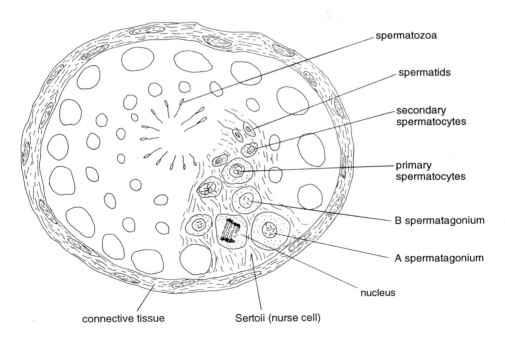

Figure 2.8   A cross-section view through a seminiferous tubule within the stallion's testis, illustrating the gradual meiotic division of spermatogonia to spermatozoa.

The whole cycle of development of a mature sperm takes approximately 50–60 days and seems to occur in waves, ensuring a continual supply of mature sperm, providing the stallion is not overused. The mean daily sperm production of a mature stallion is in the order of $7–8 \times 10^9$ sperm and it has been reported to take 8–11 days for sperm to pass from the testis to the exterior. This time, however, is a function of stallion use, as the more sperm used, the quicker they progress through the system. Total sperm production is related to testis size/volume, which can be assessed by using calipers or ultrasonically. The use of calipers in the stallion is less easy than in other livestock that have more pendulous testes. Ultrasonic assessment, therefore, gives a more accurate result (Love, Garcia, Riera and Kenney, 1991) (Graph 2.1).

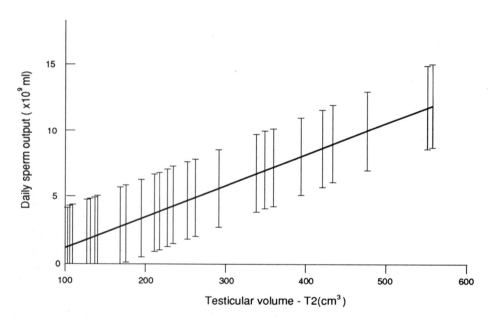

Graph 2.1   The relationship between predicted daily sperm output and testicular volume for 26 stallions (Love et al., 1991).

Sperm are structurally made up of three areas with three distinct functions: the head, the mid piece, and the tail. Figure 2.9 illustrates the characteristics of a normal sperm. During development the cytoplasm of the cell is forced along the length of the tail leaving the head mainly made up of nuclear material, containing the haploid number of chromosomes (half the normal number, 32, to allow fusion with the ova to give the normal diploid complement

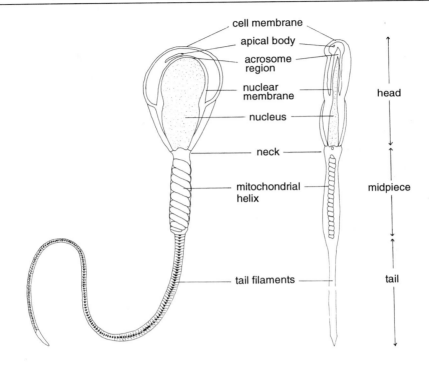

Figure 2.9   A typical stallion sperm.

of 64). The head of the sperm has a double membrane, the outer cell membrane and the inner nuclear membrane, except in the acrosome region at the top of the head, where there is an additional acrosome membrane. The importance of this membrane will become apparent when we discuss fertilisation in a later chapter, as it is responsible for the breakdown of the cell membrane and the nuclear membrane at fertilisation which allows the fusion of the male and female nuclei.

The midpiece or neck of the sperm contains a high proportion of mitochondria, organelles within the cell that produce energy. The midpiece is often termed the power plant of the sperm, providing the energy to drive the tail. The tail is made up of a series of muscle fibrils, equivalent to those found in all major muscle blocks of the body. Using the energy provided by the midpiece, the tail is whipped from side to side producing a wave-like motion.

Semen

Semen is made up of seminal fluid plus sperm, and in the stallion is a milky white gelatinous fluid. The components of seminal fluid have been discussed earlier. The sperm concentration of semen varies with the fraction examined. There are three identifiable fractions: presperm fraction, sperm rich fraction and the postsperm fraction. The presperm fraction is the initial fraction from the penis often seen prior to entering the mare as a dribble of 10 ml or so. It is secreted mainly by the bulbourethral glands and contains no sperm. Its function is to clean and lubricate the urethra prior to ejaculation. Stale urine and bacteria collect readily in the urethra and obviously can harm sperm. The high concentration of bacteria in this fraction deems it unsuitable for collection for artificial insemination.

The sperm rich fraction is the major deposit by the stallion and commences as soon as the glans penis swells to block the exit from the vagina. This fraction is normally 40–80 ml in volume and contains 80–90% of the sperm and the biochemical components of semen.

The third fraction is the post sperm or gel fraction. Its volume varies enormously from none at all to 80 ml and is dependent on libido (sexual enthusiasm), as the higher the libido, the greater the

Table 2.1   The effect of age on the seminal characteristics of stallions
(Squires, Pickett and Amann, 1979)

Characteristics	Age (years)			p
	2–3 (n=7)	4–6 (n=16)	9–16 (n=21)	
Seminal volume (ml)				
Gel	2.1	5.1	13.3	NS
Gel-free	14.2_a	26.2_b	29.8_b	<0.05
Total	16.2_a	31.4_{ab}	43.2	<0.01
Spermatozoa				
Conc. (10^6/ml)	120.4	160.9	161.3	NS
Total (10^9)	1.8_a	3.6_{ab}	4.5_b	<0.05
Motility (%)	55.0	63.1	59.9	NS
pH	7.68_a	7.64_{ab}	7.59_b	<0.05

Means in the same row with different subscripts differ
NS = no significant difference

gel fraction. Season affects the gel fraction volume, this being lower in the non-breeding season. Breed and previous use also have an effect. If a stallion is used more than once per day, the second ejaculate frequently has half the gel fraction of the first. The volume of gel-free fraction is also affected by age, as are sperm numbers and concentration, as illustrated in the data given in Table 2.1

At ejaculation the stallion secretes semen in a series of up to 9 jets, the average volume of semen and the sperm concentration decreasing with successive jets (Graph 2.2).

Motility results given in Table 2.1 illustrate that a relatively large proportion of sperm are not capable of movement. This can be used as a guide to viability of sperm and will be discussed in more detail in Chapter 20 on artificial insemination.

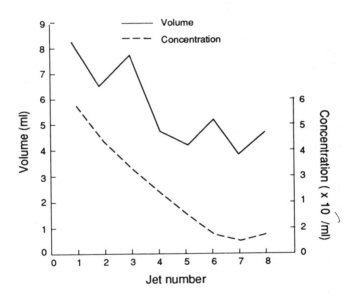

Graph 2.2   Volume and sperm concentration of successive jets (Kosiniak, 1975).

Conclusion

It can be concluded that the male reproductive tract is specifically designed for the efficient deposition of sperm within the female tract. It also ensures that sperm are deposited in a medium (seminal plasma) which is able to provide all the elements required for their survival.

References and Further Reading

Boyle, M.S., Cran, D.G., Allen, W.R. and Hunter, R.H.F. (1987) Distribution of spermatozoa in the mare's oviduct. J.Reprod.Fert.Suppl., 35, 79.

Dym, M. and Madhwa Raj, H.G. (1977) Response of adult rat Sertoli cells and Leidig cells to depletion of luteinising hormone and testosterone. Biol.Reprod., 17, 676.

Friedman, R., Scott, M., Heath, S.E., Hughes, J.P., Daels, P.F. and Tran, T.Q. (1991) The effects of increased testicular temperature on spermatogenesis in the stallion. J.Reprod.Fert.Suppl., 44,127.

Hoffman, L.S., Adam, T.E. and Evans, J.W. (1987) Circadian, circhoral and seasonal variation in patterns of gonadotrophin secretion in geldings. J.Reprod.Fert.Suppl., 35, 51.

Johnson, L. and Nguyen, H.B. (1986) Annual cycle of the Sertoli cell population in adult stallions. J.Reprod.Fert., 76, 311.

Kosiniak, K. (1975) Characteristics of successive jets of ejaculated semen of stallions. J.Reprod.Fert.Suppl., 23, 59.

Little, T.V. and Woods, G.L. (1987) Ultrasonography of accessory sex glands in the stallion. J.Reprod.Fert.Suppl., 35, 87.

Love, C.C., Garcia, M.C., Riera, F.R. and Kenney, R.M. (1991) Evaluation of measurements taken by ultrasonography and calipur to estimate testicular volume and predict daily sperm output in the stallion. J.Reprod.Fert.Suppl., 44, 99.

McDonnell, S.M., Garcia, M.C. and Kenney, R.M. (1987) Pharmacological manipulation of sexual behaviour in stallions. J.Reprod.Fert.Suppl., 35, 45.

Roberts, S.J. (1986) Veterinary Obstetrics and Genital Diseases, Theriogenology, 3rd Ed. pp. 752–893. David and Charles, North Pomfret, V.T. 1986.

Rossdale, P.D. and Ricketts, S.W. (1980) Equine Stud Farm Medicine. 2nd Ed. Ballière Tindall, London.

Squires, E.L., Pickett, B.W. and Amann, R.P. (1979) Effect of successive ejaculation on stallion seminal characteristics. J.Reprod.Fert.Suppl., 27, 7.

Thompson, D.L. Jnr., Johnson, L. and Weist, J.J. (1987) Effects of month and age on prolactin concentrations in stallion's semen. J.Reprod.Fert.Suppl., 35, 67.

Thompson, D.L. Jnr., Pickett, B.W., Squires, E.L. and Amann, R.P. (1979) Testicular measurements and reproductive characteristics in stallions. J.Reprod.Fert.Suppl., 27, 13.

CHAPTER 3

Endocrine Control of Reproduction in the Mare

THE mare is naturally a seasonal breeder showing sexual activity only during the spring, summer and autumn months. This is termed the breeding season and the non-breeding season is termed anoestrus. On average the season lasts from March until November (10–15 cycles), but type of breed has a significant effect on the exact length, as has individual variation. Ponies and the large heavier breeds, especially of the coldblooded type, tend to show shorter seasons than the finer, more hotblooded types eg the Thoroughbred and lighter riding horses. During her breeding season the mare shows a series of spontaneous oestrous cycles at regular intervals and is, therefore, termed a polyoestrus spontaneous ovulator. The rest of her breeding cycle is summarised in Figure 3.1, which illustrates the major milestones in the mare's reproductive life.

The oestrous cycles of the mare commence at puberty and last on average 21 days (range 20–22 days). Each cycle is a pattern of physiological and behavioural events under hormonal control and is divided into two periods: oestrus when the mare is sexually receptive, normally for 4–5 days, and dioestrus, when she will reject all sexual advances, normally for a period of 16 days. The exact times of these two periods varies considerably between individuals. The true length of oestrus is difficult to detect as many mares take awhile to come into a truly sexually receptive state. This tends to be towards the end of the apparent oestrus period, coinciding with ovulation, which usually occurs 24–36 hours before the end of oestrus. The length of oestrus is also affected by season, tending to be longer and less distinct during the early spring and late autumn months, as the mare goes through the transition period of coming into or going out of the breeding season.

Oestrous cycles continue throughout the mare's lifetime and only cease during the non-breeding season. The mare is an efficient breeder, showing oestrus cycles during lactation, unlike other seasonal breeders such as the ewe, and she is, therefore, capable of

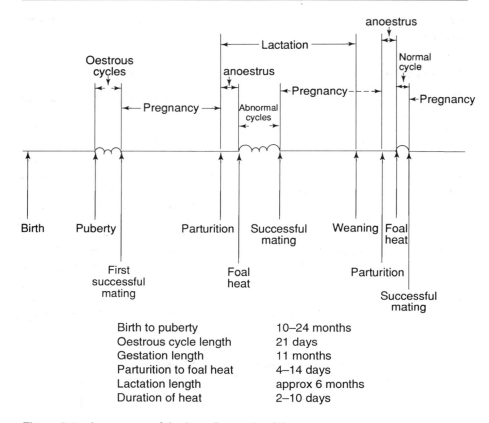

Figure 3.1   A summary of the breeding cycle of the mare.

being pregnant and lactating at the same time. The mare shows her first oestrus after foaling often within 4–10 days: this oestrus is termed her foal heat. After the foal heat the mare may start to show her regular 21 day cycles, but in many cases due to the effects of lactation, it takes awhile for the system to settle down to a regular pattern again (Mathews, Rophia and Butterfield, 1967).

For ease of understanding, the following discussion on the oestrous cycle of the mare will be divided into two sections, the associated physiological and behavioural changes.

Physiological Changes

The major physiological events associated with reproductive activity in the mare are endocrine changes, which in turn govern and drive the other physiological changes as well as her behavioural patterns.

The Endocrinological Control of the Oestrous Cycle

The endocrinological control of the oestrous cycle is governed by the hypothalamic-pituitary-gonad axis; a similar axis controls the reproductive activity of the stallion. The gonads, in the case of the mare, are the ovaries (Figure 3.2). Overriding the whole of this control mechanism is the effect of photoperiod: decreasing day length causing oestrous cycles to cease and increasing causing them to occur. Day length is perceived by the pineal gland in the brain which, by means of the hormone melatonin and subsequently prolactin, controls the activity of the hypothalamic-pituitary-ovarian axis. Melatonin is produced nocturnally by the pineal gland, and under the influence of short day lengths dominates the reproductive system of the mare, inhibiting the activity of the axis (Kilmer, Sharp, Berglund, McDowell, Grubaugh and Peck, 1982). As day length increases, inhibition of the axis is removed allowing LH and FSH production by the pituitary (Fitzgerald, Affleck, Barrows, Murdock,

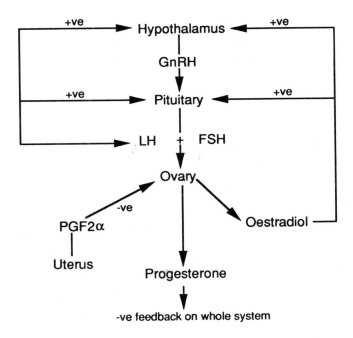

Figure 3.2   The hypothalamic-pituitary-ovarian axis that governs reproduction in the mare.

Barker and Loy, 1987). LH is released in a pulsatile manner and the frequency of these pulses is seen to increase from 0.38 to 4.74 pulses/day as the mare moves from anoestrus to the first ovulation of the season (Fitzgerald, L'Anson, Legan and Loy, 1985). The means by which melatonin controls the hypothalamus is unclear, but seems likely to involve endogenous opioids including B endorphins plus prolactin.

Opioids also seem to be responsible for changes in metabolic rate, increasing the efficiency of food conversion during the winter months, a time of food deprivation. Especially evident in the more native breeds, this demonstrates an innate ability of the equine body to anticipate environmental conditions and respond accordingly (Morley, Levine, Yim and Lowy, 1983; Argo and Smith, 1983). Prolactin, therefore, is thought to translate the seasonal changes in melatonin to pituitary inhibition/activity and concentrations are elevated during long day length and act to enhance pituitary activity (Johnson and Becker, 1987; Thompson and Johnson, 1987). The mechanism by which melatonin influences prolactin secretion is better known in other seasonal breeders than in horses (Johnson, 1987; Evans, Alexander, Irvine, Livesey and Donald, 1991).

Prolactin concentrations in the horse show the typical sinusoidal seasonal cycle with peak concentrations in summer as seen in other seasonal mammals (Johnson, 1986; Thompson, Weist and Nett, 1986; Worthy, Colquhoun, Escreet, Dunlop, Renton and Douglas, 1987). In addition, short term and diurnal fluctuations are evident (Roser, O'Sullivan, Evans, Swedlow and Papkoff, 1987). As seen, prolactin secretion is mediated via the stimulation of the pineal gland through the production of melatonin in a daily rhythm stimulated by darkness and inhibited by light. It is unclear whether alterations in prolactin secretion within the mare's system are caused by an alteration in a prolactin inhibiting factor, or a prolactin releasing factor or the amount of prolactin available for release. Evidence does suggest that the amount of prolactin available for release is primarily affected (Thompson and Johnson, 1987).

The hypothalamus, when day length is appropriate, is driven to produce gonadotrophin releasing hormone (GnRH). GnRH release is pulsatile in manner and is passed directly down a specialised portal system, the hypothalamic-pituitary portal vessels passing directly to the anterior pituitary. The level of GnRH in the mare's circulatory system is, therefore, relatively low, as its passage is directed along these specialised portal vessels. In response to GnRH the anterior pituitary produces the gonadotrophins follicle stimulating hormone and luteinising hormone, the target organ for which is the ovaries (Hughes, Stabenfeldt and Evans, 1972a, 1972b).

Follicle Stimulating Hormone

Follicle stimulating hormone (FSH), as its name suggests, is responsible for the stimulation of follicle development. It is passed into the
general circulatory system of the mare and its concentration rises
gradually, day 15 onwards, from levels of 4 ng/ml, to reach
concentrations of 9 ng/ml during oestrus. Up to 10 follicles may be
affected by the initial rise in FSH but only a select one or two develop
to a stage (diameter of 3 cm or more) that can react to the message to
ovulate (Vandeplasshe, Henry and Coryn, 1979) (Graph 3.1). This
graph, along with subsequent ones illustrating plasma hormone
concentrations, is drawn to give an appreciation of the relative, rather than absolute, hormone concentrations. The absolute
levels reported vary considerably between different scientific reports.
Where known, concentrations are discussed within the text.

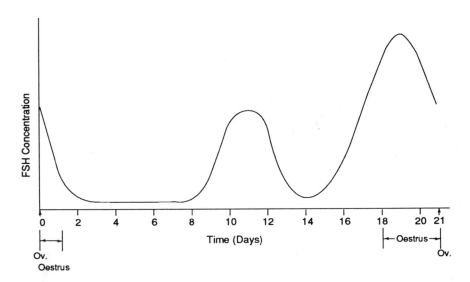

**Graph 3.1   Variations in the relative plasma concentrations of follicle
stimulating hormone in the non-pregnant mare.**

Oestradiol

As the follicles develop they secrete oestradiol 17β, an ovarian
steroidal oestrogen, produced by the granulosa cells within the
developing follicle. The granulosa cells are provided with the
precursors (building blocks) for oestradiol 17β by the follicular
theca cells (Seamens and Sharp, 1982). Oestradiol 17β is secreted
into the main circulatory system and 24–48 hours prior to
ovulation reaches a peak of 10–15 pg/ml, dropping to basal levels

immediately post oestrus. This decline in oestrogen secretion is associated with the conversion of the granulosa cells from oestradiol 17β producing cells to progesterone producing cells as part of the initial process of luteinisation (Tucker, Henderson and Duby, 1991). Oestrogen is responsible for the behavioural changes in the mare associated with oestrus and sexual receptivity.

As FSH levels rise, oestradiol levels also increase, both reaching a peak within oestrus, and thus ensuring that maximum follicular development and oestrus are synchronised (Knudsen and Velle, 1964; Garcia, Freedman and Ginther, 1979) (Graph 3.2).

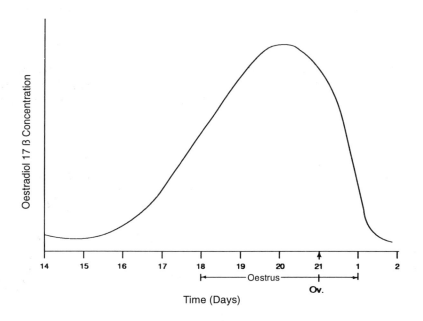

Graph 3.2   Variations in the relative plasma concentrations of oestradiol in the non-pregnant mare

Luteinising Hormone
At this stage the follicles are ready to ovulate. The message to ovulate is luteinising hormone (LH), which, secreted by the pituitary, also rises at the time of oestrus. LH rises in an episodic manner, the frequency and amplitude of pulses increasing to give this rise. Its receptors on the follicular theca cells increase in number as its concentration rises. Increasing LH thus drives additional androgen precursor production providing more oestradiol 17β building blocks, which diffuse across to the granulosa cells for conversion to

oestradiol 17β, which in turn drives oestrus behaviour. Hence, rising LH levels are associated with increasing oestradiol 17β secretion, further ensuring the synchronisation of ovulation and oestrous behaviour. LH levels begin to rise from their basal levels of <3 ng/ml with a pulse frequency of 1.4 pulses/24hrs several days before the onset of oestrus. They then reportedly reach a peak of 10–16 ng/ml after ovulation (Whitmore, Wentworth and Ginther, 1973). It has been reported by some researchers that LH not only induces ovulation but is also involved in driving final follicular development. LH drops from its peak to its low dioestrus levels within a few days of ovulation (Pattison, Chen and King, 1972; Evans, Hughes, Nealy, Stabenfeldt and Winget, 1979; Pantke, Hyland, Galloway, Maclean and Hoppen, 1991) (Graph 3.3).

Progesterone
Ovulation of a follicle results in the formation of a corpus luteum within the lumen of the follicle left after the ova and follicular fluid have been released. The luteal tissue contained within the corpus

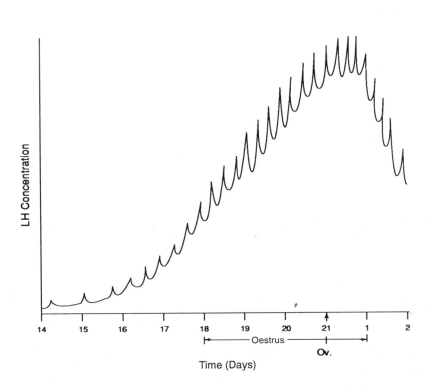

Time (Days)

Graph 3.3   Variations in the relative plasma concentrations of luteinising hormone in the non-pregnant mare

luteum secretes progesterone. Progesterone levels, therefore, rise post ovulation starting within 24–48 hours. Maximum concentrations (10 ng/ml) are reached 5–6 days post ovulation and are maintained at this level until day 15–16 of the oestrus cycle. If the mare has not conceived, progesterone levels drop dramatically 3–4 days prior to the next ovulation to give basal levels during oestrus. Oestrus cannot begin until progesterone levels have fallen below 1 ng/ml (Graph 3.4).

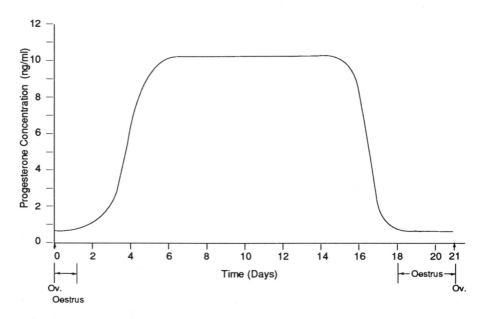

Graph 3.4   Variations in the relative plasma concentrations of progesterone in the non-pregnant mare.

Progesterone has an inhibitory effect on the release of gonadotrophins in most farm livestock. The block to gonadotrophin release in the mare is, however, not so complete. High progesterone levels do seem to have an inhibitory effect on the release of LH, as seen during dioestrus preventing any rise in LH until progesterone levels fall off at day 15–16 of the cycle. However, progesterone does not seem to have such an inhibitory effect on FSH. Indeed, unique to the mare there seems to be a second rise of FSH about 10–12 days after ovulation. This rise in FSH is thought to start the development of a first series of follicles but these then seem to become atretic (degenerate) for some unknown reason and do not

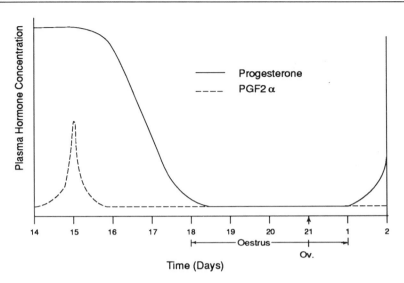

Graph 3.5 **Variations in the relative plasma concentrations of prostaglandin F2α and progesterone in the non-pregnant mare.**

provide the pool of follicles from which the ovulating ovum originates. The reason for, and significance of, this surge in FSH is unclear as it seems to be the FSH rise at the time of oestrus that provides the follicle for ovulation (Stabenfeldt, Hughes and Evans, 1972) (Graph 3.5).

If the mare fails to conceive and is not pregnant, progesterone levels fall off at around 15–16 days after ovulation, allowing the mare to return to oestrus and ovulate on day 21. In order to induce the decline in progesterone a message has to be received by the reproductive system of the mare telling her that there is no conceptus present. The messenger is the hormone prostaglandin F2α (PGF2α).

Prostaglandin F2α
PGF2α is difficult to measure in the peripheral circulatory system because of its short half-life and pulsatile manner of release. However, PGF2α has a metabolic breakdown product, prostaglandin F metabolites (PGFM), which has a longer half-life and is easier to measure in blood serum. It very closely mimics changes in PGF2α. Using levels of PGFM as a guide it can be seen that PGF2α levels rise between day 14 and 17 post ovulation, immediately before progesterone levels start to decline. In mares suffering from

retained corpora lutea or those that are pregnant, no such rise is detected. PGF2α is known to be secreted by the uterine endometrium and causes luteolysis (destruction) of the corpus luteum (CL), thus causing progesterone levels to decline. In the mare, PGF2α reaches the ovary via the main circulatory system, not by a local counter current transport system as in the ewe and cow. This can be demonstrated by hysterectomy, as in the mare removal of the uterine horn ipsilateral to (on the same side as) the CL does not result in maintenance of that CL (Ginther and First, 1971) (Graph 3.5).

This message of non-pregnancy is also thought to involve the release of oxytocin. Oxytocin is produced by the posterior pituitary and is transported via the main circulatory system to the uterus, seemingly enhancing the release of PGF2α.

The falling progesterone levels remove any inhibition on gonadotrophin release, allowing the hormone changes associated with oestrus and ovulation to begin again. Graph 3.6 summarises the major fluctuations in hormone concentrations during a single oestrous cycle of a non-pregnant mare.

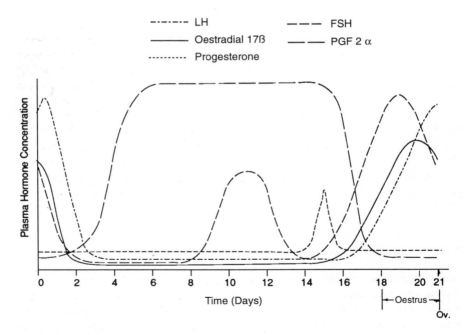

Graph 3.6   A summary of the major plasma hormone concentration changes during the oestrus cycle of the non-pregnant mare.

The following is a summary of all the major events in the mare's oestrus cycle:

DAY 0 Ovulation LH rising FSH falling Oestradiol falling	**DAY 11** FSH peak **DAY 14** FSH at basal levels **DAY 15** PGF2α peak Progesterone falling
DAY 1 LH peak	

DAY 2 Oestrus ends
Dioestrus begins
LH peaking or possibly
   declining
FSH approaching basal
   levels
Oestradiol approaching
   basal levels
	DAY 16 FSH rising
	
	DAY 18 FSH rising
	Progesterone basal
	Oestradiol rising
	Oestrus beginning
	LH rising

DAY 3 Progesterone rising	**DAY 20** Progesterone basal FSH rising LH rising Oestradiol rising
DAY 6 Progesterone at maximum	
DAY 10 FSH rising	**DAY 21** Ovulation

Variations in the above pattern within the oestrous cycle have been noted. LH has been reported to be released in an episodic fashion (1-4 ng/ml) during dioestrus, these episodes occasionally being large enough to cause ovulation mid cycle, despite high dioestrus progesterone levels. This evidence of dioestrus rises in LH and the previously discussed second FSH peak mid cycle indicates that, unlike many other species studied, progesterone does not serve to act as a complete block to gonadotrophin release in the mare. It also seems that the CL of the mare requires the action of a pituitary hormone, possibly LH, to prevent its regression during dioestrus. Both these phenomena are as yet unexplained in the mare (Short, 1973).

Physiological Changes of the Genital Tract

As well as the cyclical changes in hormone concentration, changes in the mare's reproductive tract are also seen; these are in fact driven by the fluctuations in hormone levels. In the uterus the epithelium proliferates during oestrus and early dioestrus preparing it for implantation. The epithelial cells tend to be tall and columnar during oestrus and early dioestrus, changing to cuboidal in nature during late dioestrus. The epithelial glands also change configuration and secrete more during oestrus and early dioestrus, easing firstly the passage of the sperm and then the passage of any fertilised ova. Levels of leucocytes within the uterus also vary, increasing during oestrus and so helping to prevent infection during a highly vulnerable time. This increase in leucocytes is thought to be associated with the increase in circulating oestradiol levels. In general these changes cause the uterine wall to increase in thickness and turgidity during oestrus into dioestrus, preparing for the imminent implantation of an embryo. If this does not occur, luteolysis results in a reduction of the thickness of the uterine wall and a reversal of the preparatory changes for implantation.

The cervix also shows variation within the cycle, and its appearance, as viewed by means of a vaginascope, can be used as a diagnostic aid to the detection of reproductive activity in a mare. During dioestrus the cervix is tightly closed, forming a tight seal against entry into the uterus. Its appearance is white, firm and dry. During oestrus the cervix relaxes, opening the cervical seal to allow partial entry of the penis at mating. During oestrus the cervix appears moist, red and dilated as the secretions of the uterine epithelial cells and cervical cells increase, again easing entry of the penis (Warszawsky, Parker, First and Ginther, 1972).

Variations

The mare is notorious for variations/abnormalities in her reproductive cycle. This is especially evident in mares descended from lines bred for performance rather than reproductive ability. This is in contrast to other farm livestock that have been specifically bred for centuries for their ability to reproduce rather than perform.

A wide variation in the length of oestrus is evident between mares, the extremes being cycles of between 1 and 50 days in length. In general a variation can be seen with the time of year, longer and less distinct oestrous periods being evident during the beginning and end of the breeding season. Nutritional intake also causes variation in oestrous length. When nutrition is limited oestrus tends to be longer and less distinct, making it less likely that the mare will conceive during such a non-ideal time. This effect of poor nutrition may be an additional signal to the mare indirectly indicating seasonal and, therefore, day length changes.

The length of dioestrus also varies between mares, with the extremes being 10 days to several months. This delay is normally due to one of three reasons. Firstly, a silent ovulation: ovulation occurred but it was not accompanied by oestrus, giving the impression that the mare had been in dioestrus for a prolonged period of time. Secondly, the existence of a persistent corpus luteum that has not reacted to PGF2α or has not received enough PGF2α to elicit a response. Thirdly, inactive ovaries, usually associated with the transition into or out of the non-seasonal state or true anoestrus. Two other variations within the cycle occur: they are ovulation in dioestrus, which has already been discussed, and oestrus with no ovulation, which occasionally occurs out of season.

The causes of most of these variations can be put down to managerial or environmental influences, eg nutrition, temperature, day length etc. Occasionally they are due to genetic faults, lactational effects or embryonic death. To further complicate matters it seems that in the mare there is a carryover effect from one year to the next. Hence problems, especially those associated with environmental or managerial disturbances, may not became apparent until 12–18 months later.

Covering at the foal heat is often unsuccessful as fertility rates are normally relatively low. The oestrous cycles following this foal heat tend to be disturbed, showing prolonged oestrus and dioestrus.

Behavioural Changes

Cyclical hormonal changes govern the mare's behavioural patterns associated with oestrus and dioestrus, elevated oestradiol concentrations being the major effector, stimulating the central nervous system. There are many variations between individuals in the extent and strength of behavioural changes (Munro, Renton and Butcher, 1979). Details of the signs of oestrus and their interpretation are given in Chapter 13. A summary of the major behavioural changes is as follows:

Oestrus initiated by elevated oestradiol concentrations

Signs of oestrus
- docility
- urination stance
- lengthening and eversion of the vulva
- exposure of the clitoris (winking)
- tail raised
- urine bright yellow with a characteristic odour
- acceptance of the stallion's advances

Dioestrus initiated by basal (low) oestradiol concentrations

Signs of dioestrus
- hostility
- rejection of stallion's advances

Conclusion

The prime aim behind all the control mechanisms for the female reproductive system is to synchronise the physiological and behavioural changes associated with oestrus in order to synchronise mating, ovulation and fertilisation, along with embryo and uterine development.

References and Further Reading

Argo, C.M. and Smith, J.S. (1983) The relationship of energy requirements and seasonal changes of food intake in Soay rams. J.Physiol.Lond., 343,23, Abstr.

Curlewis, J.D., Loudon, A.S.I., Milne, J.A. and McNeilly, A.S. (1988) Effects of chronic long acting bromocriptine treatment on liveweight, voluntary food intake, coat growth and breeding season in non-pregnant red deer hinds. J.Endocri., 119, 413.

Evans, M.J., Alexander, S.L., Irvine, C.H.G., Livesey, J.H. and Donald, R.A. (1991) In vitro and in vivo studies of equine prolactin secretion throughout the year. J.Reprod.Fert.Suppl., 44, 27.

Evans, J.W., Hughes, J.P., Neely, D.P., Stabenfeldt, G.H. and Winget, C.M. (1979) Episodic LH secretion patterns in the mare during the oestrous cycle. J.Reprod.Fert.Suppl., 27, 143.

Fitzgerald, B.P., Affleck, K.J., Barrows, S.P., Murdock, W.L., Barker, K.B. and Loy, R.G. (1987) Changes in LH pulse frequency and amplitude in intact mares during the transition into the breeding season. J.Reprod.Fert., 79, 485.

Fitzgerald, B.P., L'Anson, H., Legan, S.J. and Loy, R.G. (1985) Changes in patterns of luteinising hormone secretion before and after the first ovulation in the post partum mare. Biol.Reprod., 33, 316.

Freedman, L.J., Garcia, M.C. and Ginther, O.J. (1979) Influences of ovaries and photoperiod on reproductive function in the mare. J.Reprod.Fert.Suppl., 27, 79.

Garcia, M.C., Freedman, L.H. and Ginther, O.J. (1979) Interaction of seasonal and ovarian factors in the regulation of LH and FSH secretion in the mare. J.Reprod.Fert.Suppl., 27, 103.

Garcia, M.C. and Ginther, O.J. (1976) Effects of ovariectomy and season on plasma luteinising hormones in mares. Endocrinology, 98, 958.

Ginther, O.J. (1992) In: Reproductive Biology of the Mare, Basic and Applied Aspects. 2nd Ed. Pub Equiservices 4343, Garfoot Road, Cross Plains, Wisconsin 53528, U.S.A.

Ginther, O.J. and First, N.L. (1971) Maintenance of the corpus luteum in hysterectomised mares. Am.J.Vet.Res., 32, 1687.

Ginther, O.J. and Wentworth, B.C. (1974) Effect of a synthetic gonadotrophin releasing hormone on plasma concentrations of luteinising hormone in ponies. Am.J.Vet.Res., 35, 79.

Hughes, J.P., Stabenfeldt, G.H. and Evans, J.W. (1972a) Estrous cycle and ovulation in the mare. J.Am.Vet.Med.Ass., 161, 1367.

Hughes, J.P., Stabenfeldt, G.H. and Evans, J.W. (1972b) Clinical and endocrinological aspects of the estrus cycle of the mare. Proc.18th.Ann. Conv.Am.Ass.Equine Practnrs., 119.

Johnson, A.L. (1986) Serum concentrations of prolactin, thyroxine and triiodothyronine relative to season and the oestrous cycles of the mare. J.Anim.Sci., 62, 1012.

Johnson, A.L. (1987) Seasonal and photoperiodic induced changes in serum prolactin and pituitary responsiveness to thyrotrophin-releasing hormone in the mare. Proc.Soc.Exp.Biol.Med., 184, 118.

Johnson, A.L. and Becker, S.E. (1987) Effects of physiologic and pharmacologic agents on serum prolactin concentration in the non-pregnant mare. J.Anim.Sci., 65, 1292.

Kilmer, D.N., Sharp, D.C., Berghund, L.A., Grubaugh, W., McDowell, K.J. and Peck, L.S. (1982) Melatonin rhythms in pony mares and foals. J.Reprod.Fert.Suppl., 32, 303.

Knudsen, O. and Velle, W. (1964) Ovarian oestrogen levels in the non-pregnant mare. Relationship to histological appearance of the uterus and its clinical status. J.Reprod.Fert., 2, 130.

Lowe, J.E., Foote, R.H., Baldwin, B.H., Hillman, R.B. and Kallfeiz, F.A. (1987) Reproductive patterns in cyclic and thyroidectomised mares. J.Reprod.Fert.Suppl., 35, 281.

Mathews, R.G., Rophia, R.T. and Butterfield, R.M. (1967) The phenomenon of foal heat in mares. Aust.Vet.J., 43, 579.

Morley, J.E., Levine, A.S., Yim, G.K. and Lowy, M.J. (1983) Opioid modulation of appetite. Neurosci.Behav.Rev., 3, 155.

Munro, C.D., Renton, J.P. and Butcher, R.A. (1979) The control of oestrous behaviour in the mare. J.Reprod.Fert.Suppl., 27, 217.

Pantke, P., Hyland, J., Galloway, D.B., Maclean, A.A. and Hoppen, H.O. (1991) Changes in Luteinising hormone bioactivity associated with gonadotrophin pulses in the cycling mare. J.Reprod.Fert.Suppl., 44,13.

Pattison, M.L., Chen, C.L. and King, S.L. (1972) Determination of LH and oestradiol-17β surge with reference to the time of ovulation in mares. Biol.Reprod., 7, 136.

Pineda, M.H., Ginther, O.J. and McShan, W.H. (1972) Regression of corpus luteum in mares treated with an antiserum against an equine pituitary fraction. Am.J.Vet.Res., 33, 1767.

Roser, J.F., O'Sullivan, J., Evans, J.W., Swedlow, J. and Papkoff, H. (1987) Episodic release of prolactin in the mare. J.Reprod.Fert.Suppl., 35, 687.

Seamens, K.W. and Sharp, D.C. (1982) Changes in equine follicular aromatase activity during the transition from winter anoestrous. J.Reprod. Fert.Suppl., 32, 225.

Sharp, D.C., Vernon, M.W. and Zavy, M.T. (1979) Alteration of seasonal reproductive patterns in mares following superior cervical ganglionectomy. J.Reprod.Fert.Suppl., 27, 87.

Short, R.V. (1973) Role of hormones in sex cycles. In: Reproduction in Mammals. 3. Hormones in Reproduction, Ed. C.R. Austin and R.V.Short, p.42. Cambridge University Press, Cambridge.

Stabenfeldt, G.H., Hughes, J.P. and Evans, J.W. (1972) Ovarian activity during the oestrus cycle of the mare. J.Endocri., 90, 1397.

Stewart, F. and Allen, W.R. (1979) Binding of FSH, LH and PMSG to equine gonadal tissue. J.Reprod.Fert.Suppl., 27, 431.

Strauss, S.S., Chen, C.L., Kalra, S.P. and Sharp, D.C. (1979) Localisation of gonadotrophin releasing hormone (GnRH) in the hypothalamus of ovarectomised pony mares by season. J.Reprod.Fert.Suppl., 27, 123.

Tetzke, T.A., Ismail, S., Mikuckis, G. and Evans, J.W. (1987) Patterns of oxytocin secretion during the oestrous cycle of the mare. J.Reprod.Fert.Suppl., 35, 245.

Thompson, D.L. and Johnson, L. (1987) Effects of age, season and active immunisation against oestrogen on serum prolactin concentrations in stallions. Dom.Anim.Endocri., 4, 17.

Thompson, D.L., Weist, J.J. and Nett, T.M. (1986) Measurement of equine prolactin with an equine-canine radioimmunoassay, seasonal effects of prolactin response to thyrotrophin releasing hormone. Dom.Anim.Endocri., 3, 247.

Tucker, K.E., Henderson, K.A. and Duby, R.T. (1991) In vitro steroidogenesis by granulosa cells from equine preovulatory follicles. J.Reprod.Fert.Suppl., 44, 45.

Vandeplassche, M., Henry, M. and Coryn, M. (1979) The mature mid-cycle follicle in the mare. J.Reprod.Fert.Suppl., 27, 157.

Warszawsky, L.F., Parker, W.G., First, N.L. and Ginther, O.J. (1972) Gross changes of internal genitalia during the estrous cycle in the mare. Am.J.Vet.Res., 33, 19.

Whitmore, H.L., Wentworth, B.C. and Ginther, O.J. (1973) Circulating concentrations of luteinising hormone during estrous cycles of mares as determined by radioimmunoassay. Am.J.Vet.Res., 34, 631.

Worthy, K. Colquhoun, K., Escreet, R., Dunlop, M., Renton, J.P. and Douglas, T.A. (1987) Plasma prolactin concentrations in non-pregnant mares at different times of the year and in relation to events in the cycle. J.Reprod.Fert.Suppl., 35, 269.

CHAPTER 4

Endocrine Control of Reproduction in the Stallion

THE stallion, like the mare, is a seasonal breeder but tends to show a less distinct season than the mare, and unlike her, if given enough encouragement, is capable of breeding all the year around. Sperm production, unlike ova production, is a continual process not governed by cyclical hormonal changes. However, season does have an effect upon seminal volume, sperm concentration in the gel free fraction, sperm per ejaculate, the number of mounts per ejaculate and reaction time to the mare (Pickett, Faulkner and Sutherland, 1970; Pickett and Voss, 1972; Pickett, Faulkner and Voss, 1975; Johnson, 1991). Graphs 4.1 to 4.5 demonstrate the effect of these factors on reproductive components.

As with the mare, reproductive activity commences at puberty and continues for the rest of a stallion's lifetime, though semen quality tends to decline in stallions over 20 years of age. The exact timing of puberty is unclear and varies with breed and development. Various researchers have used histological changes within the testis, especially in association with the Leidig cells, to indicate the timing of puberty. Ages of 1 to 2.2 years have been suggested using such parameters (Cornwall, 1972; Naden, Amann and Squires, 1990). However, work suggests the colt may be up to 5 years old before full adult reproductive ability is reached. The characteristics tested were testicular weights, daily sperm production, testosterone concentrations, Leidig and Sertoli cell number and volume (Johnson and Neaves, 1981; Thompson and Honey, 1984; Berndston and Jones, 1989). Stallions of 3 years of age have been reported to have reached puberty and to be spermatogenically active and perfectly capable of fertilising mares covered; however, they have a limited sperm producing capacity. By 5 years of age they are capable of producing adequate numbers of spermatozoa to mate as many mares as an adult stallion (Johnson, Varner and Thompson, 1991).

As with the mare, the reproductive activity of the stallion can be

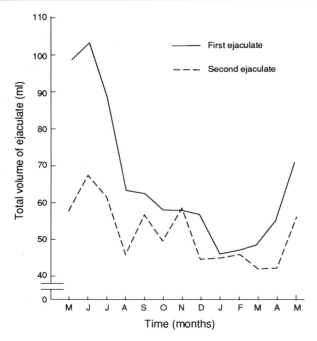

Graph 4.1 **Total seminal volume produced throughout the year** (Pickett and Voss, 1972).

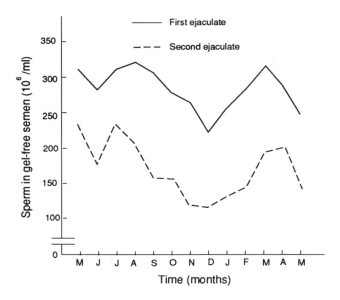

Graph 4.2 **Number of sperm in gel free semen throughout the year** (Pickett and Voss, 1972).

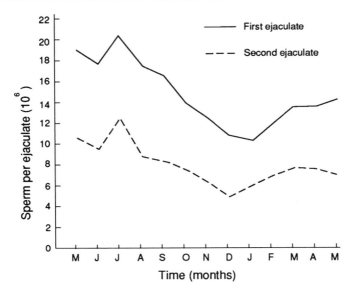

Graph 4.3   Number of sperm per ejaculate throughout the year (Pickett and Voss, 1972).

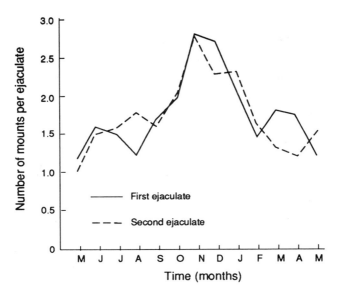

Graph 4.4   Mean number of mounts required per ejaculate throughout the year (Pickett and Voss, 1972).

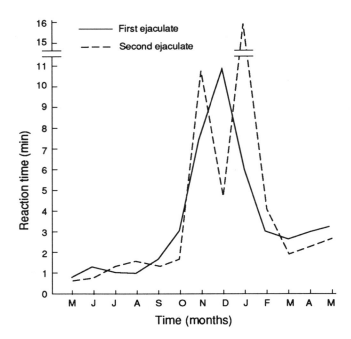

Graph 4.5   The effect of season on sexual behaviour as measured by reaction time (Pickett and Voss, 1972).

divided into physiological and behavioural changes and will be detailed in turn.

Physiological Changes

Hormone patterns are the major physiological changes associated with stallion reproductive activity and indeed govern the remaining physiological and behavioural changes.

The Endocrinological Control of Stallion Reproduction

Control of stallion reproduction is governed by the hypothalamic-pituitary-gonad axis as seen in the mare, except the testis fulfils the function of the gonads (Amann, 1981a, b). This axis is detailed in Figure 4.1.

Environmental stimuli in the form of day length and temperature have an overriding effect on the above axis. Season is governed by the secretion of melatonin from the pineal gland in response to day

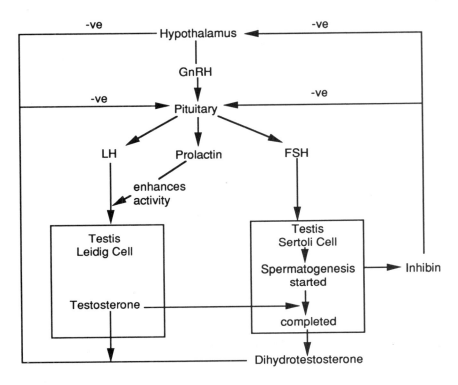

Figure 4.1 The hypothalamic-pituitary-testis axis that governs reproduction in the stallion.

length. In the stallion, as in the mare, melatonin is secreted during the hours of darkness and hence increases with shortening days, that is as winter approaches (Thompson, Pickett, Bernditson, Voss and Nett, 1977; Burns, Jawad, Edmunson, Cahill, Boucher, Wilson and Douglas, 1982; Clay, Amann and Pickett, 1987). In addition to this, prolactin concentrations increase as day length increases – 4–5 µg/l in summer compared to 1.7 µg/l during winter – and it is thought that via these changes in prolactin concentrations melatonin inhibits the frequency and amplitude of testosterone production by an inhibitory effect on the hypothalamus as seen in rams (Lincoln and Short, 1980; Evans, Alexander, Irvine, Livsey and Donald, 1991). Both melatonin and photoperiod can be manipulated, as in the mare, to alter the timing of the breeding season, but overstimulation with artificial long days for a prolonged period of time results in refractoriness (Clay, Squires, Amann and Pickett, 1987; Cox, Redhead and Jawad, 1988; Argo, Cox and Grey

1991). Photoperiod also controls coat growth and weight gain though its effect may be overridden by environmental temperature (Tucker and Wetterman, 1976).

As day length reduces, the inhibition of the axis is removed, allowing the hypothalamic activity of the breeding season, and as a result GnRH (gonadotrophin releasing hormone) is produced. Directly connecting the hypothalamus and the anterior pituitary gland is a series of hypothalamic-pituitary portal vessels down which GnRH is directly passed. GnRH acts, therefore, directly on the pituitary and does not enter the general circulatory system in any significant concentration.

The anterior pituitary is stimulated to produce FSH (follicle stimulating hormone) and LH (luteinising hormone) (Irvine, 1984). The testes, which are the target organs for LH and FSH, consist of two major cell types, Leidig and Sertoli cells. The plasma concentrations of LH and FSH in the sexually active stallion are in the order of 3–4 ng/ml and 7–7.5 ng/ml respectively (Seamens, Roser, Linford, Lui and Hughes, 1991).

Leidig cells make up the intertubular spaces or interstitial tissue of the testis and are responsible for the production of testosterone. These cells and, therefore, testosterone secretion are controlled by LH (Amann, 1981). LH is produced in a pulsatile fashion, and so it seems is testosterone. Such pulsatile release of testosterone can mean that a single blood sample for the hormone can give erroneous results, and a hormone profile taken over a period of time and averaged out is a much more accurate indication of true testosterone levels.

Sertoli cells line the seminiferous tubules and act as nurse cells to developing spermatids, nourishing and protecting them during their development. These cells are driven by both FSH and testosterone. FSH is known to start the process of spermatogenesis, developing spermatogonia to secondary spermatocytes. Testosterone then completes their development from secondary spermatocytes to spermatozoa ready for passage to the epididymis for maturation. In addition, testosterone controls the development of male genitalia, testis descent in the foetus or neonate, pubertal changes and accelerated growth, plus the maintenance and function of the accessory glands. It is also responsible for male libido and sexual behaviour by stimulation of the central nervous system (CNS), plus development of personality, stallion behaviour and muscular development. Testosterone also feeds back negatively on pituitary function to reduce the release of LH and FSH and hence its own production (Irvine, Alexander and Turner, 1986; Flink, 1988). A derivative of testosterone, dihydrotestosterone, also feeds back negatively on the pituitary and has a limited effect on other testosterone-driven stallion characteristics. This negative feedback ensures that the system does not overrun itself. Testosterone may also be produced by a limited population of Leidig cells occasionally found in the wall of the vas deferens; this is evident in some geldings that have been successfully gelded but still show unacceptably high levels of stallion characteristics. Testosterone is also produced by the adrenal glands in both the stallion and the gelding and it is this testosterone that is responsible for the continued, but reduced, male characteristics of the gelding.

Not only is the production of testosterone dependent upon season (Graph 4.6), but there is also evidence of a diurnal rhythm of testosterone secretion in the stallion.

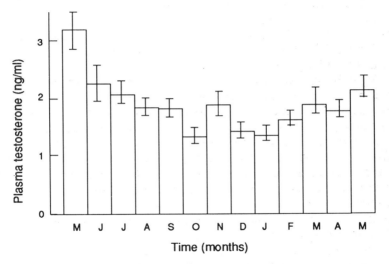

Graph 4.6   Concentration (mean + SE) of testosterone in the peripheral plasma of mature stallions over a 13 month period (Berndston, Pickett and Nett, 1974).

Testosterone concentrations have been reported to be higher at 06.00 and 18.00 hours. It has been postulated that in the wild this ensures that mating activity is greatest at dawn and dusk, times of least risk to stallions and mares from predators.

Two other hormones also thought to be involved in the control of male reproduction are inhibin and prolactin. Inhibin is also produced by the testis and is thought to have an additional negative feedback effect on hypothalamic and pituitary function. Prolactin is produced by the anterior pituitary and is thought to enhance the activity of LH in some way. It is known that if low levels of prolactin are in the system, then the effect of LH on testosterone production is greater. These relatively low levels of prolactin are not to be confused with the very high levels of prolactin that in many species, including the horse, seem to be associated with the control of reproductive activity in the breeding and non-breeding season. The importance of inhibin and prolactin in the horse is as yet unclear.

The stallion's testis contains a higher concentration of oestrogens and oestrones (150–200 pg/ml) than testes of other mammals. The control and significance of such testicular oestrogen is unclear, and present evidence suggests that it is in fact non-bioactive (Raeside, 1969; Seamens et al., 1991).

Behavioural Changes

Testosterone drives male behaviour, especially that associated with reproduction. There is much variation between individuals, but in summary the following behaviour is controlled by testosterone (Pickett, Faulkner and Sutherland, 1975). On sight of a mare the frequency and amplitude of GnRH and hence LH and FSH release increases, driving testosterone production and hence producing the characteristic stallion behaviour (Irvine and Alexander, 1991). It is also thought that GnRH acts directly on higher brain centres in addition to the pituitary (Merchenthaler, Setalo, Csontos, Petruz, Flerko and Negro-Vilar, 1989).

As far as behavioural patterns are concerned, a stallion will fix his eyes upon the mare, draw himself up to his full height, arch his neck and stamp or paw the ground. He will often show the characteristic flehman behaviour of drawing back his top lip as if tasting the air and will roar. If the mare seems receptive, he will approach her from the front, muzzle to muzzle, and if there is still no hostility he will work his way over her neck, back and rump to her perineum and vulva. If the mare still stands with no objection he will turn and approach her from behind and to one side, nearly always the left-hand side. He will then mount her from that side, possibly after a few dummy mounts first to test her reaction and give him confidence that she will not object and lash out at him.

Ejaculation follows shortly after entry into the mare and is signalled by the rhythmical flagging of his tail. Ejaculation is normally in the form of 6–9 jets emptying the contents of the epididymis, vas deferens, accessory glands and finally urethra into the top of the vagina and bottom of the cervix. There is a decreasing volume of semen with each successive jet; 70% of sperm and the biochemical components of semen are deposited in the first 3 jets, the latter jets being richer in gel. Such mating behaviour is directly affected by circulating testosterone concentrations. At the beginning and the end of the season reaction time to a mare is longer and the number of mounts per ejaculate is greater, these being direct indications that when testosterone levels are declining sexual enthusiasm is also waning. Graphs 4.1–4.5 illustrate this effect.

Ejaculation is followed by a period of quiescence before the stallion dismounts the mare. This time allows the glans penis to return to its normal size prior to withdrawal.

Conclusion

The control of reproduction in the stallion is designed to ensure continual reproduction activity, not cyclic as in the mare. The only constraint on the stallion's reproductive activity is season, a limitation which ensures that offspring are born at a time of year most appropriate to their survival.

References and Further Reading

Amann, R.P. (1981a) A review of the anatomy and physiology of the stallion. J.Equine Vet.Sci., 1, 3, 83.

Amann, R.P. (1981b) Spermatogenesis in the stallion, A review. J.Equine Vet.Sci., 1,4, 131.

Argo, C.M., Cox, J.E. and Gray, J.L. (1991) Effect of oral melatonin treatment on the seasonal physiology of pony stallions. J.Reprod. Fert.Suppl., 44, 115.

Berndston, W.E. and Jones, L.S. (1989) Relationship of intratesticular testosterone content to age, spermatogenesis, Sertoli cell distribution and germ cell : sertoli cell ratio. J.Reprod.Fert., 85, 511.

Berndston, W.E., Pickett, B.W. and Nett, T.H. (1974) Reproductive physiology of the stallion and seasonal changes in the testosterone concentrations of peripheral plasma. J.Reprod.Fert.Suppl., 39, 115.

Blue, B.J., Pickett, B.W., Squires, E.L., McKinnon, A.O., Nett, T.M., Amann, R.P. and Shiner, K.A. (1991) Effect of pulsatile or continuous administration of GnRH on reproductive function of stallions. J.Reprod.Fert.Suppl., 44, 145.

Boyle, M.S., Skidmore, J., Zhang, J. and Cox, J.E. (1991) The effects of continuous treatment of stallions with high levels of a potent GnRH analogue. J.Reprod.Fert.Suppl., 44, 169.

Burns, P.J., Jaward, M.J., Edmunson, A., Cahill, C., Boucher, J.K., Wilson, E.A. and Douglas, R.H. (1982) Effect of increased photoperiod on hormone concentrations in Thoroughbred stallions. J.Reprod.Fert., 32, 103.

Clay, C.M., Squires, E.L., Amann, R.P. and Pickett, B.W. (1987) Influences of season and artificial photoperiod on stallions: testicular size, seminal characteristics and sexual behaviour. J.Anim.Sci., 64, 517.

Clay, C.M., Squires, E.L., Amann, R.P. and Nett, T.M. (1989) Influences of season and artificial photoperiod on stallion's pituitary and testicular responses to exogenous GnRH. J.Anim.Sci., 67, 763.

Cornwall, J.C. (1972) Seasonal variation in stallion's semen and puberty in the Quarter horse colt. MS Thesis, Louisiana State University.

Cox, J.E., Redhead, P.H. and Jawad, N.N.A. (1988) The effect of artificial photoperiod at the end of the breeding season on plasma testosterone concentrations in stallions. Aust.Vet.J., 65, 239.

Cox, J.E. and Williams J.H. (1975) Some aspects of the reproductive endocrinology of the stallion and cryptorchid. J.Reprod.Fert.Suppl., 23, 75.

Dowsett, K.F., Pattie, W.A., Knott, L.M., Jackson, A.E., Hoskinson, R.M., Rigby, R.P.G. and Moss, B.A. (1991) A preliminary study of immunological castration in colts. J.Reprod.Fert.Suppl., 44, 183.

Evans, M.J., Alexander, S.L., Irvine, C.H.G., Livesey, J.H. and Donald, R.A.S. (1991) In vitro and in vivo studies of equine prolactin secretion throughout the year. J.Reprod.Fert.Suppl., 44, 27.

Flink, G. (1988) Gonadotrophin secretion and its control, In: The physiology of reproduction, pp 1349–1377. Eds. E. Knobil and J.Neill. Raven Press, New York.

Ganjam, V.K. (1978) Episodic nature of the 4 and 5 steroidogenic pathways and their relation to adreno-gonadal steroidogenic axis in stallions. J.Reprod.Fert.Suppl., 27, 67.

Ganjam, V.K. and Kenney, R.M. (1975) Androgens and oestrogens in normal and cryptorchid stallions. J.Reprod.Fert.Suppl., 23, 67.

Gordon, I. (1983) Controlled breeding of farm animals. Pergamon Press, Oxford.

Irvine, C.H.G. (1984) Gonadotrophin releasing hormone. J.Equine Vet.Sci., 3, 4168.

Irvine, C.H.G. and Alexander, S.L. (1991) Effect of sexual arousal on gonadotrophin-releasing hormone, luteinising hormone and follicle stimulating hormone secretion in the stallion. J.Reprod.Fert.Suppl., 44, 135.

Irvine, C.H.G., Alexander, S.L. and Turner, J.E. (1986) Seasonal variation in the feedback of sex steroid hormones on serum LH concentration in the male horse. J.Reprod.Fert., 76, 221.

Johnson, L. (1991) Seasonal differences in equine spermatogenesis. Biol.Reprod., 44, 284.

Johnson, L. and Neaves, W.B. (1981) Age related changes in Leidig cell populations, seminiferous tubules and sperm production in stallions. Biol.Reprod., 24, 703.

Johnson, L. and Tatum, M.E. (1989) Temporal appearance of seasonal changes in numbers of Sertoli cells, Leidig cells and germ cells in stallions. Biol. Reprod., 40, 994.

Johnson, L., Varner, D.D. and Thompson, D.L. (1991) Effect of age and season on the establishment of spermatogenesis in the horse. J.Reprod.Fert.Suppl., 44, 87.

Lincoln, G.A. and Short, R.V. (1980) Seasonal breeding, Natures contraceptive? Recent progress Horm. Res. 36, 1.

Magistrini, M., Chanteloube, Ph. and Palmer, E. (1987) Influence of season and frequency of ejaculation on the production of stallion semen for freezing. J.Reprod.Fert.Suppl., 35, 127.

Merchenthaler, I., Setalo, G., Csontos, C., Petruz, P., Flerko, B. and Negro-Vilar, A. (1989) Combined retrograde tracing and immunocytochemical identification of luteinising hormone-releasing hormone and somatostatin containing neurons projecting into the median eminence of the rat. Endocrinology, 125, 2812.

Naden, J., Amann, R.R. and Squires, E.L. (1990) Testicular growth, hormone concentrations, seminal characteristics and sexual behaviour in stallions. J.Reprod.Fert., 88, 167.

Pickett, B.W., Faulkner, L.C. and Sutherland, T.N. (1970) Effect of month and stallion on seminal characteristics and sexual behaviour. J.Anim.Sci., 31, 713.

Pickett, B.W., Faulkner, L.C. and Voss, J.L. (1975) Effect of season on some characteristics of stallion semen. J.Reprod.Fert.Suppl., 23, 25.

Pickett, B.W. and Voss, J.L. (1972) Reproductive management of stallions. Proc. 18th. Ann.Conv.Am.Ass.Equine Practnrs., 501.

Raeside, J.I. (1969) The isolation of oestrone sulfate and oestradiol 17β sulfate from stallion testis. Can.J.Biochem., 47, 811.

Roser, J.F. and Hughes, J.P. (1991) Prolonged pulsatile administration of gonadotrophin-releasing hormone (GnRH) to fertile stallions. J.Reprod.Fert.Suppl., 44, 155.

Rossdale, P.D. and Ricketts, S.W. (1980) Equine Stud Farm Medicine. 2nd Ed. Ballière Tindall, London.

Seamens, M.C., Roser, J.F., Linford, R.L., Liu, I.K.M. and Hughes, J.P. (1991) Gonadotrophin and steroid concentrations in jugular and testicular venous plasma in stallions before and after GnRH injection. J.Reprod.Fert.Suppl., 44, 57.

Swierstra, E.E., Gebauer, M.R. and Pickett, B.W. (1974) Reproductive physiology of the stallion. I Spermatogenesis and testis composition. J.Reprod.Fert., 40, 113.

Thompson, D.L. and Honey, P.G. (1984) Active immunisation of prepubertal colts against estrogens, hormonal and testicular response after puberty. J.Anim.Sci., 51, 189.

Thompson, D.L.Jr., Pickett, B.W., Bernditsone, W.E., Voss, J.L. and Nett, T.H. (1977) Reproductive physiology of the stallion. VIII Artificial photoperiod, collection interval and seminal characteristics, sexual behaviour and concentration of LH and testosterone in serum. J.Anim.Sci., 44, 656.

Thompson, D.L.Jr., St George, R.L., Jones, L.S. and Garza, F.Jr. (1985) Patterns of secretion of LH, FSH and testosterone in stallions during the summer and winter. J.Anim.Sci., 60, 741.

Tucker, H.A. and Wetterman, R.P. (1976) Effects of ambient temperature and relative humidity on serum prolactin and growth hormone in heifers. Proc.Soc.Exp.Biol.Med., 151, 623.

CHAPTER 5

Anatomy and Physiology of Pregnancy in the Mare

T HE anatomy of pregnancy in the mare can be divided into four main sections for ease of consideration: fertilisation, early embryo development, placentation and organ development.

Fertilisation

The ovum released by the follicle is wafted down the fallopian tube towards the isthmus, where it meets the utero-tubular junction and awaits the arrival of the sperm. It is unable to pass through this junction until it has been fertilised.

The sperm, having been ejaculated into the cervix, make their way up through the uterus and uterine horns to the utero-tubular junction. They move by means of contractions of the female tract and the driving action of their own tails. It is thought that they are attracted towards the ovum by chemical attractants produced by the ovum awaiting fertilisation. On arrival at the utero-tubular junction they pass through to the fallopian tube and if the timing is correct meet a waiting ovum. As the sperm pass up through the female tract they come in contact with uterine secretions which induce a capacitation response in the sperm heads. A sperm has to be capacitated before it is capable of fertilising an ovum, as capacitation activates the enzymes in the acrosome region of the sperm head in readiness for penetration of the ovum.

Sperm, once in the vicinity of the ovum, stick to its outer gelatinous layer. This outer gelatinous layer replaces the corona radiata cells prior to ovulation. Figure 5.1 shows the ovum's structure prior to ovulation.

Sperm, by means of the whipping action of their tails, force themselves through the outer gelatinous layer to the zona pellucida. They then penetrate the zona pellucida using the enzyme acrosin released from the sperm head of capacitated sperm. This

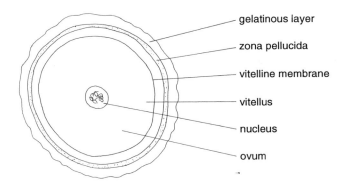

gelatinous layer

zona pellucida

vitelline membrane

vitellus

nucleus

ovum

Figure 5.1   The equine ovum prior to ovulation.

enzyme digests a pathway through the zona for the sperm. If an ovum is examined microscopically immediately after fertilisation, tracks through the zona pellucida of all the sperm attempting to fertilise the ovum can be identified.

As the head of the sperm meets the vitelline membrane of the ovum the two fuse. This fusion initiates the final meiotic division resulting in three polar bodies and the single ovum. The nuclei of the sperm and the egg (often termed the pronuclei) unite, their haploid complement (32) of chromosomes joining together to give the full diploid (64) of the new individual. All the characteristics of the new individual are dictated by this newly combined genetic material.

There is some variation in the reported length of time that the equine ovum remains viable, figures varying between 4 and 36 hours having been reported. After fertilisation the ovum is known to actively control, possibly via the localised secretion of oestrogens, its passage through the utero-tubular junction to the uterine horn, overtaking on its way any unfertilised ova from that or any previous ovulation (Betteridge and Mitchell, 1975; Onuma and Ohnami, 1975; Flood, Jong and Betteridge, 1979). Any ova not fertilised may take several months to degenerate.

In order to ensure the successful fusion of one male pronucleus and one female pronucleus it is essential to ensure that only one sperm penetrates the vitelline membrane of the ovum. Polyspermy (penetration by more than one sperm) is prevented by an instantaneous block set up as soon as one sperm touches the vitelline membrane. This instantaneous response involves a chemical reaction within the vitelline membrane forcing a gap between itself and the zona pellucida. No sperm can cross this gap and hence there is an instantaneous block to polyspermy.

Embryology

Twenty-four hours after mating, the fertilised ovum, now termed a zygote, has divided by mitosis (growth by cell division) into 2 cells. At this stage the outer gelatinous layer is lost and the fertilised ovum, still within the zona pellucida, continues to divide into 4, 16, 32 cells etc. At 4 days old it is a bundle of cells, still contained within its zona pellucida, but surrounded by a new outer smooth coat. It is now termed a morula (Figure 5.2).

At this stage the total volume and external size of the bundle of cells has not changed from the 2 cell zygote stage. The cytoplasm of

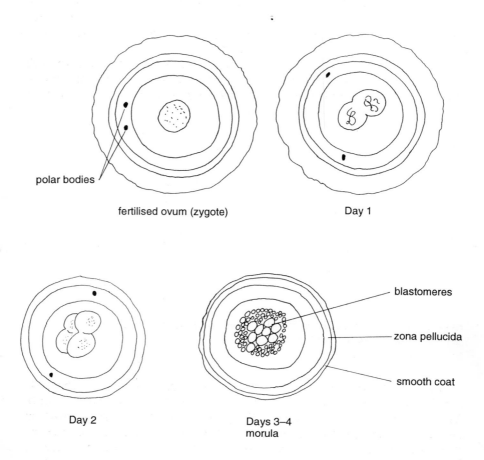

polar bodies

fertilised ovum (zygote)   Day 1

Day 2   Days 3–4
                        morula

blastomeres

zona pellucida

smooth coat

Figure 5.2   The developmental stages from fertilised ovum to morula in the equine conceptus, illustrating the loss of the gelatinous outer layer by Day 2 and the formation of a smooth coat by Day 4.

the original ovum has either been divided up between all the cells in the morula or used for energy. Nevertheless the amount of genetic material has dramatically increased, giving a full identical complement to all cells of the morula.

As the cells continue to divide, the morula makes its way by anticlockwise rotational swimming to the uterus. It arrives here at Day 6, when the total size of the morula starts to increase, this helps to force a break in the zona pellucida, through which it hatches, leaving it surrounded by an outer capsule. At this time it starts to derive nutrients for its growth and cell division from its surrounding uterine secretions, as by this stage it has used up all its own reserves. The provision of such additional nutrients allows a further increase in size. The morula is now free and floating in the uterus, deriving all its nutritional requirements from uterine histotroph (milk), a secretion designed to match exactly the requirements of the developing conceptus.

At Day 8 the cells of the morula become differentiated (organised) and three distinct areas can be identified: the embryonic mass (shield), the blastocoel and the trophoblast (Figure 5.3). The morula is now termed a blastocyst.

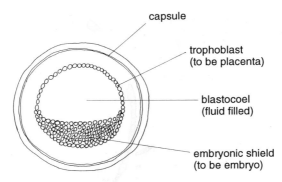

capsule

trophoblast
(to be placenta)

blastocoel
(fluid filled)

embryonic shield
(to be embryo)

Figure 5.3   The equine blastocyst at Day 8 post fertilisation, showing the differentiation of areas.

These three areas go to form the embryo proper (embryonic shield), the yolk sack (blastocoel) and the placenta (trophoblast). This cell differentiation marks the beginning of the switching on and off of various genes, cells then becoming destined to pursue set lines of development. Prior to this differentiation all cells in theory were capable if extracted from the morula of each developing into a new individual as none of its genes had been switched off. After differentiation this is no longer possible as certain cells have been given the message to pursue only set lines of development. The mechanism behind this switching on and off of genes and its trigger are unknown in the horse. It is important to note that at this differentiation stage the conceptus is very susceptible to external physical effects eg drugs, chemicals, disease etc. These can easily disrupt the differentiation process, resulting in deformities, abnormalities and a high risk of abortion.

From this stage on further differentiation takes place. Day 9 marks the differentiation of two germ layers (cell layers), the ectoderm consisting of the outer blastocyst cell layers and any zona pellucida remains and the endoderm consisting of the inner cell lining (Figure 5.4).

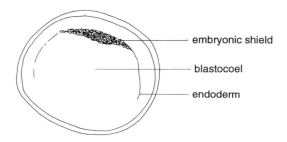

Figure 5.4   The equine conceptus at Day 9 post fertilisation illustrating the differentiation of the ectoderm and endoderm layers.

The endoderm grows and develops, working its way around the inside of the trophoblast to give a complete inner layer. This endoderm and the ectoderm together form the yolk sack wall and provide the means by which the embryonic shield receives its nourishment from the uterine secretions. The blastocoel, or fluid filled centre, is now termed the yolk sack (Figure 5.5).

At Day 14 when the embryo has reached 1.3 cm in diameter the mesoderm or third germ cell layer begins to develop. It becomes progressively evident between the ectoderm and endoderm, in the centre of the yolk sack wall, again working its way down from the embryonic shield to enclose the whole blastocyst. These three germ cell layers are the cell layers from which all placental and embryonic tissue development originates. In the case of the placenta the ectoderm forms the outer cell layers nearest the uterine epithelium, the mesoderm forms the blood vessels and nutrient transport system and the endoderm forms the inner cell lining that will become the allantoic sack (Figure 5.6).

At Day 16 folds appear in the outer cell layers and the beginnings

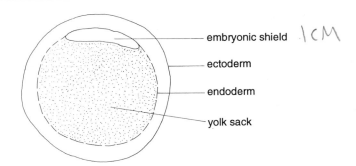

Figure 5.5 *The equine conceptus at Day 12 post fertilisation, illustrating the yolk sack, which at this stage provides the nutrients required by the developing conceptus.*

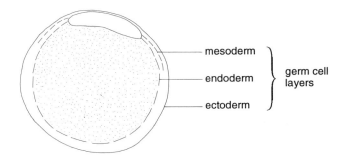

Figure 5.6 *The equine conceptus at Day 14 post fertilisation, illustrating the developing mesoderm which forms the blood vessels and nutrient transport system of the conceptus.*

of the protective layers surrounding the embryo become evident. The ectoderm folds over the top of the embryonic shield taking the mesoderm with it. The outer layer of these folds is now made up of the ectoderm plus a mesoderm layer and is termed the chorion. These two folds fuse, producing a fluid-filled protective space for the embryonic shield; this is the amniotic sack containing the amniotic fluid (Figure 5.7).

The amnion protects the embryo through the rest of its life in utero. Initially it is visible as a clear fluid-filled bubble surrounding the embryo. As pregnancy progresses it tends to collapse and lie close to the foetus. Throughout its life in utero the amniotic sack provides a clean environment in which the embryo can develop. The source of its surrounding amniotic fluid is not clear. However, its composition is very much like blood serum, and the exchange of

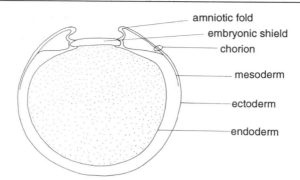

amniotic fold
embryonic shield
chorion

mesoderm

ectoderm

endoderm

Figure 5.7   The equine conceptus at Day 16 post fertilisation, illustrating the formation of the amniotic folds over the embryonic shield.

fluids between the amniotic sack and the kidneys, intestine and respiratory tract is known to occur. The foetus in later stages seems to breath in and swallow its surrounding amniotic fluid. The volume of amniotic fluid surrounding the foetus is about 0.4 litres at 100 days post fertilisation and increases to 3.5 litres at full term.

At Day 16 when the amnion is first evident the mesoderm has not yet spread to enclose the whole of the yolk sack. The area over which the mesoderm has spread, and which, therefore, has the three layers ectoderm, mesoderm and endoderm and which is nearest to the embryonic shield, is called the trilaminar omphalopleur. The area into which the mesoderm has not yet spread and, therefore, has only ectoderm and endoderm covering it is termed the bilaminar omphalopleur. The junction of these two areas, ie the line delineating the limit of mesoderm migration, is called the sinus terminalis (Figure 5.8).

From about Day 18 onwards embryology can be dealt with in two sections: placentation and organ development.

Placentation

The placenta has two major functions, first protection and secondly regulation of foetal environment, in the form of nutrient intake and waste output. The placenta develops from the extra embryonic membranes, the trophoblast of the blastocyst. The first source of nutrients and, therefore, a form of primitive placenta is the yolk sack or blastocoel. This provides both a temporary store of nutrients and a transport system for nutrients derived from uterine secretions to the embryonic disc.

Day 16 sees the first evidence of blood vessels developing in the

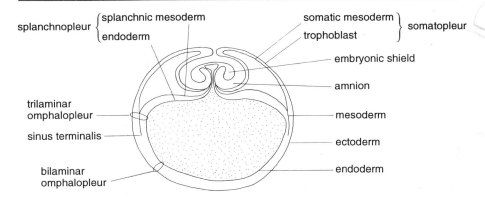

Figure 5.8   *The equine conceptus at Day 18 post fertilisation, illustrating the*
near-completion of the amniotic sack surrounding the embryonic shield.
The trilaminar omphalopleur is also shown, nearest the embryo,
consisting of the endoderm, mesoderm and ectoderm, and the
bilaminar omphalopleur, into which the mesoderm has not yet spread.

centre of the yolk sack wall in the mesoderm. This will become the
blood system of the placenta. By Day 18 the vitelline artery,
carrying blood towards the mother, and the vitelline vein, carrying
blood away from the mother, are identifiable.

On Day 20 an outpushing of the embryonic hindgut can be seen
immediately below the placenta. This is termed the allantois or
chorionic sack and continues to grow with the embryo. This sack is
filled with allantoic fluid which is partly secreted by the chorio-
allantoic membrane, lining the underside of the placenta, plus
urinary fluid from the foetus excreted from the foetal bladder via
the urachus within the umbilical cord (Figures 5.9 and 5.10).

The volume of the allantois at Day 45 is approximately 110 ml
increasing to 8.5 litres by Day 310, a considerably larger volume
than seen in the amniotic sack. The allantoic fluid increases in
volume as the foetus grows, producing more urinary fluid to be
stored. During the first trimester (three-four months) it is clear
yellow in colour, changing to brown/yellow as pregnancy develops.

This developing allantoic sack moves over the top of the embryo as
its contents increase, forcing the embryo down to the bottom of the
blastocyst, reducing as it goes the extent of the yolk sack, until the
yolk sack is hardly visible. As the allantoic sack increases in size the
umbilical cord becomes evident. The attachment point of the umbilical cord normally corresponds to the position of initial implantation.
It consists of two vitelline arteries, one vitelline vein and the urachus,
plus some supporting and connective tissue. The arteries and veins
are responsible for blood transfer from the placenta to the foetal

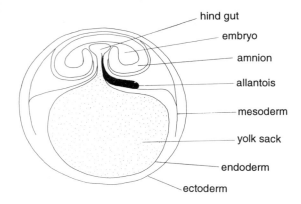

Figure 5.9   The development of the equine placenta at Day 20 post fertilisation.

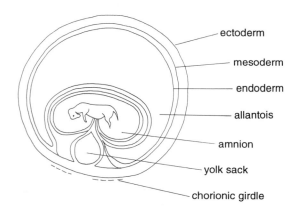

Figure 5.10   The development of the equine placenta at Day 40 post fertilisation.

system and the urachus transfers waste products from the bladder to the allantois; as such it extends no further than the allantois and does not reach the placenta (Figure 5.11).

As the foetus develops, its nutrient demand increases. The nutrients provided via the yolk sack are soon not enough to meet the demand; thus a more intimate relationship needs to develop between the mother and the embryo, and hence it begins to implant. This occurs as a gradual process from Day 20 onwards and until this point the yolk sack is fully functional.

The first identifiable attachment between mother and foetus occurs around Day 25, at the chorionic girdle (Figure 5.12). This is a temporary attachment and normally occurs at the junction between the uterine body and the uterine horn. The chorionic girdle

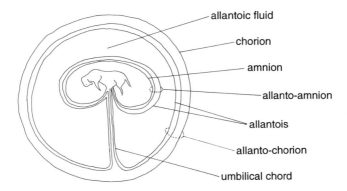

Figure 5.11   The placental arrangement of the equine conceptus near term.

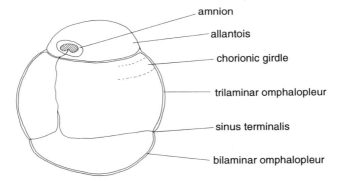

Figure 5.12   The development of the placenta of the equine conceptus at Day 25 post fertilisation, illustrating the position of the chorionic girdle attachment.

is a band of shallow folds encircling the chorio-allantoic membrane at a point near the junction between the uterine horn and uterine body. Cells within this girdle elongate and invade the uterine epithelium, engulfing some of the epithelial cells. These form a temporary attachment between foetus and mother at this localised area. At Day 38 the foetal cells then migrate into the maternal endometrium and detach from the allanto-chorionic. The conceptus is then released and can migrate within the uterus; however, it normally returns to the junction of the uterine horn and body for final implantation. The invading foetal cells form endometrial cups in a band around the inside of the uterus, at the junction of the body and the horn, at the original position of the chorionic girdle. These endometrial cups secrete pregnant mare serum gonadotrophin (PMSG), which is essential for the maintenance of pregnancy in the early stages. PMSG will be discussed in detail in Chapter 6.

Around Day 90 the endometrial cups begin to degenerate and slough off from the uterine endometrium. The reason for this seeming rejection is not fully understood, but may be a maternal rejection of the 'foreign' foetal tissue. The remains of these sloughed-off endometrial cups may be reabsorbed by the foetus during the remainder of the pregnancy or they may be seen in the placenta at birth as invaginations or pouches in the allanto-chorion.

As the endometrial cup attachment at the chorionic girdle is lost, the rest of the foetal allanto-chorion begins to attach to the uterine epithelium. This attachment becomes gradually firmer over the next 100 days, being fully attached by Day 150. At Day 45–70 the allanto-chorion takes on a velvety appearance created by fine microvilli all over its surface; hence the equine placenta is termed diffuse. These microvilli organise themselves into discrete microscopic bundles or tufts that invade into receiving invaginations in the uterine epithelium. These bundles of microvilli are termed microcotyledons, and their attachment develops over a period of time, being fully complete and functional by Day 150 (Figure 5.13).

Between the foetus and the mother is formed a strong fixed attachment consisting of six cell layers, three on the foetal side (endoderm, mesoderm and ectoderm) and three on the maternal side (epithelium, endometrium and blood vessel wall). The equine placenta is therefore termed epitheliochorial. The presence of microcotyledons increases the surface area of the placenta and also, therefore, the area for nutrient and gas exchange. Within each microcotyledon the maternal and foetal blood supply system come in very close proximity, allowing efficient diffusion.

However, the thickness of the placental attachment prevents the diffusion of any larger protein molecules across the placental barrier to the foetus. As immunoglobulins are large protein molecules, the attainment of passive immunity in the foal by diffusion across the placenta is thus limited. Passage of immunoglobulins via colostrum is, therefore, of utmost importance in the mare, as will be discussed in further detail in Chapter 15 and 16. The thickness of the placenta varies in other mammals, but in all cases, the thicker the placenta, the less efficient the transfer of passive immunity and hence the increased reliance upon colostrum. However, a thicker placenta as seen in the mare has the advantage of providing extra protection to the foetus from harmful maternal blood-borne factors. The whole of the allanto-chorionic surface is attached to the uterus except at the cervix, termed the cervical star, and at the entry to the fallopian tubes (Figure 5.14).

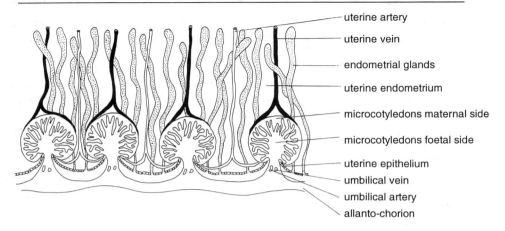

Figure 5.13   Equine placental microcotyledons in the fully developed placenta.

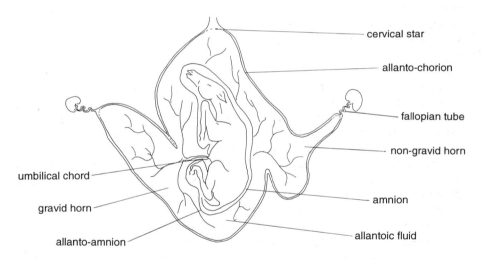

Figure 5.14   The equine foetus and placenta near term.

Placental Efficiency

Nutrient and gaseous exchange across the mare's placenta is relatively efficient when compared with other farm livestock, particularly the ewe. However, it must be remembered that measurements taken on placental efficiency involve the acute catheterisation of the umbilical arteries, and hence the technique used may affect the results obtained. Silver, Steven and Comline (1973) demonstrated the relative efficiency of the mare's placenta, as changes in the mare's blood oxygen and glucose concentrations were mimicked more closely by changes in the foetal blood concentration than in sheep.

Free fatty acids and lactate also follow the same pattern. It may well be deduced, therefore, that malnutrition in the mare will have a greater effect on the foetus than is evident in ruminants, though such an association has yet to be confirmed.

As pregnancy progresses the maternal epithelial stretches as the uterus increases in size. As a result the placenta tends to get thinner and hence the resistance to gas and nutrient exchange decreases, and the placenta becomes more efficient as the demands of the foetus increase. By full term the placenta may weigh in the region of 4 kg. Its surface area is approximately 14,000 cm² and it is in the order of 1 mm thick. The foal's birth weight is directly proportional to the surface area of the placenta, as this is the limiting factor controlling nutrient and gas exchange and hence their availability to the developing foetus. The surface area of a placenta may be restricted for several reasons. In the case of twins, the area of the uterus available for each placenta is restricted by the presence of the other foetus. If the division of the uterine surface area available to each twin is equal, then both twins have a chance of survival but their birth weights will be reduced. If the division is unequal, then the smaller one may well abort, but the placenta of the larger one cannot expand into the uterus surface originally occupied by the now dead foetus. At term, therefore, a single foal will be born but with a reduced birth weight due to the restriction of its placental size (Figure 5.15).

Placental Blood Supply

As mentioned previously the mesoderm of the blastocyst surrounding the yolk sack forms the first blood supply to the foetus. Early on, a clear network of blood vessels can be identified surrounding the trilaminar omphalopleur of the conceptus, with two major vessels connecting the network to the rudimentary heart. Additional pathways develop to feed areas of considerable growth. Hence when the yolk sack degenerates and the nutrient supply of the foetus is taken over by the allanto-chorionic placenta, a well-formed network of blood vessels lining the allanto-chorion is already present. This fine network enlarges and invaginates the microcotyledons of the placenta. Each microcotyledon is supplied on the foetal side by several arteries, but exit back to the foetal heart is via a single vein. This arrangement slows down the blood flow through the microcotyledons and encourages efficient diffusion and gas exchange. The oxygenated and nutritionally replenished

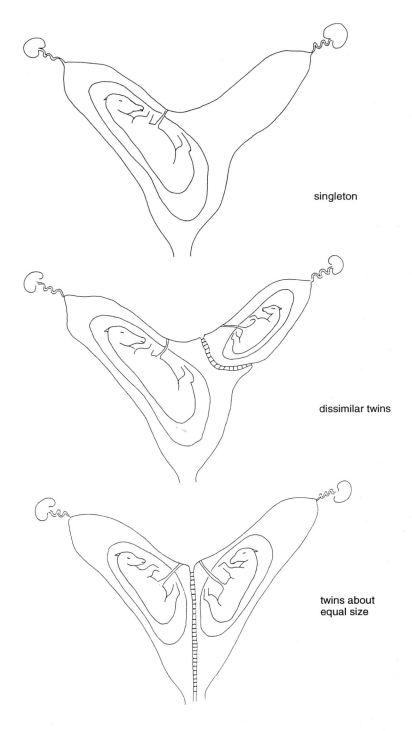

singleton

dissimilar twins

twins about
equal size

Figure 5.15   The three types of twin placentations seen most commonly in the equid.

73

blood returns to the foetal heart by the umbilical vein (Figure 5.16).

It should be remembered that in the foetus, because of the bypass of the non-functional foetal lungs, deoxygenated blood is carried to the placenta in arteries and oxygenated in the veins.

On the maternal side a very similar arrangement exists with the blood supply to the maternal side of the microcotyledons. Oxygenated and nutritionally enriched blood approaches the microcotyledons in a fine network of arteries, but the drainage back to the maternal system is also via a single vein, again slowing down the passage of blood and increasing the efficiency of nutrient and gas exchange. This transfer across the utero-foetal placental barrier can be compared in many ways to the gaseous exchange within the mammalian lung.

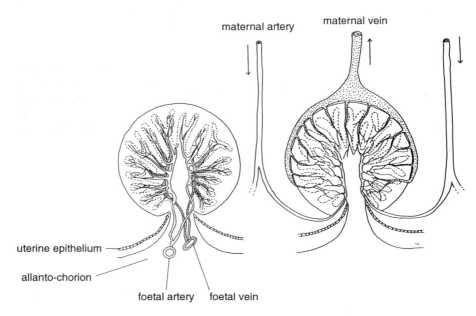

Figure 5.16   The venous and artery blood supply to the microcotyledons within the equine placenta, allowing the transfer of nutrients and waste products from the maternal (right) to the foetal (left) system and vice versa.

Organ Development

Organ development in the equid can be divided into two basic sections, gastrulation and neuralation. The former can be subdivided into segregation, delamination and involution.

Gastrulation

Gastrulation is defined as the organisation of the embryo into three germ layers: ectoderm, mesoderm and endoderm. The first stage of gastrulation is termed segregation, during which the central blastomeres or cells of the embryonic disc organise themselves into smaller outer cells and larger inner cells (Figure 5.17).

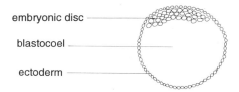

embryonic disc

blastocoel

ectoderm

Figure 5.17   Day 9 segregation in the equine conceptus, illustrating the larger inner blastomeres and smaller outer blastomeres within the embryonic disc.

The larger cells collect underneath the disc and migrate in two directions. Firstly they migrate to line the remaining ectoderm of the blastocyst and go to form the endoderm. Secondly they migrate within the embryonic disc, creating at Day 11 the first asymmetry within it, the thicker area at one end being termed the caudal and the thinner, the cranial end (Figure 5.18).

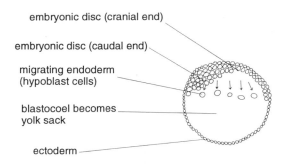

embryonic disc (cranial end)

embryonic disc (caudal end)

migrating endoderm
(hypoblast cells)

blastocoel becomes
yolk sack

ectoderm

Figure 5.18   Day 11 segregation in the equine conceptus, indicating the migration of the large hypoblast cells from the lower part of the embryonic disc to form the ectoderm.

The next stage of gastrulation is termed delamination. This commences at Day 12 and marks the first evidence of epiblast cells, hypoblast cells and the primitive gut (Figure 5.19). The epiblast cells are those of the embryonic disc, which is now termed the neural plate. The hypoblast cells are the migrating endoderm and

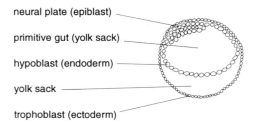

neural plate (epiblast)

primitive gut (yolk sack)

hypoblast (endoderm)

yolk sack

trophoblast (ectoderm)

Figure 5.19   Day 12 delamination in the equine conceptus, showing the future development of the endoderm layer.

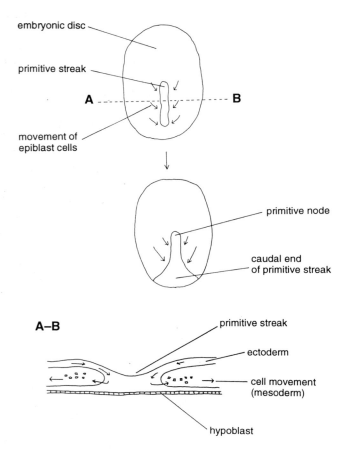

embryonic disc

primitive streak

A   B

movement of epiblast cells

primitive node

caudal end of primitive streak

A–B

primitive streak

ectoderm

cell movement (mesoderm)

hypoblast

Figure 5.20   Day 14 involution of the equine conceptus. A bird's-eye view of the embryonic shield along with a cross-section view through A–B, illustrating the passage of ectoderm cells through the primitive streak to reappear between the ectoderm and endoderm, forming the mesoderm. Further cell movement results in the flattening of the caudal end of the primitive streak.

76

within this ring of hypoblast is the yolk sack or primitive gut. At Day 14 a change in the neural plate becomes evident and this is the beginning of the primitive streak formed within the epiblast cells. At this stage it is about 1 cm in length.

Involution is the third stage of gastrulation, when the epiblast cells move inwards to the centre of the caudal end of the disc (Figure 5.20). These moving epiblast or ectoderm cells reappear as mesodermal cells between the ectoderm and the hypoblast or endoderm. As the cells move through to the lower level they leave a depression in the upper surface of the epiblast. These migrating epiblast cells tend to move in greater concentrations at the caudal end of the primitive streak, making it wider. The primitive streak makes the future longitudinal axis of the embryo.

At Day 15 epiblast cell movement tends to slow down and the slight indentation along the longitudinal axis of the primitive streak becomes deeper as cells continue to move out from underneath to form the mesoderm and are not replaced by migrating epiblast cells above. This deep groove is now termed the primitive groove. The cells associated with the primitive groove are termed node cells to differentiate them from the cells of the remainder of the blastocyst. At Day 15 these node cells can be identified as precursors of future body organs. The ectoderm node cells form the neural plate running the length of the top of the primitive groove, the cranial end of which goes to form the head. The spreading mesoderm in the immediate vicinity of the neural plate goes to form the somites or body trunk and the mesoderm immediately below the primitive groove goes to form the notochord (spine and central nervous system). Finally the wide caudal end forms the tail end of the foetus (Figure 5.21).

The process of gastrulation is now completed, the major cell blocks are identifiable and the longitudinal axis of the embryo is determined.

Neuralation

The next stage, termed neuralation, involves the development of the central nervous system (CNS), gut and heart. Day 16 sees three major changes. Firstly the ectoderm near the neural plate thickens and two neural folds develop either side of the neural plate. The neural plate becomes depressed and the neural folds fold over, join and then fuse to enclose a hollow tube, the spine and CNS to be (Figure 5.22).

Secondly, the mesoderm either side of the neural plate organises itself into 14 somites. Thirdly at the cranial (cephalic or head) end of the neural plate an increase in cell growth above the surface becomes apparent, with an accompanying increase in the length of

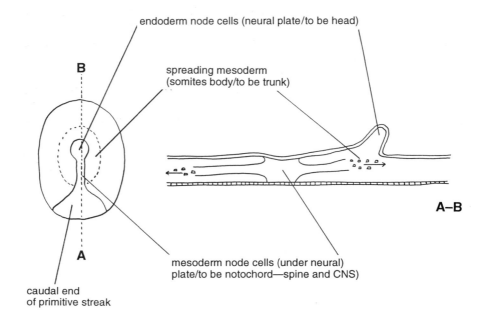

endoderm node cells (neural plate/to be head)

B

spreading mesoderm
(somites body/to be trunk)

A–B

A

mesoderm node cells (under neural)
plate/to be notochord—spine and CNS)

caudal end
of primitive streak

*Figure 5.21   Day 15 completed gastrulation in the equine conceptus. A bird's-eye
view and cross-section view through A–B of the embryonic shield. The
formation of the head from the cranial end of the neural plate is
illustrated, along with the somites or body trunk formation from the
mesoderm in the immediate vicinity of the neural plate and the spine
and CNS formation from the mesoderm immediately below the
primitive groove.*

the neural plate. This is the first step in the formation of the
embryo body. This cell growth folds over to form the head process,
heart and pharynx.

By Day 18 lateral folds are beginning to develop either side of the
head process. As cells move into this area and cell division
increases, the cranial end of the neural plate lifts away from the
remaining tissue (Figures 5.23 and 5.24).

These lateral folds move down from the cranial end to the caudal
end, lifting the whole body away from the underlying tissue (Figure
5.24).

This lifting away from the remaining tissue leaves just one
attachment point in the centre, the first evidence of the umbilical
cord. The embryo continues to lift off the underlying tissue and the
head and tail processes fold back down to give the embryo its
characteristic C shape configuration. At this stage two more
somites are evident, making 16 in total.

The gut tube also begins to develop from the pharynx fold by closure of the endoderm folds, in a way similar to that by which the neural tube was formed from folds in the ectoderm. The hind gut of the foetus extends out into the blastocoel to form the allantois as illustrated in Figures 5.9 and 5.10.

The embryo now lies away from the underlying tissue, the placental tissue, and is connected directly to the mother only by the umbilical cord, which contains a blood system derived from the mesoderm along with supporting connective tissue. The embryo now has an identifiable neural tube, the forerunner of the CNS, and a head process with enlarged neural tube, the brain to be. Its

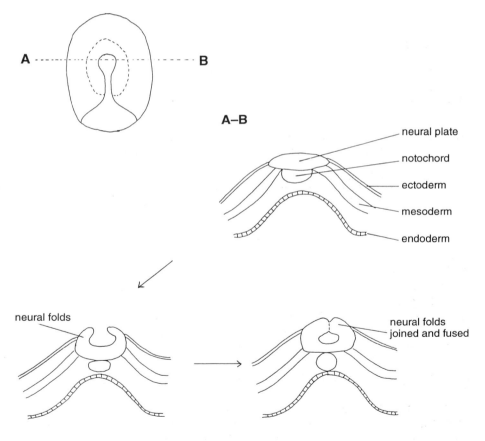

Figure 5.22   Day 16–17 neuralation of the equine conceptus. The ectoderm near the neural plate thickens and two neural folds develop either side of the neural plate and join to enclose a hollow tube, the spine and CNS to be.

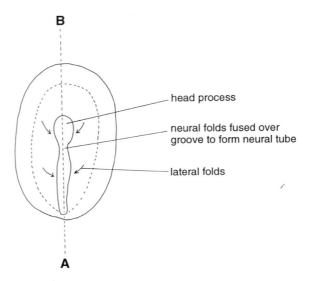

Figure 5.23   *Day 18 neuralation of the equine conceptus. A bird's-eye view of the embryonic shield illustrating cell movement in towards the cranial end of the neural plate, which subsequently lifts away from the underlying tissue.*

Figure 5.24   *A cross-section view (A–B) of Figure 5.23, illustrating the gradual appearance of the head and tail processes during neuralation in the 19-day-old equine conceptus. The embryo now begins to take up the characteristic C shape.*

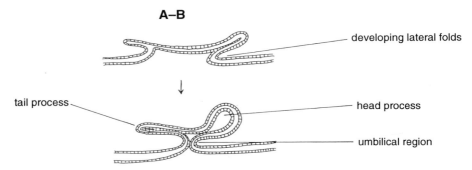

pharynx and gut tube are also present, as are the somites, or body muscle blocks. Therefore, by Day 23 all the basic bodily structures are evident though only in a rudimentary state.

Fine Differentiation and Growth

From Day 23 onwards development is in the form of fine differentiation and organ growth. By Day 40 all the main body features are evident, eg limbs, tail, nostrils, pigmented eyes, ears, elbow and stifle regions, eyelids, etc, and the embryo is now termed a foetus. Day 39–45 heralds sexual differentiation and evidence of external genitalia. The weight of the foetal gonads reaches a maximum at Day 180–200 (Douglas and Ginther, 1975), the weight of foetal testes and ovaries being equivalent and developing to the following pattern:

Day 80   1.4 g
Day 140   18.7 g
Day 200   48.0 g
Day 320   31.4 g

The increase and decrease in size is due to a proliferation and degeneration of interstitial cells (Walt, Stabenfeldt, Hughes, Neely and Bradbury, 1979).

At this stage most of the development is complete and an increase in growth now occurs (Graph 5.1). At Day 60 the eyelids close and finer eye development occurs, teats are present and the palate fused. Day 160 sees the first evidence of hair, around the

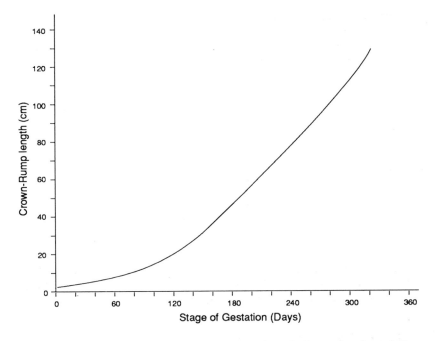

Graph 5.1   The increase in foetal crown-rump length throughout gestation (Roberts, 1971; Evans and Sack 1973; Ginther, 1992).

Table 5.1   Summary of foetal development

Day of gestation	Major milestones
1	Zygote, 2 cells
4	Morula, 16 cells plus
6	Hatching of morula
8	Blastocyst, differentiated into embryonic mass, blastocoel and trophoblast
9	Ectoderm and endoderm germ layers evident, gastrulation begins
11	Segregation giving first embryonic asymmetry, caudal and cranial ends evident
12	Delamination epiblast cells, hypoblast cells and primitive gut evident
14	Mesoderm evident, primitive streak appearing, involution commencing
15	Primitive streak now evident as a groove
16	Neuralation starts, folds leading to the formation of the amnion seen, first blood vessels evident in mesoderm, chorionic vesicle round (2–4 cm)
18	Vitelline artery and vein identifiable, foetus begins to take on characteristic C shape
20	Allantois forming from outpushings of the foetal hind gut, chorionic vesicle oval in shape (2.5–4.5 cm), eye vesicle and ear present
21	Amnion complete
23	All basic bodily structures evident though in rudimentary state
26	Forelimb bud seen, three branchial arches present, eye visible
30	Genital tubercle present, eye lens seen
36	Rudimentary 3 digits seen on hoof, facial clefts closing, eyes pigmented and acoustic groove forming
40	Endometrial cups forming, ear forming, nostrils seen, eyelids seen, all limbs evident and elbow and stifle joint areas identifiable, chorionic vesicle oval in shape (4.5–7.5 cm)
42	Ear triangle in shape, mammary buds along ridge seen
45	External genitalia evident, allantoic sac volume 110 ml
47	Palate fused
49	Mammary teats evident
55	Ear covers acoustic groove, eyelids closing
60	Chorionic vesicle 13.3 × 8.9 cm
63	Eyelids fused, fine eye development occurring, hoof, sole and frog areas of hoof evident
75	Female clitoris prominent
80	Scrotum clearly seen
90	Endometrial cups degenerate, chorionic vesicle 14 × 23 cm
95	Hoof appears yellow in colour
112	Tactile hairs on lips growing
120	Fine hair on muzzle, chin and eyelashes beginning to grow, eye prominent and ergot evident
150	Full attachment of placental microcotyledons, eyelashes clearly seen, enlargement of mammary gland forming
180	Mane and tail evident
240	Hair of poll, ears, chin, muzzle and throat evident
270	Whole of body covered with fine hair, longer mane and tail hair clearly seen
310	Allantoic sack volume 8.5 litres
320	Testis may drop from this time onwards
320–340	Birth of fully developed foetus

eyes and muzzle, and by Day 222 hair has begun to develop at the tip of the tail and the beginnings of a mane are evident (Colour plate 16). By Day 270 hair covers the whole of the body surface.

From Day 150 onwards the hippomane, an accumulation of waste minerals within the allantois, becames apparent. The hippomane increases in size as pregnancy develops (Colour plates 16–19).

From Day 320 onwards the testis in the male foetus may drop through the inguinal canal; however, this does not occur in all colt foetuses as some drop neonatally.

The main milestones in equine foetal development are summarised in Table 5.1.

Full term, normally at 320 days in ponies and up to two weeks later in Thoroughbred and riding type horses, heralds the birth of a very well-developed foetus capable of all basic bodily functions including walking within 15 minutes of birth. Details of the foal's adaptation to the extrauterine environment are given in Chapter 15.

Conclusion

Our understanding of embryo development specific to the equid is largely incomplete. Continued widening of our knowledge is essential if we are to understand and hence minimise embryo mortality, a significant cause of apparent infertility in the mare. As the environmental factors affecting embryo mortality are understood, so our management of the equid can be geared to minimising such losses.

References and Further Reading

Allen, W.R. (1974) Palpable development of the conceptus and fetus in Welsh pony mares. Equine Vet.J., 6, 69.

Allen, W.R. and Moor, R.M. (1972) The origin of the equine endometrial cups. 1. Production of PMSG by fetal trophoblast cells. J.Reprod.Fert., 29, 313.

Amoroso, E.C. (1952) Placentation. In: Marshall's Physiology of Reproduction, Ed. A.S. Parkes, Vol. 2 p.127. Longmans Green, London.

Betteridge, K.J., Eaglesome, M.D. and Flood, P.F. (1979) Embryo transport through the mare's oviduct depends upon cleavage and is independent of the ipsilateral corpus luteum. J.Reprod.Fert.Suppl., 27, 387.

Betteridge, K.J. and Mitchell, D. (1975) A surgical technique applied to the study of tubal eggs in the mare. J.Reprod.Fert.Suppl., 23, 519.

Bruck, I. and Hyland, J.H. (1991) Measurements of glucose metabolism in single equine embryos during early development. J.Reprod.Fert.Suppl., 44, 419.

Clegg, M.T., Boda, J.M. and Cole, H.H. (1954) The endometrial cups and allantochorionic pouches in the mare with emphasis on the source of equine gonadotrophin. Endocrinology, 54, 448.

Comline, R.S. and Silver, M. (1970) pO_2, pCO_2 and pH levels in the umbilical and uterine blood of the mare and ewe. J.Physiol.Lond., 209, 587.

Douglas, R.H. and Ginther, O.J. (1975) Development of the equine fetus and placenta. J.Reprod.Fert.Suppl., 23, 503.

Enders, A.C. and Lui, I.K.M. (1991) Lodgement of the equine blastocyst in the uterus from fixation through endometrial cup formation. J.Reprod.Fert.Suppl., 44, 427.

Evans, H.E. and Sack, W.O. (1973) Prenatal development of domestic and laboratory mammals. Growth curves, external features and selected references. Anat.Histol.Embryol., 2, 11.

Flood, P.F., Betteridge, K.J. and Irvine, D.S. (1979) Oestrogens and androgens in blastocoelic fluid, and cultures of cells from equine conceptuses of 10–22 days gestation. J.Reprod.Fert.Suppl., 27, 413.

Flood, P.F., Jong, A. and Betteridge, K.J. (1979) The location of eggs retained in the oviducts of mares. J.Reprod.Fert., 57, 291.

Flood, P.F. and Marrable, A.W. (1975) A histochemical study of steroid metabolism in the equine fetus and placenta. J.Reprod.Fert.Suppl., 23, 495.

Ginther, O.J. (1971) Maintenance of the corpus luteum in hysterectomised mares. Am.J.Vet.Res., 32, 1687.

Ginther, O.J. (1992) Reproductive Biology of the Mare, Basic and Applied Aspects. 2nd Ed. Equiservices, 4343, Garfoot Road, Cross Plains, Wisconsin 53528, USA.

Gunzel, A. and Merkt, H. (1979) Oestrus and fertility following progestagen treatment of mares showing clinical evidence of early pregnancy failure. J.Reprod.Fert.Suppl., 27, 453.

Hawkins, D.L., Neely, D.P. and Stabenfeldt, G.H. (1979) Plasma progesterone concentrations derived from the administration of exogenous progesterone to ovarectomised mares. J.Reprod.Fert.Suppl., 27, 211.

Holtan, D.W., Nett, T.M. and Estergreen, V.L. (1975) Plasma progestagens in pregnant mares. J.Reprod.Fert.Suppl., 23, 419.

Holtan, D.W., Squires, E.L., Lapin, D.R. and Ginther, O.J. (1979) Effect of ovariectomy on pregnancy in mares. J.Reprod.Fert.Suppl., 27, 457.

Onuma, and Ohnami, Y. (1975) Retention of tubal eggs in mares. J.Reprod.Fert.Suppl., 23, 507.

Oriol, J.G., Donaldson, W.L., Dougherty, D.A. and Antczak, D.F. (1991) Molecules of the early equine trophoblast. J.Reprod.Fert.Suppl., 44, 455.

Palmer, E., Bezard, J., Magistrini, M. and Duchamp, G. (1991) In vitro fertilisation in the horse. A retrospective study. J.Reprod.Fert.Suppl., 44, 375.

Pruitt, J.A., Forrest, D.W., Burghardt, R.C., Evans, J.W. and Kraemer, D.C. (1991) Viability and ultrastructure of equine embryos following culture in a static or dynamic system. J.Reprod.Fert.Suppl., 44, 405.

Roberts, S.J. (1971) Veterinary Obstetrics and Genital Disease. Edwards, Ann Arbor, Mich.

Short, R. (1969) Implantation and maternal recognition of pregnancy. In: Ciba Foundation Symposium on Foetal Autonomy. Ed. G.E.W. Wolstenholme and Maeve O'Connor, pp. 2–31. Churchill, London.

Silver, M., Steven, D.H. and Comline, R.S. (1973) Placental exchange and morphology in ruminants and mares. In: Foetal and Neonatal Physiology. Ed. R.S. Comline, K.W. Cross, G.S. Davies and P.W. Nathanielsz, p.245. Cambridge Univ. Press, Cambridge.

Van Niekerk, C.H. and Allen, W.E. (1975) Early embryonic development in the horse. J.Reprod.Fert.Suppl., 23, 495.

Walt, M.L., Stabenfeldt, G.H., Hughes, J.P., Neely, D.P. and Bradbury, R. (1979) Development of the equine ovary and ovulation fossa. J.Reprod.Fert.Suppl., 27, 471.

CHAPTER 6

Endocrine Control of Pregnancy in the Mare

WHEN examining the endocrinological control of pregnancy in the mare, gestation can be divided into two stages, early (fertilisation to Day 150) and late (Day 150 to full term).

Early Pregnancy

By Day 5 the conceptus has migrated to the uterus and exists by deriving nutrients from the uterine hystotroph or secretions. No major changes from the non-pregnant cycle are evident as yet. However, by Day 15 a message has to be received by the reproductive system of the mare if it is to continue in a pregnancy mode, blocking the drop in progesterone, and hence allowing progesterone levels to remain elevated as essential for the initial maintenance of pregnancy. If the mare is pregnant then ovarian progesterone production, from a corpus luteum (CL), has to be maintained until at least Day 75. If there is a failure of the functional corpus luteum during this time then the mare will abort.

The importance of this decision day, Day 15, is seen in experiments with early pregnant mares. If the embryo is removed from a pregnant mare prior to Day 15 then she will return to oestrus at her normal time (21 days after the last). If, however, the embryo is removed at Day 16 or later, then the mare will not return to oestrus as expected and will show a prolonged dioestrus, due to the persistence of the corpus luteum. The length of the delay will depend to a certain extent on the age of the embryo at removal.

Experiments show quite clearly that 'D' day as far as the hormonal control of pregnancy is concerned is Day 15 (Hersham and Douglas, 1979). The exact nature of the message informing the mare of the presence of a conceptus is unclear, but in some mammals there is evidence that the message is via oestrogens produced by the very young conceptus. These act locally on the uterine epithelium to inform the mother of the presence of a foetus. This is the likely mechanism in the mare, as it has been demonstrated that equine conceptuses are capable of synthesising oestrogens as early as Day 12 (Heap, Hamon and Allen, 1982). The concentration of oestrogens increases but does not become evident in the dam's circulation until Day 70 post coitum (Flood, Butteridge and Irvine, 1979; Stewart, Stabenfeldt, Hughes and Meagher, 1982; Daels, Stabenfeldt, Hughes, Odensvik and Kindahl, 1990). By whatever means the message is delivered, the result is the maintenance of the CL beyond Day 15 (Ball, Altschul, McDowell, Ignotz and Currie, 1991). In the non-pregnant mare the CL is knocked out at about Day 15 by prostaglandin F2α (PGF2α), allowing the cyclical changes associated with oestrus and ovulation to begin. Therefore, in the pregnant mare this action of PGF2α must be blocked.

Again, the mechanism is unknown but there are several hypotheses. Firstly, it has been suggested that PGF2α binding to the CL is reduced. However, doubt has been placed on this hypothesis as it has been reported that the concentration of PGF2α receptors on CL is high during the period Day 16–18 post ovulation. The second hypothesis is that the secretion of PGF2α is reduced, or thirdly that an alternative component is produced by the uterus which competitively binds with the PGF2α receptors on the CL. A candidate for such a competitor is prostaglandin E (PGE), very similar in structure to PGF2α but inactive (Allen, 1970; Heap, 1972; Vernon, Strauss, Simonelli, Zavy and Sharp, 1979).

From Day 15 onwards maternal progesterone and foetal oestrogens are the dominant hormones in the system and are very important in the production and composition of uterine hystotroph and pregnancy-specific proteins, collectively termed uterine milk. The composition of uterine milk in the mare is very important, as the conceptus survives in a free-living form for a relatively long period of time, implantation not occurring until Day 38 onwards.

If we consider the hormonal changes individually, then we can build up a picture of how the mare maintains a developing conceptus.

Progesterone

Between days 6 and 14 the concentration of progesterone is 8–15 ng/ml, similar to that seen during dioestrus of the non-pregnant oestrus cycle. Day 15 is decision time, after which the concentrations of progesterone in pregnant and non-pregnant mares diverge. Progesterone levels in pregnant mares decline slightly after Day 16 but not to the extent seen in non-pregnant mares. Levels continue to decline slowly to reach concentrations of approximately 6 ng/ml by Day 30. Levels subsequently rise again to reach 8–10 ng/ml by Day 45–55 and remain at this level or possibly fall slightly until Day 150 (Holton, Nett and Estergreen, 1975). Experiments in the mare indicate that ovarian progesterone, ie that produced by a CL, is essential for the maintenance of all pregnancies at least until Day 75 and in some cases up to Day 150. After Day 150 placental progesterone is adequate to take over and maintain a pregnancy. All pregnant mares ovariectomised (the ovaries and hence the functional CL are removed) before Day 75 will abort. If mares are ovariectomised in the period Day 75–150, different reports show differing abortion rates. After Day 150 ovariectomy has no effect, with all mares successfully carrying foetuses to full term (Holtan, Squires, Lapin and Ginther, 1979). This demonstrates that ovarian progesterone is essential prior to Day 75 in all mares; in the period Day 75–150 placental progesterone gradually takes over from the ovarian progesterone and by Day 150 ovarian progesterone is not needed. Indeed at this time the CL on the ovary can be seen to have regressed.

Ovarian progesterone is not secreted continually by a single CL. In the mare an increase in ovarian activity is evident between Days 20 and 30 post coitum: follicles develop, driven by follicle stimulating hormone (FSH) surges similar to those seen during dioestrus. A dominant follicle becomes apparent and ovulates between Days 40 and 60, forming a secondary CL. This secondary CL is unusual and relatively unique to the mare. During the period Day 40–70 the secondary CL gradually takes over the production of progesterone; the primary CL regresses and becomes non–functional between Days 75 and 150. From Day 70 onwards the placenta begins to produce progesterone, gradually taking over the function of the secondary CL (Allen and Hadley, 1974; Evans and Irvine, 1975; Squires and Ginther, 1975).

Pregnant Mare Serum Gonadotrophin

As discussed in Chapter 5, Day 40 marks the appearance of endometrial cups, which are responsible for the secretion of pregnant mare serum gonadotrophin (PMSG). PMSG is secreted into the mare's circulatory system and reaches a maximum concentration between Days 50 and 70 post coitum. It is secreted by the foetal tissue within the endometrial cups and the maximum concentrations achieved vary considerably, different reports giving levels of between 10 and 100 iu/ml (Allen and Moor, 1972). Levels are known to be affected by genotype; maximum levels reached and the duration of these high levels are greater in mares mated to close relatives, eg brother to sister matings (Stewart, Allen and Moor, 1977). The parous state of the mare also has an effect on PMSG levels, multiparous mares having lower levels than first-timers. Levels always tail off and normally reach basal levels by Day 100–120 (Allen, 1969a and 1969b).

The importance of PMSG and the reason why it is only secreted for a short period of pregnancy are unclear; however, several hypotheses suggest its involvement in the prevention of foetal immunological rejection by the mother. An immunological response is also implicated in the reduction in PMSG secretion. It seems that the PMSG-secreting foetal sides of the endometrial cups are slowly rejected by the maternal system and as a result PMSG secretion declines until Day 100 when all cups have regressed. This hypothesis of an immunological rejection of PMSG-secreting foetal cells is supported in work done on sister-brother matings. The foetus from such a mating has a relatively similar genetic make-up to its mother; in addition PMSG is secreted much longer than in non-related matings and endometrial cups also regress much later, presumably because the relatively similar genetic make-up of the foetus does not elicit the same rejection response as in non-related matings.

The exact role of PMSG is also uncertain, but recent evidence indicates that it may be involved in follicular development in readiness for the formation of the secondary CL. PMSG is known to have FSH-like properties and also some LH-like properties. As such it is used as a superovulating agent in farm livestock other than the horse. It does not seem to be solely responsible for driving follicular development in pregnant mares but works as a synergist with FSH surges. PMSG is also thought to act as a luteinising agent in the formation of the secondary CL and it may be implicated as an agent responsible in part for the maintenance of the primary CL (Allen, 1975; Nett, Holtan and Estergreen, 1975).

It has also been suggested that PMSG is involved in preventing maternal rejection of the developing conceptus. The uterus is a privileged site, the only place in the body that, under the influence of the hormone changes associated with pregnancy, will tolerate a foreign body, the conceptus. The genetic make-up of the conceptus is only half that of its mother and, therefore, is foreign and under normal conditions such an invasion would set up an immunological response and the foreign body (the conceptus) would be rejected. For some reason this rejection is prevented in the pregnant mare and all other mammals that carry foetuses in utero. It seems that some sort of immunological barrier is set up between the foetus and its mother blocking the immunological response. PMSG may be involved in the prevention of rejection from Day 40 onwards. However, as it is not secreted prior to Day 40, some other component must be responsible during early pregnancy.

Equine Chorionic Gonadotrophin

It has been suggested that equine chorionic gonadotrophin (eCG) may be involved in the maintenance of early pregnancy. ECG is secreted by the trophoblast cells and can in some mammals be isolated, forming a barrier between the foetal and maternal tissue in early pregnancy. It may also be involved as a luteotrophic agent actively maintaining the CL of early pregnancy. In the mare this immunological barrier is then possibly taken over by PMSG from Day 40 onwards, though it is reported that eCG continues to be produced and may then act primarily as a luteotrophic agent (Daels, De Moraes, Stabenfeldt, Hughes and Lasley, 1991).

Maternal Oestrogens

The concentration of maternal oestrogens also varies within pregnancy. Between Days 0 and 35 levels remain very similar to those seen during the non-pregnant dioestrus period, but then rise sharply to reach 3–5 ng/ml by Day 40. In the period Day 40–45 they decline slightly and subsequently remain at this constant level until Day 60–70, after which they slowly rise again (Darenius, Kindahl and Madej, 1988; Stabenfeldt, Daels, Munro, Kindahl, Hughes and Lasley, 1991).

These rising levels of maternal oestrogens between Days 35 and 40 are thought to be secreted by the developing follicles prior to the formation of the secondary CL in much the same way as oestrogens are produced by developing follicles prior to ovulation and oestrus in the normal oestrus cycle (Daels et al., 1991). These rising oestrogens, however, do not result in oestrous behaviour. Evidence for the ovarian origin of these oestrogens is their absence in pregnant mares after ovariectomy prior to Day 40 and the delayed decline in concentrations seen after foetal death not accompanied by immediate CL regression (Daels, et al., 1991; Stabenfeldt, et al., 1991). However, the second rise in oestrogen at Day 60–70 is not affected by ovariectomy but is affected by induced or spontaneous foetal death (Darenius et al., 1988) and can, therefore, be assumed to originate from the foeto-placental unit. The precursors for such production are foetal gonad in origin. By Day 85 oestrogens in the mare's peripheral blood system are higher than those detected in non-pregnant mares, and the level can be used as a pregnancy test. The continuing rise in oestrogens after Day 80 is due to increased foeto-placental production of equilin and equilenin, two oestrogens unique to the pregnant mare.

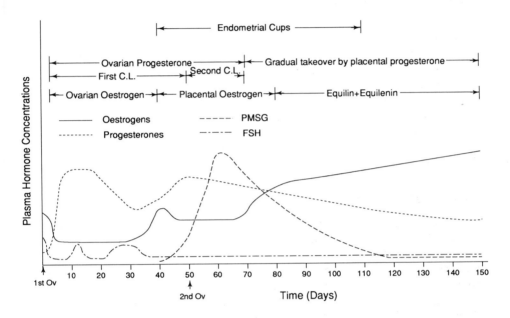

Graph 6.1   A summary of the plasma hormone concentrations during early pregnancy in the mare (Day 0–150).

Late Pregnancy

As far as hormone control is concerned, late pregnancy can be classified as Day 150 onwards.

Progesterone

Progesterone levels, which prior to Day 150 were elevated but slowly declining, were once thought to remain at a steady 1–3 ng/ml until 20–30 days prior to parturition and then rise to a peak at parturition and subsequently fall off dramatically to levels less than 1 ng/ml within a few hours of delivery. However, this rise seems in fact to be caused by three 5α pregnane metabolities of progesterone rather than progesterone per se. These progesterone metabolities have a very similar structure to progesterone but are inactive. As such, high levels of pregnanes cause the reproductive system to assume low levels of progesterone as the pregnane molecules attach to the progesterone receptors (Barnes, Nathanielsz, Rossdale, Comline and Silver, 1975; Moss, Estergreen, Becker and Grant, 1979; Holtan, Houghton, Silver, Fowdey, Ousey and Rossdale, 1991).

Oestrogens

Oestrogens within the maternal system continue to rise in late pregnancy reaching a peak, at Days 210–280, of approximately 8 ng/ml, the two main oestrogens being the equine-specific equilin and equilenin. Oestrogen levels then decline as parturition approaches, reaching the order of 2 ng/ml at parturition.

It is interesting to note that initially the hormone changes during late pregnancy in the mare, high progesterone and low oestrogen, were seemingly opposite to those detected in other mammals. The significance of this and possible explanations are detailed in Chapter 8 (Nett, Holtan and Estergreen, 1973; Nett, Holtan and Estergreen, 1975; Holtan and Silver, 1992).

Prostaglandin

During the major part of late pregnancy, prostaglandin remains at low levels equivalent to those seen during early pregnancy. How-

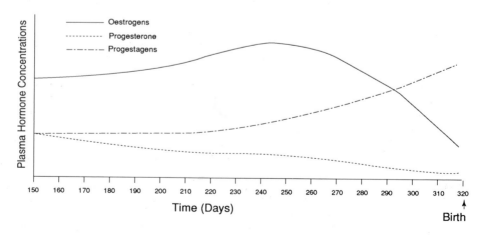

Graph 6.2   A summary of the plasma hormone concentrations in the mare during late pregnancy (Day 150–parturition).

ever, near to term levels do increase slightly, but significant elevations in their concentration are not detected until labour has started, when they play a vital role in myometrial contraction (Barnes, Comline, Jeffcote, Mitchell, Rossdale and Silver, 1978).

Conclusion

It is evident that the control of pregnancy in the mare follows a similar pattern to that seen in other mammals, with the exception of progesterone and oestrogen concentrations in late pregnancy. The significance of this apparent departure from the norm, if indeed it is present, is yet to be elucidated.

References and Further Reading

Allen, W.R. (1969a) A quantitative immunological assay for pregnant mare serum gonadotrophin. J.Endocr., 43, 581.

Allen, W.R. (1969b) The immunological measurement of pregnant mare serum gonadotrophin. J.Endocr., 43, 593.

Allen, W.R. (1970) Endocrinology of early pregnancy in the mare. Equine Vet.J., 2, 64.

Allen, W.R. (1975) Ovarian changes during early pregnancy in pony mares in relation to PMSG production. J.Reprod.Fert.Suppl., 23, 425.

Allen, W.R. and Hadley, J.C. (1974) Blood progesterone concentrations in pregnant and non-pregnant mares. Equine Vet.J., 6, 87.

Allen, W.R. and Moor, R.M. (1972) The origin of the equine endometrial cup. 1. Production of PMSG by foetal trophoblastic cells. J.Reprod.Fert., 29, 313.

Amoroso, E.C. (1955) Endocrinology of pregnancy. Br.Med.Bull., 11, 117.

Ball, B.A., Altschul, M., McDowell, K.J. Ignotz, G. and Currie, W.B. (1991) Trophoblastic vesicles and maternal recognition of pregnancy in mares. J.Reprod.Fert.Suppl., 44, 445.

Barnes, R.J., Comline, R.S., Jeffcote, L.B., Mitchell, M.D., Rossdale, P.D. and Silver, M. (1978). Fetal and maternal concentrations of 13, 14-dihydro-15-oxo-prostaglandin F in the mare during late pregnancy and at parturition. J.Endocr., 78, 2, 201.

Barnes, R.J., Nathanielsz, P.W., Rossdale, P.D., Comline, R.S. and Silver, M. (1975) Plasma progestagens and oestrogens in fetus and mother in late pregnancy. J.Reprod.Fert.Suppl., 23, 617.

Daels, P.F., Jorge De Moraes, M., Stabenfeldt, G.H., Hughes, J.P. and Lasley, B. (1991) The corpus luteum a major source of oestrogen in early pregnancy in the mare. J.Reprod.Fert.Suppl., 44, 502.

Daels, P.F., Stabenfeldt, G.H., Hughes, J.P., Odensvik, K. and Kindahl, H. (1990) The source of oestrogen in early pregnancy in the mare. J.Reprod.Fert. 90, 55.

Darenius, K., Kindahl, H. and Madej, A. (1988) Clinical and endocrine studies in mares with a known history of repeated conceptus losses. Theriogenology, 29, 1215.

Evans, M.J. and Irvine, C.H.G. (1975) Serum concentrations of FSH, LH and progesterone during the oestrus cycle and early pregnancy in the mare. J.Reprod.Fert.Suppl., 23, 193.

Flood, P.F., Betteridge, K.J. and Irvine D.S. (1979) Oestrogens and androgens in blastocoelic fluid and cultures of cells from equine conceptuses of 10–22 days gestation. J.Reprod.Fert.Suppl., 27, 413.

Hamilton, W.J. and Day, F.T. (1945) Cleavage stages of the ova of the horse, with notes on ovulation. J.Cleavage., 79, 127.

Heap, R.B. (1972) Role of hormones in pregnancy. In: Reproduction in Mammals. 3. Hormones in Reproduction, Ed. C.R. Austin and R.V. Short, p.73. Cambridge University Press, Cambridge.

Heap, R.B., Hamon, M., Allen, W.R. (1982) Studies on oestrogen synthesis by pre-implantation equine conceptus. J.Reprod.Fert., 32, 343.

Hersham, L. and Douglas, R.H. (1979) The critical period for maternal recognition of pregnancy in mares. J.Reprod.Fert.Suppl., 27, 395.

Holtan, D.W., Houghton, E., Silver, M., Fowden, A.L., Ousey, J. and Possdale, P.D. (1991) Plasma progestagens in the mare, fetus and new born foal. J.Reprod.Fert.Suppl., 44, 517.

Holtan, D.W., Nett, T.M. and Estergreen, V.L. (1975) Plasma progestagens in pregnant mares. J.Reprod.Fert.Suppl., 23, 419.

Holtan, D.W. and Silver, M. (1992) Readiness for birth; another piece of the puzzle. Equine Vet.J., 24, 5, 336.

Holtan, D.W., Squires, E.L., Lapin, D.R. and Ginther, O.J. (1979) Effects of ovariectomy on pregnancy in the mare. J.Reprod.Fert.Suppl., 27, 457.

Moss, G.E., Estergreen, V.L., Becker, S.R. and Grant, B.D. (1979) The source of 5αpregnanes that occur during gestation in mares. J.Reprod. Fert.Suppl., 27, 511.

Nett, T.M., Holtan, D.W. and Estergreen, V.L. (1973) Plasma estrogens in pregnant mares. J.Anim.Sci., 37, 962.

Nett, T.M., Holtan, D.W. and Estergreen, V.L. (1975) Oestrogens, LH, PMSG and prolactin in serum of pregnant mares. J.Reprod.Fert.Suppl., 23, 457.

Spincemaille, J., Bouters, R., Vandeplassche, M. and Bonte, P. (1975) Some aspects of endometrial cup formation and PMSG production. J.Reprod.Fert.Suppl., 23, 415.

Squires, E.L. and Ginther, O.J. (1975) Follicular and luteal development in pregnant mares. J.Reprod.Fert.Suppl., 23, 429.

Stabenfeldt, G.H., Daels, P.F., Munro, C.J., Kindahl, H., Hughes, J.P. and Lasley, B. (1991) An oestrogen conjugate enzyme immunoassay for monitoring pregnancy in the mare: Limitations of the assay between Days 40 and 70 of gestation. J.Reprod.Fert.Suppl., 44, 37.

Stewart, D.R., Stabenfeldt, G.H., Hughes, J.P. and Meagher, D.M. (1982) Determination of the source of elaxir. Biol.Reprod., 27, 17.

Stewart, F., Allen, W.R. and Moor, R.M. (1977) Influence of foetal genotype on FSH:LH ratio of PMSG. J.Endocr., 73, 415.

Stewart, F. and Maher, J.K. (1991) Analysis of horse and donkey gonadotrophin genes using southern blotting and DNA hybridization techniques. J.Reprod.Fert.Suppl., 44, 19.

Vernon, M.W., Strauss, S., Simonelli, M., Zavy, M.T. and Sharp, D.C. (1979) Specific PGF2α binding by the corpus luteum of the pregnant and non-pregnant mare. J.Reprod.Fert.Suppl., 27, 421.

CHAPTER 7

Physical Process of Parturition

PARTURITION is the active expulsion of the foetus along with its associated fluid and placental membranes. The length of gestation on average in the mare is 320–335 days, tending to be up to 2 weeks shorter in ponies than in Thoroughbreds and horses of the larger riding type.

There are several signs that indicate parturition is approaching. These may become evident at any time in the last few weeks of pregnancy. It must be remembered that the signs which will be detailed below should not be used in isolation and that there is much variation between individuals and between successive pregnancies. Therefore, when watching for imminent parturition, a combination of the following signs should be looked for. It is also useful to have information to hand on a mare's previous pregnancies as general behavioural patterns may be characteristic to a particular mare.

Signs of Imminent Parturition

Changes in the appearance of the udder are one of the first signs of parturition. During the last month as lactogenesis (milk production) gets underway, the udder size increases as colostrum is produced and stored. The udder also may feel relatively warm to the touch. This temperature increase is due to the associated increase in cell metabolism. At this time the udder may seem to increase in size at night, especially if the mare is kept in, and decrease during the day when she is turned out and able to move about. When there is no apparent change between day and night in udder size, parturition is imminent, as at this stage the udder is so full of milk that exercise does not affect its size. How much the udder increases in size is obviously dependent upon the size of the mare and to a certain extent on whether or not she has foaled previously. Colour

96

plate 20 (facing page 100) shows the udder of a Welsh Cob Section D mare 5 days prior to parturition.

The teats also start to change. Initially they become shorter and fatter as the udder becomes full and the bases of the teats are stretched. As the time for parturition becomes imminent the teats fill as milk produced in the udder increases; they then elongate and become tender to touch. Some mares may even start to milk themselves as the milk production by the udder gets too great for its storage capacity. If a mare does start to milk herself it is very important to minimise any milk loss. The milk at this stage is in fact colostrum, high in antibodies responsible for passive immunity development in the foal, and as there is a finite amount of colostrum, if a mare is seen to milk herself, or habitually does so, it is a good idea to milk her out a bit and store the collected colostrum for feeding to the foal immediately post partum. Colostrum can be very successfully frozen for use at a later date.

Many mares 'wax up', a term given to the clotting of colostrum at the end of the teat (Colour plate 21). This is a good sign of imminent parturition. However, the lack of wax is not indicative that parturition is a long time off, as these plugs of colostrum can easily be dislodged, especially in active mares turned out during the day.

Changes in the birth canal also become apparent as parturition approaches. About 3 weeks prior to parturition a hollowness may appear either side of the tail root as the muscles between the hip and the buttock relax. The whole pelvic area may seem to sink as the muscles and ligaments relax to allow expansion of the birth canal for passage of the foal. If the area either side of the tail root is felt daily in the last three to four weeks of pregnancy it may be possible to detect a change as the muscle tone relaxes (Colour plate 22). As parturition becomes more imminent the cervix begins to dilate, especially during first stage labour.

Changes in the mare's abdomen may also be detected in late pregnancy. As the foetus increases in size the abdomen expands correspondingly, becoming large and pendulous. However, in the final stages of pregnancy the abdomen appears to shrink as the foal moves up into the birth passage ready for delivery (Colour plate 23).

As parturition gets closer still, the mare becomes restless and agitated, especially as she enters first stage labour. Some restlessness may also be apparent in late gestation; and naturally, at this stage, the mare will move to the periphery of the herd in readiness to move away completely for labour itself. During first stage labour she may show signs very similar to those indicative of colic, eg walking in circles, swishing her tail, looking around at her sides, kicking her abdomen etc. If a mare does show signs of colic in late gestation it is pertinent to consider that it may in fact be first stage labour, and so, her eating, drinking and defecating should be checked. As first stage labour progresses she may well sweat profusely.

As discussed, not all mares show all these symptoms but a combination of one or two will give an accurate prediction that foaling is imminent. Some commercial products have been produced to aid in the diagnosis. These make use of some of the mare's natural signals, reacting to movement of the mare, temperature increase or an increase in sweating. They normally produce a signal transmitted to an audio receiver. Closed circuit television is also quite popular to observe late pregnant mares prior to expected parturition.

Process of Parturition

Parturition in most mammals involves three distinct stages. The first stage involves the positioning of the foal and preparation of the internal structures for delivery. The second stage is the actual birth of the foal and the third stage is the expulsion of the afterbirth. All three stages involve considerable myometrial activity, mainly within the uterus itself, but with some involvement of the abdominal muscles as an extra force.

First Stage of Labour

This stage involves uterine myometrial contractions and lasts between 1–4 hours, positioning the foal ready for birth. Figure 7.1 illustrates the forces involved in this stage.

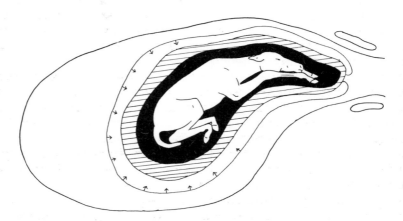

Figure 7.1   The forces involved in the first stage of labour are provided by contractions of the uterine myometrium, as indicated by the arrows.

The uterine muscles contract in mild waves from the uterine horn towards the cervix. These contractions, helped by the movement of the mare and, to a certain extent, those of the foal, result in the repositioning of the foal and passage into the birth canal. Throughout late pregnancy the foal lies in a ventral position (its vertebrae lying along the line of the mother's abdomen) with its forelimbs flexed. During first stage labour it rotates into a dorsal position with its forelimbs, head and neck fully extended, becoming engaged in the birth canal (Figure 7.2). This movement occurs normally three to four hours prior to the commencement of labour, or right at the beginning of labour. In some cases a mare may show signs of first stage labour and then cools off only to show further signs several hours later.

If parturition is to continue other changes in addition to the engaging of the foal also occur. The cervix gradually dilates. This is encouraged during the later part of first stage labour by the pressure of the allanto-chorionic membrane and the foal's forelimbs against the utero side of the cervix. During the birth of a dead foetus, dilation of the cervix does not occur to such an extent nor does it proceed so rapidly. It is, therefore, thought that dilation of the cervix is actively encouraged by the movement of the foal.

The vulva continues to relax and vaginal secretions increase (Colour plate 24). The exact mechanism by which parturition myometrial activity is induced and controlled is unclear. However, active movement of the foal against the cervix and within the birth canal has been shown to increase PGFM, and hence by inference PGF2α, levels. Such an increase in PGF2α is not seen during the birth of a dead foal.

The cervix contains a high concentration of collagen. The relative amount of collagen to muscle fibres progressively increases from the uterine horn through the uterus to the cervix. It is thought that PGF2α affects this collagen, causing it to relax. A similar effect has been shown in the ewe.

At the end of first stage labour the foetus's forelegs and muzzle push their way through the dilating cervix taking with it the allanto-chorionic membrane. The placental membrane at the cervix, termed the cervical star, is one of the three small sites of the placenta that have no microcotyledons and, therefore, no attachment to the mother. The other two are at the entry to each fallopian tube. The cervical star is the thinnest area of the placenta and hence at the end of the first stage of labour it ruptures as the pressure of the myometrial contractions against the placental fluids increases, forcing them and the foetus through the cervix. The subsequent release of the allantoic fluid (breaking of the waters) is the signal for the beginning of the second stage of labour.

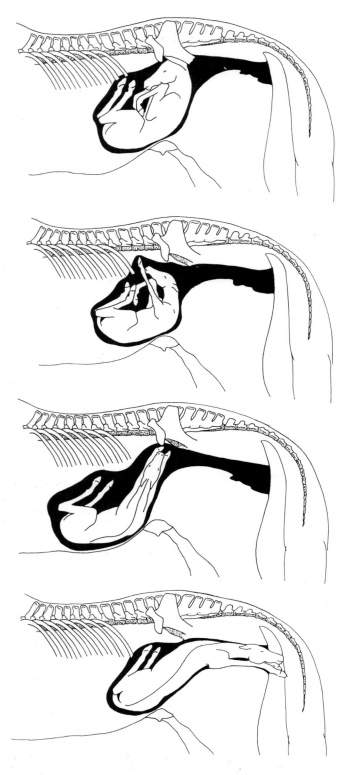

Parturition

▲
20 The udder of a Section D mare 5
    days prior to parturition.

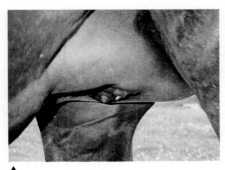

▲
21 One of the signs of imminent
    parturition is the accumulation
    of dried colostrum on the teats
    of the mare, termed waxing up.

22 A further sign of imminent
    parturition is a hollowing of the
    back above the pelvis as a
    result of a relaxation of the birth
    canal.

▼

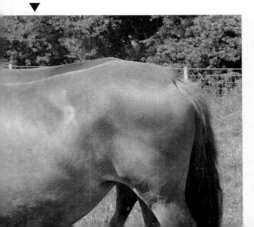

23 One of the obvious signs of
    pregnancy is the increase in
    size of the abdomen.
    Immediately prior to parturition
    the size of the abdomen appears
    to shrink as the foal moves up
    into the birth canal.

▼

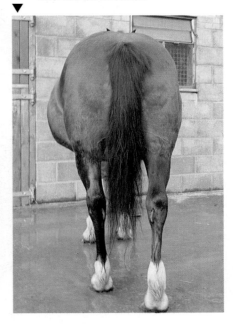

24 During first stage labour
    continued relaxation of the
    vulva is seen along with an
    increase in vaginal secretions.

▼

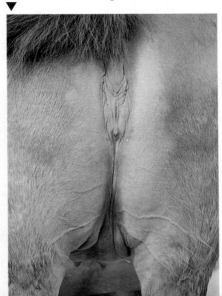

▲
25 *The angle of the birth canal ensures that the foal is delivered in a curved manner, being expelled down towards the mare's hind legs.*

▲
26 *In order to reduce the diameter of the foal's thorax passing through the birth canal, the foal is delivered with one foot slightly in advance of the other.*

27 *The end of the second stage of labour is marked by the foal lying with its hind legs still within the vulva of its mother.*
▼

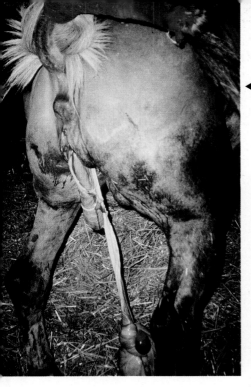

28 *The uterine contractions of third stage labour plus vasoconstriction of the placental blood vessels result in the expulsion of the placenta.*

29 *The amniotic sack, the white membrane in the background, and the placenta proper, in the foreground, after expulsion during the third stage of labour. As can be seen the placenta is expelled inside out, the red velvety outer allanto-chorion being inside and the smooth inner allanto-chorion being outermost.*

30 *A close up of the placenta in its inside-out state, illustrating the innermost red allanto-chorion and the outermost smooth white allanto-chorion.*

▼

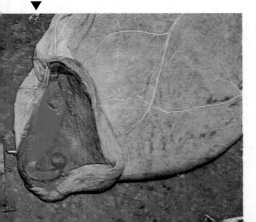

▲

31 *The hippomane often found within the placenta of the mare. (This example is 11 cm wide.)*

Foaling

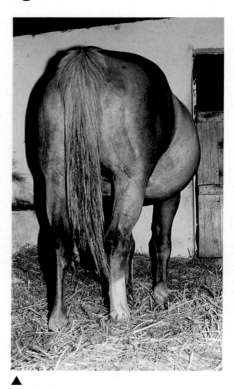

▲
32 Foaling mares outside is becoming increasingly more popular, especially with native type mares and multiparous mares foaling later on in the season.

▲

33 The first signs of labour in the mare are periods of restlessness in which she stands up and lies down, grimaces, looks at her flanks, digs up her bedding etc.

34 During second stage labour the mare invariably takes up a recumbent position, the most efficient for straining.
▼

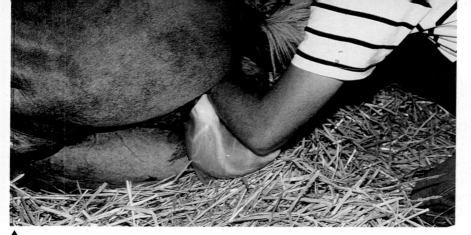

▲
35 At the start of the second stage of labour a brief internal examination may be made to ascertain whether or not the foal is lying in the correct position.

36 The allantoic sack should be seen as a white membrane protruding from the mare's vulva.

▶

37 If all is well the mare should be left to foal unaided and observed from a discreet distance.

▼

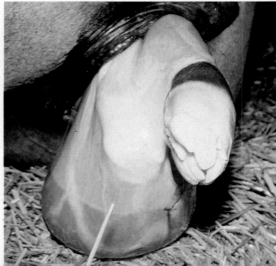

◄ **38** Gentle traction may be used to aid a mare in the final stages of delivery, especially if she is showing signs of exhaustion.

39 The umbilical cord should be left to break naturally. This normally occurs as the foal makes its first attempts to rise to its feet. ►

40 The foal may be dried off immediately after delivery. At the same time the heart rate of the foal can be checked by placing a hand on the foal's thorax.

▼

41 *The amnion may be tied up to prevent the mare standing on it and ripping it out prematurely.*

42 *The mare should be encouraged to remain recumbent for a while after birth in order to reduce the inspiration of air, and, therefore, bacteria in through the still relaxed vulva.*

43 *As soon as it has been expelled the placenta should be examined for completeness.* ▶

◀ **44** *During and after third stage labour the mare and foal should be left undisturbed to start the process of bonding. There is no need to drag the foal around to the head of the mare, as in time she will invariably reach around herself and move towards it without assistance.*

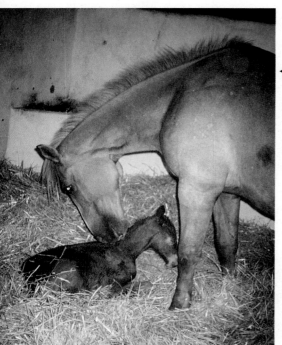

45 *Luminal cysts within the uterus are generally fluid filled and of about 3 cm in diameter. They do not normally cause excessive problems unless evident in large numbers, in which case they may interfere with implantation.*

▼

Second Stage of Labour

The release of allantoic fluid at the start of stage two lubricates the vagina and is thought to induce the stronger uterine contractions of the second stage. These strong contractions continue until the birth of the foal. Second stage labour involves stronger contractions of the uterine myometrium, supplemented by abdominal muscle contractions. The cause of the induction of these stronger contractions is again unclear but it is associated with the release of allantoic fluid. Oxytocin is also thought to be involved primarily with the second stage of labour, blood concentrations being significantly higher than during first stage.

The supplementary force provided by the abdominal muscles is termed voluntary straining. During voluntary straining the mare breaths in deeply holding the rib cage and diaphragm at maximum inspiration. As a result the rib cage and abdominal muscles are induced to contract against the pressure, and this pressure is transferred to the uterine contents, adding extra impetus to the expulsion of the uterine contents (Figure 7.3).

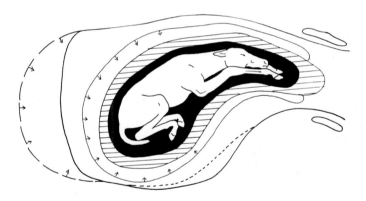

Figure 7.3   Second stage labour involves stronger contractions of the uterine myometrium, supplemented by contractions of the abdominal muscles, as indicated by the arrows.

As discussed previously, at the end of first stage labour the foal lies in a dorsal position with extended forelimbs within the birth canal. The shape of the birth canal is governed by the soft tissue and bones surrounding it. The pelvis contains the sides and bottom

Figure 7.2   During the first stage of labour the foal is gradually rotated and positioned within the birth canal in readiness for expulsion during the second stage contractions.

of the canal and the sacral and coccygeal vertebrae the top. The diameter of the entry into the birth canal (20–24 cm) is slightly larger than the exit diameter (15–20 cm), and slightly higher, nearer the mare's vertebrae than the exit. Hence, the foal is funnelled through the birth canal in a curved manner, being expelled down towards the mare's hind legs (Colour plate 25).

The foal is delivered forelegs first followed by the head lying extended between the forelegs parallel with the knees. The two forelegs are not delivered aligned: normally one leg is slightly in advance of the other, the fetlock of one being in line with the other hoof (Colour plate 26). This lack of alignment of the forelegs reduces the cross-section diameter of the foal's thorax, which is the widest part of the foal and any reduction of which reduces foal trauma at birth and eases foaling. Once the thorax and shoulders have passed through the birth canal the remainder of the process is relatively easy. At the end of stage two labour the foal lies with its head near the mare's back legs, with its own hind limbs still within the mare (Colour plate 27).

Third Stage of Labour

Stage three of labour is normally completed within three hours of the end of stage two. Uterine contractions continue at a level similar to that evident during stage one labour, again originating at the uterine horns and passing down in waves to the cervix (Colour plate 28). At the same time the placental membranes begin to shrink, as the blood vessels vasoconstrict and draw away from the uterine endometrium. This releases the remaining attachments between the allanto-chorion and the uterine epithelium and forces the placenta to be expelled inside out. The placenta is delivered with its red velvety outer allanto-chorion inside and the smooth inner allanto-chorion outermost (Colour plates 29 and 30). The contractions of third stage labour also expel any remaining fluids and help in uterine involution (recovery).

Passed out either during second stage labour or along with the placenta during third stage labour is a small, brown, often leathery, structure termed the hippomane (Colour plate 31). It is found within the allantoic fluid and is an accumulation of waste salts and minerals collected throughout pregnancy. It has a high concentration of calcium, hexosamine, magnesium, nitrogen, phosphorous and potassium, and is first evident at around Day 85 of pregnancy. It has been associated with much folklore, including attributed aphrodisiac properties and responsibility for keeping the foal's mouth open. There is, however, no evidence to support these claims.

Foaling Abnormalities

The vast majority of foalings, up to 90%, go according to plan and require no outside interference from man. Occasionally, however, things do go wrong and it is as well to be prepared for and have an understanding of such eventualities. In such cases prompt action can often save the life of both foal and mare. Causes of foaling problems can be divided into three main categories: foetal dystocia, maternal dystocia and abnormal maternal conditions.

Dystocia can be classified as a stress or complication associated with parturition that prevents a natural birth. It can be divided into foetal dystocia as a result of foetal complications or maternal dystocia as a result of maternal problems. Dystocia often causes a delay in parturition, which may result in a reduction in the oxygen intake by the foal due to either partial placental breakdown or constriction of the umbilical blood supply. This may result in the birth of a weak foal requiring careful intensive post natal care or more permanent problems, including brain damage in the more extreme cases. Delay in parturition due to dystocia may result in two other complications. The uterus may close around the foetus as it dries out and the lubricating effect of the allantoic fluid may diminish. As a result manipulation of the foal becomes increasingly difficult. Further drying out of the tract due to a delay hinders passage through the birth canal and thus makes any successful manipulation harder.

If dystocia does occur there are three main actions that can be taken. These are manipulation or mutation of the foal, involving manual correction of the foetal position within the uterus allowing the foal to be pulled out. Pulling a foal or traction normally follows manipulation, as in such cases the mare is often too exhausted to push the foal out herself without help. Another action that may be taken, especially in cases where extensive manipulation would be needed, is a caesarean. There is a higher risk to both mare and foal in a caesarean as it must be conducted under some form of anaesthetic. The mortality rates for foals born by caesarean are approximately 10%. The mortality rates for mares delivering by caesarean are, however, lower than this, especially if the mare is operated upon before labour has progressed too far.

A caesarean involves a ventral midline incision if the mare is in a recumbent position or a flank incision if she is standing. Apart from the mortality risk, mares that have undergone caesareans have the added risk of developing adhesions. Adhesions is the term given to the adhering of internal structures due to their drying out when exposed to air during the operation. If the adhered structures include parts of the reproductive tract, there is a risk of inter-ference with their function and therefore a detrimental effect on the mare's subsequent reproductive performance. Infection is another potential problem, though with the advances in modern medicine this is nowhere near the risk it used to be.

The final and most drastic alternative in response to severe cases of dystocia is fetotomy. This is the dividing up of the foal's body in utero and is performed on foals that have died in utero during parturition or late pregnancy. A caesarean is a possibility in such cases but with it runs a higher mortality risk to the mare. Therefore, if the foal is already known to be dead, in utero fetotomy is the better option.

Foetal Dystocia

Foetal dystocia is caused almost exclusively by malpresentations in utero that make normal unaided birth impossible. The foal in utero may be found in one of many positions, and the ease of correction and likely outcome depend on its position and the skill of the person manipulating it. Some of the more common conditions and their treatment are detailed below.

Forward presentations

One or more forelegs can be flexed or folded under the foal whose head is in the normal position within the birth canal. As discussed previously, the widest cross-section presented is the thorax area; therefore, in this position it includes the flexed forelegs, making passage impossible (Figure 7.4).

The head of the foal may be flexed back. In this position the nose of the foal will not be felt in the birth canal. This again presents a much wider maximum cross-section than the mare can deliver naturally (Figure 7.5).

The foal's legs may be more misaligned than the normal hoof to fetlock alignment. One leg may become caught on the pelvic brim at the point of the elbow, preventing the foal's delivery (Figure 7.6).

The foal's forelegs may lie over its head and be lodged behind its ears. This presents a cross-section too large for natural passage (Figure 7.7).

All these four positions can be corrected reasonably easily by pushing the foal back into the mare and manipulating it so it presents a normal position, followed by traction to aid the mare in its delivery. It must be remembered that traction has to be in a curved manner to ease the foal along the curved passageway dictated by the pelvic anatomy.

Other more complicated positions are seen which need veterinary

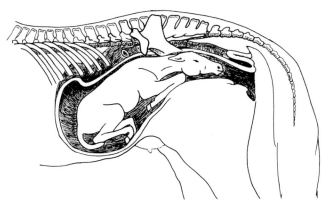

Figure 7.4   *Flexion of one or both forelegs at delivery significantly increases the cross-section across the foal's thorax and, therefore, makes the passage of the foal very difficult.*

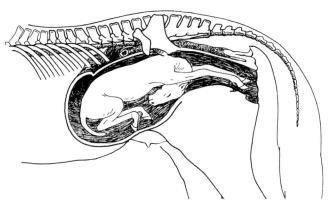

Figure 7.5   *If only the forefeet are presented in the birth canal the head may be flexed back, again presenting such a wide cross-section of foal that it is very difficult for the mare to foal naturally.*

assistance and in the worst scenario a caesarean may be required, especially if straining by the mare has caused damage to the cervix, vagina or uterus. The foal may have both its head and neck turned back, presenting just two forelegs (a more extreme version of Figure 7.5), or the presentation of the head and forelegs may be correct but the hind legs are also presented (Figure 7.8), putting four legs in the birth canal. Four legs are also presented first in the crosswise position (Figure 7.9).

Finally the foal may be presented in a ventral position, in which case the arch of the foal's back bone does not allow expulsion in the

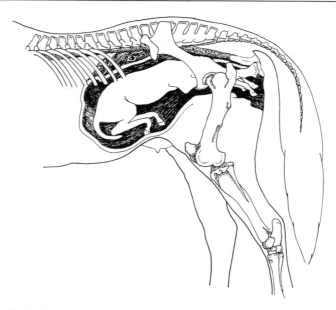

Figure 7.6   *If the forefeet presented are more misaligned than the normal hoof to fetlock misalignment, this may be due to one elbow being flexed and becoming lodged at the pelvic brim.*

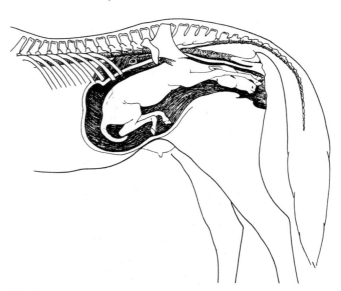

Figure 7.7   *The forelegs may be positioned over the foal's head. Not only does this increase the cross-sectional diameter of the foal but it also creates the risk of a rectal vaginal fissure.*

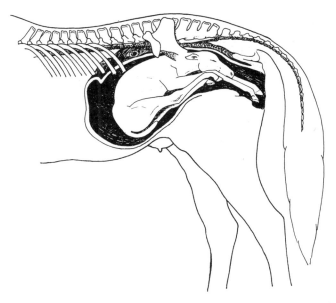

Figure 7.8   *If both the forefeet and hindfeet are felt within the birth canal, veterinary assistance should be called immediately as this position can prove very difficult to correct, especially if the mare has progressed far into labour.*

Figure 7.9   *A crosswise presentation also presents four feet first. This can be corrected by pushing the hindlegs back into the uterus.*

required curved manner. This position must be corrected by rotation of the foal before delivery is possible (Figure 7.10).

Backward presentations

Backward positions may also be seen when the back legs are presented first in the birth canal, a position that need not be manipulated in utero but must be carefully watched during labour. In such a case there is danger that the umbilical cord may easily become trapped between the abdomen of the foal and the pelvic brim, starving the foal of oxygen. Foals in this position must, therefore, be pulled quickly and the amniotic sack removed from the muzzle to allow breathing as soon as possible.

A more complicated backward position is indicated when no back legs are presented and only the tail can be felt. This position is known as a breech and veterinary assistance will be required to deliver such a foal (Figure 7.11).

Other positions, variations of the above, may be found, but those discussed are the most common.

Maternal Dystocia

Maternal dystocia is the failure of natural delivery as a result of a maternal complication. These complications may be found in

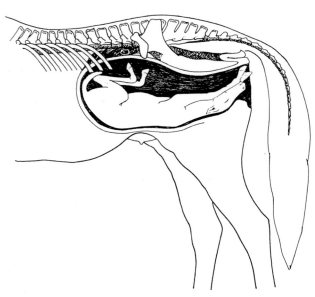

Figure 7.10   Ventral position of the foal. The foal must be rotated into a normal position before delivery is possible.

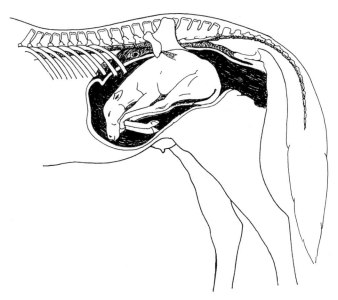

Figure 7.11   *If no legs can be felt within the birth canal and only the tail and rump are presented, this is a true breech position and will require veterinary assistance.*

isolation or in combination. Some of the more common ones will be discussed below.

Placenta previa is the passage out of the birth canal of the placental membranes before the forelegs of the foal. Usually the red allanto-chorionic membrane is seen protruding out from the mare's vulva, due to its failure to rupture at the cervical star. Its failure to rupture may be due to extra-thick placental membranes, in which case manual breakage of the protruding allanto-chorion will allow parturition to progress normally. However, failure to rupture at the cervical star may be because the membranes have ruptured elsewhere. Rupture away from the cervical star may be an indication that premature separation of the placenta from the maternal epithelium has occurred. As this separation is one of the triggers for the foal's first breath, there is a danger of suffocation in such cases.

Small pelvic openings may also cause problems. Restriction in this area may be the result of an accident or fracture or caused by malnutrition during the mare's early life when her bone structures were still developing. In such cases a caesarean is often the only course of action that can be taken in order to deliver a live foal. It is questionable as to whether such mares should be bred from.

Uterine problems may also cause dystocia. One of the more rare conditions is uterine torsion, where the uterus has become twisted, often as the result of a fall. The twist, which is often towards the cervix end of the uterus, prevents any delivery through the birth canal. The prognosis for such cases is normally poor but some success has been reported using manual correction via an incision in the flank or via the rectum; normally a caesarean is required.

Uterine inertia is a more common condition, and may often accompany foetal dystocia. In such cases the uterine myometrium becomes exhausted due to excessive straining, especially during the second stage of labour. The condition can be accentuated by age or previous uterine infections and may result in rupture of the uterus or a retained placenta. Uterine inertia may not occur to the same extent across all the myometrium. In such cases tight rings of muscle contraction may occur, and as a result there is a danger that the foal may be crushed or strangled. In most cases of uterine inertia, traction to aid the mare, along with an oxytocin injection to encourage muscle contraction, is all that is needed; however, occasionally a caesarean may be required.

Abnormal maternal conditions

Abnormal conditions associated with parturition do not normally directly prevent normal delivery but may indirectly cause concern. These conditions include problems that become evident prior to labour, such as ruptured prepubic tendon and uterine prolapse. The prepubic tendon is the tendon sheet supporting both the uterus and the abdomen. Its rupture means that these two areas no longer have the support they need and death is the normal outcome. Uterine prolapse results in the inversion of the vagina and uterus through the vulva due to incompetent uterine support, possibly associated with old age plus the additional pressure within the abdominal cavity. In such cases the uterus can be replaced manually after the administration of an anaesthetic or relaxant to the mare (Brewer and Kleist, 1963). This reduces the mare's straining and allows the uterus to be replaced and the vulva sutured to prevent a reoccurrence.

Hypocalcaemia is not common in mares; if it does occur it can be quite successfully treated by administration of calcium boro-gluconate post partum. The incidence of hypocalcaemia is higher in mares suffering from dystocia and also post caesarean section. Delayed involution may be apparent after parturition. Under normal conditions within a few hours of parturition the uterus should have shrunk to one-quarter of its fully expanded state. By Day 7 post partum it should be only two to three times the size evident in a barren mare, returning fully to its normal size by

Day 30 post partum. Evidence of a dilated cervix or enlarged uterus at Day 30 indicates problems, possibly associated with retention of placental membranes or dystocia. Uterine involution can be encouraged in such mares by daily infusion of oxytocin and antibiotics plus gentle exercise.

Some conditions occur during parturition, for example, uterine or intestinal rupture. Both of these are caused by excessive straining especially during second stage labour or as a result of weakness from previous damage to these organs. Mares fed large infrequent meals in late pregnancy are more predisposed to intestinal rupture than those fed little and often. Finally, some conditions may not become evident until after parturition but are a result of the forces of delivery. For example, haemorrhage, both internal, that can prove fatal, and the less drastic external, may be a result of labour but not be detected until after delivery. Rectal vaginal fissures or perineal lacerations may also cause problems when the foal's hoof punctures the roof of the vagina and passes through the floor of the rectum opening up a cloaca as the foal is delivered (Figure 7.12).

Finally, retention of the placental membranes may cause serious complications. Normally the placenta reduces in size gradually as its blood flow reduces; it therefore shrinks and separates away from the uterus and is delivered by the final contractions of the uterus in

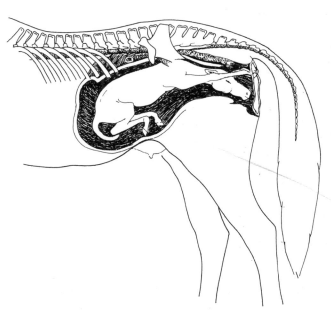

Figure 7.12   During foaling the foal's feet may pass through the vaginal wall and up into the rectum, forming a rectal vaginal fissure.

third stage labour. Retention of the placental membranes will result in infection that if left untreated will prove fatal. Retention of the afterbirth for more than 10 hours indicates problems and the mare should be seen to immediately. Retention is normally in the previously non-gravid horn, especially after dystocia or caesarean section. It may be due to a hormonal imbalance resulting in inadequate oxytocin and, therefore, reduced muscle activity.

Treatment can be in one of several ways, including manual removal and/or oxytocin treatment. Manual removal must be carried out with great care and if attempted it must be certain that all the placenta is removed. After initial attempts an antibiotic pessary can be inserted and the rest of the membranes removed a few days later. It is imperative that all the placenta is removed as soon as possible. If the whole of the placenta has been retained then iodine solution can be pumped in through the cervical star to fill up the allantoic sack. It may take 9–11 litres for an effect to be obtained. After about 5 minutes the mare will be seen to strain against the filled placenta and so aid expulsion. This method can be used in conjunction with oxytocin treatment, though it itself induces an elevation in naturally produced oxytocin. Oxytocin may be used alone if partial retention of the placenta is suspected. This will aid the final contractions of third stage labour and hence expulsion.

References and Further Reading

Bowen, J.M. (1976) Non-surgical correction of a uterine torsion in the mare. Vet. Rec. 99, 495.

Brewer, R.L. and Kliest, G.J. (1963) Uterine prolapse in the mare. J.Am. Vet.Med.Ass., 142, 1118

Challis, J.R.G. and Olson, D.M. (1988) Parturition. In The Physiology of Reproduction. Ed. S.E. Knobil, et. al. New York, Raven Press, pp 61–129.

Forsyth, I.A., Rossdale, P.D. and Thomas, C.R. (1975) Studies on milk composition and lactogenic hormones in the mare. J.Reprod.Fert.Suppl., 23, 631.

Hillman, R.B. (1975) Induction of parturition in mares. J.Reprod.Fert. Suppl., 23, 641.

Hillman, R.B. and Lesser, S.A. (1980) Induction of parturition in mares. Vet.Clin.No.Am. (large Anim.Pract.) 2, 333.

Jeffcote, L.B. and Rossdae, P.D. (1979) A radiographic study of the foetus in late pregnancy and during foaling. J.Reprod.Fert.Suppl., 27, 563.

Ley, W.B. (1989) Daytime foaling management of the mare. 2. Induction of parturition. J.Equine Vet.Sci., 9, 95.

Littlejohn, A. and Ritchie, J.D.S. (1975) Rupture of caecum at parturition. J.S.Afr.Vet.Ass. 46, 87.

Peaker, M., Rossdale, P.D., Forsyth, I.A. and Falk, M. (1979) Changes in mammary development and composition of secretion during late pregnancy in the mare. J.Reprod.Fert.Suppl., 27, 555.

Rooney, J.R. (1964) Internal haemorrhage related to gestation in the mare. Cornell Vet., 54, 11.

Rossdale, P.D., Pashan, R.L. and Jeffcote, L.B. (1979) The use of prostaglandin analogue (fluprostenol) to induce foaling. J.Reprod.Fert.Suppl., 27, 521.

Silver, M. (1990) Prenatal maturation, the timing of birth, and how it may be regulated in domestic animals. Exper. Physiol., 75, 285.

Voss, J.L. (1969) Rupture of caecum and ventral colon of mares during parturition. J.Am.Vet.Med.Ass., 155, 745.

Wheat, J.D. and Meagher, D.V. (1972) Uterine torsion and rupture in mares. J.Am.Vet.Med.Ass., 160, 881.

CHAPTER 8

Endocrine Control of Parturition

PARTURITION in the mare occurs at approximately 11 months (310–374) days after conception. The exact length of gestation, however, varies with the type of horse. Ponies tend to foal on average 2 weeks before Thoroughbred mares at about 320 days compared to 335 days (Kenneth and Ritchie, 1953; Rophia, Mathews, Butterfield, Moss and McFalden, 1969).

Within these averages many factors have an influence on the exact timing of parturition. These include the season of mating: mares mated early in the season tend to have longer gestations than those mated later on. This is presumably nature's way of compensating for early and late matings, trying to ensure that all mares foal down at the most optimum time of the year for foal survival, that is during spring. The difference is only in the order of days, but in nature such adjustment will eventually bring a mare's parturition back in line with the average during spring.

The genotype of the offspring also affects the length of its gestation, and this can be demonstrated by comparing the pregnancy lengths of various crosses within the equine species. A stallion cross mare foetus has an average gestation of 340 days; a stallion cross jennet foetus, is 350 days; a jack cross mare foetus, is 355 days; and a jack cross jennet, 365 days.

Multiple births also tend to have a shorter gestation than singles, and colt foals tend to have pregnancies on average 2.5 days shorter than filly foals.

Low nutritional intake in the final trimester of pregnancy is also reported to cause early parturition. Finally the mare herself has control over her time of delivery. As anyone having worked with breeding mares will know, they are notorious for foaling just when you turn your back or are least expecting it! The majority of mares foal at night, as illustrated in Graph 8.1.

In this study 90% of mares foaled between the hours of 1900 and

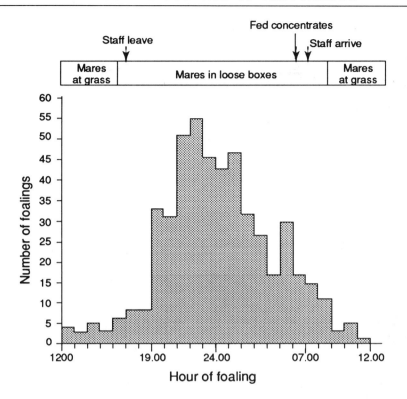

Graph 8.1   The time of foaling in the Thoroughbred mare (Rossdale and Short, 1967).

0800 when the mares were stabled in; only 10% foaled during the daylight hours while they were turned out.

Initiation of Parturition

Birth involves the expulsion of the foetus plus all associated placental membranes and fluid and is achieved by myometrial activity within the uterus and surrounding structures. In most mammals myometrial activity is inhibited by high levels of progesterone and low concentrations of oestrogen, characteristics of pregnancy in most mammals. At term the ratio of progesterone to oestrogen decreases, removing any inhibition and allowing myometrial activity to be actively driven by high PGF2α and oxytocin levels, both of which can be seen to rise. Efficient expulsion of the foetus and placental tissue is due to simultaneous contraction of the whole uterine myometrium. Therefore, there must be immediate activation of all muscle cells and efficient induction from cell to cell. This message transfer is effected by the concentration of hormones such as progesterone. Elevated progesterone levels reduce the spread of muscle cell induction (Liggins, 1979). The exact mechanism for the initiation of parturition in the horse or any other equids is as yet unclear. However, in other mammals two alternative mechanisms are apparent.

Firstly, as seen in the ewe, nanny goat, sow and cow the foetus itself actively controls the initiation of its own parturition. Towards the end of gestation the foetus comes under considerable stress due to hypoxia, a shortage of oxygen, and its physical restriction within the uterus. These increasing stress levels cause the production of corticosteroids by the foetal adrenals and pituitary gland. These pass across the placenta and change the metabolic pathways within the placental tissue, converting progesterone to oestrogen (Challis and Olson, 1988). As a result the characteristic rise in oestrogens and fall in progesterone near term in these animals is achieved. This change in progesterone metabolism also seems to drive local PGF2α production, as well as enhancing the transfer of electrical impulses through the uterine myometrium, thus initiating the first myometrial activity within the uterine wall at the start of first stage labour (Flint, Ricketts and Craig, 1979).

The second apparent method by which parturition is initiated is as seen in the human. In this case the production of corticosteroids is by the placenta and is initiated by a genetically controlled maturation signal within the foetal membranes of the placenta. This initiates, by a local response, an increase in oestrogen and decrease in progesterone which, acting at a systemic or cellular level, initiates the production of PGF2α, which again induces the first myometrial contractions of the uterine wall. In this case the foetal adrenals have very little effect on the induction of parturition (Liggins, Ritterman and Forster, 1979).

Though there is no conclusive evidence as to which mechanism occurs in the mare, most of the evidence available points to a system similar to that seen in the ewe (Silver, 1990). High corticosteroid levels have not been reported in the equine foetus during late pregnancy, mainly due to the difficulties encountered in catheterising such foetuses. However, elevated corticosteroid levels have been reported in allantoic fluid at term and it has been demonstrated that cortisol is essential in foals for the final maturation of several internal organs, including the respiratory system and digestive tract. Pashan and Allen (1979) presented evidence that parturition in the equid is influenced by an interaction between foetal and placental size and that foetal constriction may be a trigger. This hypothesis is further supported by Rossdale, McGladdery, Ousey, Holdstock, Grainger and Houghton (1992), who demonstrated that treatment of foetuses in utero with ACTH resulted in an increase in corticosteroid production by the foetal adrenals that caused premature parturition.

Endocrine Concentrations

Oestrogen

Oestrogen concentrations in the maternal blood system fall quickly over the last 30 days of gestation, and as they originate from the foetal-placental unit reach basal levels within hours of parturition. Oestradiol 17β concentrations are approximately 6 ng/ml at 30 days prior to parturition and fall to less than 2 ng/ml after parturition (Nett, Holtan and Estergreen, 1973).

Progesterone

In previous research, plasma progesterone levels were reported to change in the opposite fashion to oestrogens, rising to levels of 10–15 ng/ml in the last 20 days of gestation. This rise peaked within 3–4 days of delivery and decreased in the last 24–48 hours prior to parturition (Ousey, Rossdale, Cash and Worthy, 1987; Rossdale et al., 1992). This is the complete opposite to that seen in most mammals. These concentrations would seem on the face of it to inhibit myometrial activity.

However, the rise in progesterone is now thought to be mainly due to an increase in 5α pregnane metabolites rather than progesterone itself, maternal progesterone at this stage being undetectable (Hamon, Clarke, Houghton, Fowden, Silver, Rossdale, Ousey and Heap, 1991; Holtan, Houghton, Silver, Fowden, Ousey, and Rossdale, 1991). These metabolites represent an alteration in placental metabolism during late pregnancy and it is possible that 5α pregnanes occupy the binding sites normally taken up by progesterone and hence their increase is at the expense of progesterone. The ratio of oestrogen to progesterone, rather than the absolute levels, is known to be important; therefore, even though oestrogen levels decline it is possible that their decline is significantly less than that of progesterone and therefore in relative terms the ratio of oestrogen:progesterone increases, allowing myometrial activity to begin (Purvis, 1972; Moss, Estergreen, Becker and Grant, 1979).

Prostaglandin F2α

As discussed previously PGFM, a metabolite of PGF2α, is used as an indicator of PGF2α concentrations. From measurements of PGFM it can be deduced that PGF2α concentrations rise sharply in the peripheral plasma of the mare at term, mainly during the second stage of labour. Foetal PGF2α levels as seen in catheterised foetuses are shown to increase more gradually over the final weeks (40 days) of pregnancy (Barnes, Comline, Jeffcote, Mitchell, Rossdale and Silver, 1978; Cooper 1979; Silver, Barnes, Comline, Fowden, Clover and Mitchell, 1979; Pashan and Allen, 1980).

Oxytocin

The actions of PGF2α and oxytocin are closely linked and, therefore, they tend to show the same pattern of release. Oxytocin levels in the maternal system rise sharply at parturition, especially during the second stage, as seen with PGF2α. PGF2α seems to be primarily responsible for cervical dilation and first and second stage labour. Oxytocin is also involved in the strong contractions of second stage labour and also controls the third stage of labour (Hillman and Ganjam, 1979).

Cortisol

Cortisol concentrations do not change in the maternal system during pregnancy, though they rise, due to stress, at parturition. As far as the foetal system is concerned, as explained earlier, no sudden increase in cortisol in late pregnancy has been reported. However, a rise in cortisol within the amniotic fluid just prior to parturition (Day 250 onwards), has been reported as well as a rise in foetal blood concentrations immediately post partum (Rossdale, Silver, Comline, Hall and Nathanielsz, 1973; Nathanielsz, Rossdale, Silver and Comline, 1975; Liggins, Ritterman and Forster, 1979). This increase is gradual rather than sudden as observed in the sheep and goat and is known to be associated with final organ maturation, e.g the respiratory system, in the equine foetus (Rossdale, Silver, Comline, Hall and Nathenielsz, 1973; Alm, Sullivan and First, 1975).

Conclusion

It is, therefore, evident that the mechanism for the initiation of parturition in the mare is unclear, but seems probably to be due to foetal stress. Confirmation of this hypothesis is largely hampered by the limitation of experimental techniques to date (Graph 8.2).

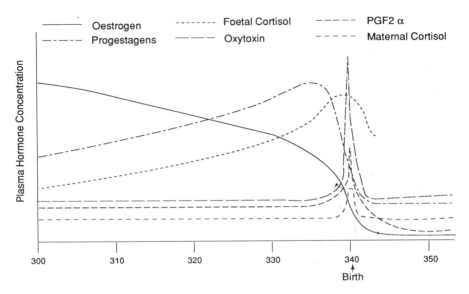

Graph 8.2   A summary of the main changes in hormone concentration evident prior to parturition.

References and Further Reading

Alm, C.C., Sullivan, J.J. and First, N.L. (1975) The effect of corticosteroid (dexamethasone), progesterone, oestrogen and prostaglandin F"α on gestational length in normal and ovariectomised ewes. J.Reprod.Fert.Suppl., 23, 637.

Barnes, R.J., Comline, R.S., Jeffcote, L.B., Mitchell, M.D., Rossdale, P.D. and Silver, M. (1978) Fetal and maternal concentrations of 13, 14–dihydro-15–oxo-prostaglandin F in the mare during late pregnancy and at parturition. J.Endocri. 78, 2, 201.

Cooper, (1979) Clinical aspects of prostaglandins in equine reproduction. Proc.2nd.Equine Pharm.Symposium, 225.

Flint, A.P.F., Ricketts, S.W. and Craig, V.A. (1979) The control of placental steroids synthesis at parturition in domestic animals. Anim.Reprod.Sci., 2, 239.

Hamon, M., Clarke, S.W., Houghton, A.L., Fowden, M., Silver, P.D., Rossdale, J.C., Ousey, J.C. and Heap, R.B. (1991) Production of 5 α dihydroprogesterone during late pregnancy in the mare. J.Reprod.Fert. Suppl., 44, 529.

Hillman, R.B. and Ganjam, V.K. (1979) Hormonal changes in the mare and foal associated with oxytocin induction of parturition. J.Reprod.Fert.Suppl., 27, 541.

Holtan, D.W., Houghton, E., Silver, M., Fowden, A.L., Ousey, J. and Rossdale, P.D. (1991) Plasma progestagens in the mare, fetus and newborn foal. J.Reprod.Fert.Suppl., 44, 517.

Kenneth, and Ritchie. (1953) Gestation periods, 3rd Ed. Commonwealth Agricultural Bureaux, Slough.

Liggins, G.C. (1979) Initiation of parturition. Br. Med. Bull., 35, 145.

Liggins, G.C., Forster, C.S., Grieves, S.A. and Schwartz, A.L. (1977) Control of parturition in man. Biol.Reprod., 16, 39.

Liggins, G.C., Ritterman, J.A. and Forster, C.S. (1979) Foetal maturation related to parturition. Anim.Reprod.Sci., 2, 193.

Moss, G.E., Estergreen, V.L., Becker, S.R. and Grant, B.D. (1979) The source of 5 pregnanes that occur during gestation in mares. J.Reprod. Fert.Suppl., 27, 511.

Nathanielsz, P.W., Rossdale, P.D., Silver, M. and Comline, R.S. (1975) Studies on foetal, neonatal and maternal cortisol metabolism in the mare. J.Reprod.Fert.Suppl., 23, 625.

Nett, T.M., Holtan, D.W. and Estergreen, V.L. (1973) Plasma oestrogens in pregnant mares. J.Anim.Sci., 37, 962.

Ousey, J.C., Rossdale, P.D., Cash, R.S.G. and Worthy, K. (1987) Plasma concentrations of progestagens, oestrone sulphate and prolactin in pregnant mares subject to natural herpes virus-1. J.Reprod.Fert.Suppl., 35, 519.

Pashan, R.L. and Allen, W.R. (1979) Endocrine changes after foetal gonadectomy and during normal and induced parturition in the mare. Anim.Reprod.Sci., 2, 271.

Pashan, R.L. and Allen, W.R. (1980) Endocrine changes after foetal gonadectomy and during normal and induced parturition in the mare. In: Animal Reproduction Science, 2, 271–288.

Purvis, A.D. (1972) Electric induction of labour and parturition in the mare. Proc.18th Ann.Conv.Amer.Ass.Equine Practnrs., San Francisco, 113.

Rophia, R.T., Mathews, R.G., Butterfield, R.M., Moss, G.E. and McFalden, M. (1969). The duration of pregnancy in Thoroughbred mares. Vet.Rec., 84, 552.

Rossdale, P.D., McGladdery, A.J., Ousey, J., Holdstock, N., Grainger, C. and Houghton, E. (1992) Increase in plasma progestagen concentrations in the mare after foetal injection with CRH, ACTH or beta methasone in late gestation. Equine Vet.J, 24 (5), 347.

Rossdale, P.D., Ousey, J.C., Cottrill, C.M., Chavatte, P., Allen, W.R. and McGladdery, A.J. (1991) Effects of placental pathology on maternal plasma progestagen and mammary secretion calcium concentrations and on neonatal adrenocortical function in the horse. J.Reprod.Fert.Suppl., 44, 579.

Rossdale, P.D. and Short, R.V. (1967) The time of foaling of Thoroughbred mares. J.Reprod.Fert., 13, 341.

Rossdale, P.D., Silver, M., Comline, R.S., Hall, L.W. and Nathanielsz, P.W. (1973) Plasma cortisol in the foal during the late foetal and early neonatal period. Res.Vet.Sci., 15, 395.

Silver, M., Barnes, R.J., Comline, R.S., Fowden, A.L., Clover, L. and Mitchell, M.D. (1979). Prostaglandins in maternal and foetal plasma and the allantoic fluid during the second half of gestation in the mare. J.Reprod. Fert.Suppl., 27, 531.

CHAPTER 9

Anatomy and Physiology of Lactation

M AMMARY glands are modified apocrine sweat glands: both develop in utero from a common precursor. They are situated along the ventral midline in all mammals in a varying number of pairs. The mare has four glands, two pairs. In most mammals each gland exits via its own teat; however, in the mare each pair of glands on either side of the midline exits via a single teat.

Anatomy

The mammary glands of the mare are situated in the inguinal region between the hind legs, protected by a layer of skin and hair that covers the surface of the skin of all the gland except for the teats. The whole of the skin surface is supplied with nerve endings, the concentrations of which are increased in the teat area, enhancing the response to touch in this area. Each of the four mammary glands is completely independent with no passage of milk from one quarter to another. They are separated and supported by the medial suspensory ligament running along the mare's midline. Further support is given by the lateral suspensory ligaments running over the surface of the mammary gland under the skin and by laminae developing off the suspensory ligament penetrating the mammary tissue in sheets (Figure 9.1).

Each udder half, either side of the midline, is made up of two quarters, and the openings from these two quarters exteriorise via a single teat (Figure 9.2).

The mammary tissue itself is made up of millions of alveoli and connecting ducts. This arrangement can be compared to a bunch of grapes, each alveolus being equated to a grape and the ducts to the branches (Figure 9.3).

The alveoli are grouped together in lobules and then the lobules into lobes. These lobes join together via a branching duct system

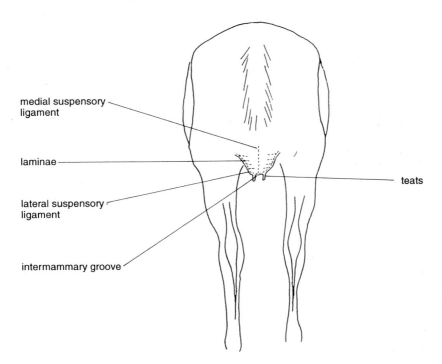

medial suspensory
ligament

laminae

lateral suspensory
ligament

intermammary groove

teats

Figure 9.1 A rear view of the mare's udder illustrating the suspensory apparatus of
the mammary gland.

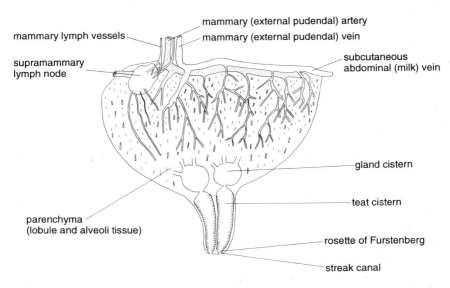

mammary (external pudendal) artery
mammary (external pudendal) vein

mammary lymph vessels

supramammary
lymph node

subcutaneous
abdominal (milk) vein

gland cistern

teat cistern

parenchyma
(lobule and alveoli tissue)

rosette of Furstenberg

streak canal

Figure 9.2 A cross-section through the mammary gland of the mare illustrating the
exit of two quarters via a single teat.

123

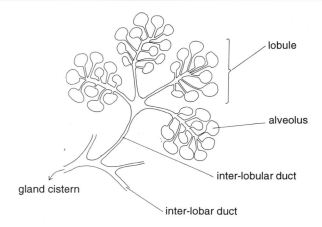

Figure 9.3   A single lobe of the mammary gland made up of several lobules which, in turn, are made up of numerous alveoli.

which eventually leads to the gland cistern. Each quarter has its own gland cistern draining into a teat cistern and on to the streak canal. In the mare each teat is fed by two streak canals, one from each quarter on that side. At the end of each teat is the rosette of Furstenberg, a tight sphincter that prevents the leakage of milk between sucklings. This sphincter can withstand a considerable build-up of milk pressure, though occasionally it may be breached, as in the case of mares that milk themselves in the late stages of pregnancy.

The alveoli, which are the milk-secreting structures, are lined by a single layer of lactating cells, surrounding a central cavity or lumen. This alveoli lumen is continuous with the mammary duct system. Milk is secreted by the lactating cells into the alveoli lumen across the luminal or apical membrane. Surrounding each alveolus is a basket network of myoepithelial cells. These muscle cells also surround the smaller ducts and their contraction is activated as part of the milk ejection reflex. Surrounding these myoepithelial cells is a capillary network supplying the alveoli with milk precursors and also a series of lymph vessels. The alveoli also have a nervous supply which is responsible for the activation of the myoepithelial basket cells as well as vasodilation and constriction of the capillary supply network (Figure 9.4).

The mammary gland as a whole is supplied with blood via two mammary or external pudendal arteries, one on each side of the midline. Blood is drained from the mammary glands via two mammary or external pudendal veins and two subcutaneous

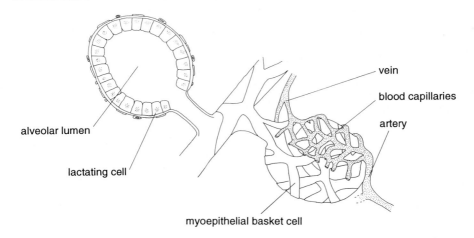

Figure 9.4   *The mammalian alveolus. On the left, a cross-sectional view illustrating the lactating cells surrounding the alveolar lumen which is continuous with the mammary duct system. On the right, an alveolus illustrating the myoepithelial basket cells and alveolar blood supply.*

abdominal (milk) veins. The external pudendal artery and vein run into the body at the inguinal region. The subcutaneous abdominal vein, which runs along the belly of the mare, can be seen more clearly in lactating mares and is hence sometimes referred to as the milk vein. The mammary gland also has two supramammary lymph nodes, one either side of the midline, and connecting to the main circulatory lymph system and that of the mammary gland itself.

Mammogenesis

Mammogenesis or mammary development is first evident in the embryo. Glands develop along either side of the midline in the inguinal region. Cells in this region proliferate to form nodules that develop to form mammary buds. At birth teats are present in males and females along with a few short branching ducts in the connective tissue associated with each teat.

From birth to puberty in both colts and fillies mammary gland growth is isometric with (at the same rate as) body growth. Most of this prepubertal growth is an increase in fat and connective tissue rather than duct development. In colts mammary growth stays isometric with body growth throughout the remainder of their lives, reaching a maximum when body growth ceases at approximately five years of age. Puberty in fillies marks a change, as mammary development becomes allometric or increases relative to body growth. This allometric growth continues past puberty, increasing and decreasing in cycles corresponding to the oestrous cycles. In all mammals the amount of mammary development within these cycles depends on the length of the oestrous cycle, as elevated progesterone levels during dioestrus are responsible for mammary development. In the mare the length of dioestrus is relatively long and hence during dioestrus lobulo-alveolar development takes place in addition to the normal increase in the branching of the duct system.

During pregnancy elevated progesterone levels cause significant lobulo-alveolar development, especially in the last trimester of pregnancy when progesterone levels are still elevated. In the last two to four weeks of pregnancy, lactogenesis, milk production, also commences (Leadon, 1984; Ousey, 1984). During lactation mammogenesis continues as cell division increases in line with milk production to satisfy the increasing demands of the foal. Cell division then decreases after the maximum yield has been reached (week five onwards). At the same time the size of the mammary gland slowly decreases until it returns to its normal non-lactating size post weaning (Graph 9.1).

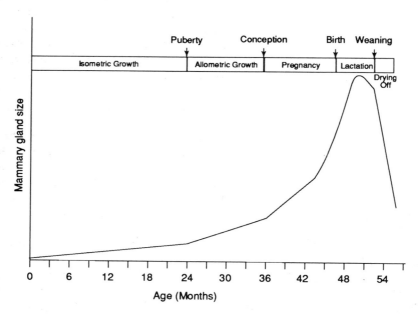

Graph 9.1   Equine mammary development from birth to 54 weeks, including development during a mare's first pregnancy.

Lactation Curve and Milk Quality

There has been nowhere near as much research conducted into the lactation of the mare as for other livestock, especially the cow. Except in a very few cultures, mare's milk is of indirect rather than direct commercial importance, its value being assessed via the standard of foal reared rather than directly by milk yield. The following discussion will concentrate on conclusions drawn either directly from experiments using mares or extrapolated from other species studied.

Lactation Curve

There is much variation in the lactation curve obtained from different individuals, largely due to man's interference and early weaning. As a general trend, however, milk yield in mares tends to increase during the first 2 months post partum. Initial levels in the first 2 weeks are in the order of 4–8 litres/day for Thoroughbreds and 2–4 litres/day for ponies. Milk production reflects demand and the size of the foal, so it continues to rise as the foal grows until about 2–3 months post partum when maximum levels of 10–18 litres/day in Thoroughbreds and 8–12 litres/day for ponies are reached. After 3 months the foal's demand for nourishment from its mother decreases as it starts increasingly to investigate grass or hay and its mother's hard feed. As the weaning process progresses towards full weaning the lactational yield drops off further with decreasing demand. As will be discussed later, milk quality also drops off at this time, further encouraging the foal to seek nourishment elsewhere and hasten the weaning process.

Lactation lasts nearly a full year naturally, the mare drying up completely a few weeks before she is due to foal the following year. However, in today's equine industry, man normally determines the length of lactation by weaning foals at about 6 months, when the yield is less than that immediately post partum. The foal is, therefore, obtaining little of its nourishment from its mother and hence weaning at 6 months will have little long term effect on its development (Graph 9.2).

The total milk yield of a Thoroughbred or one of the larger riding type horses is 2000–3000 kg of milk per lactation. As a rough guide, in these larger horses the natural daily milk yield averaged out over the whole lactation is 2–3 kg/100 kg body weight. The corresponding equation for ponies is 5 kg/100 kg body weight. The foal normally suckles 16–35 times per day taking approximately 250 g/suckle in the larger riding type horses. The number of

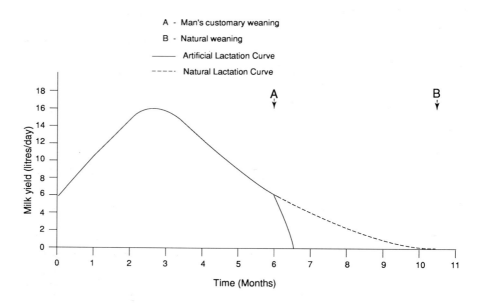

Graph 9.2   **The average lactation curve for a mare, illustrating the natural extent of lactation along with that customarily imposed by man.**

suckles per day and the amount of milk taken per suckle reduce from peak lactation towards weaning.

Milk Quality and Composition

The quality of all milk reflects the requirement of the young of that particular species and provides the foal with the energy and building blocks needed for growth throughout lactation, with the addition of immunoglobulins during the first stages of lactation (Table 9.1) (Ullrey, Struthers, Hendricks and Brent, 1966; Peaker, Rossdale, Forsyth and Falk, 1979).

Colostrum, the first milk, contains a relatively high concentration of proteins, which are immunoglobulins. Protein concentration in colostrum is in the order of 13.5% compared to 2.7% in the main lactational milk. The protein immunoglobulins in mare's colostrum are globulin, gamma, beta, alpha and albumin, in their descending order of relative concentration. This high protein concentration is at the expense of sugars and fats, in the form of lipids, that are both present in relatively low concentrations. However, within 12 hours protein levels fall dramatically and lactose and lipids levels rise. The relative concentrations within milk then stabilise at

Table 9.1 Comparative milk compositions of several species, expressed as percentages (Jennes and Sloane, 1970).

Species	Total solids	Fat	Casein	Whey protein	Lactose
Human	12.4	3.8	0.4	0.6	7.0
Cow	12.7	3.7	2.8	0.6	4.8
Goat	13.2	4.5	2.5	0.4	4.1
Sheep	19.3	7.4	4.6	0.9	4.8
Horse	11.2	1.9	1.3	1.2	6.2

levels equivalent to those reported for the remainder of lactation (Graph 9.3).

Mineral concentrations also vary with the stage of lactation. Potassium and sodium in colostrum tend to be high at 1200 mg/kg and 500 mg/kg respectively, dropping to 600 mg/kg and 300 mg/kg within 5 weeks. Calcium tends to be slightly raised in colostrum but then falls away slightly within hours, only to rise again to a peak of 1200 mg/kg at 3 weeks post partum. Magnesium levels are also

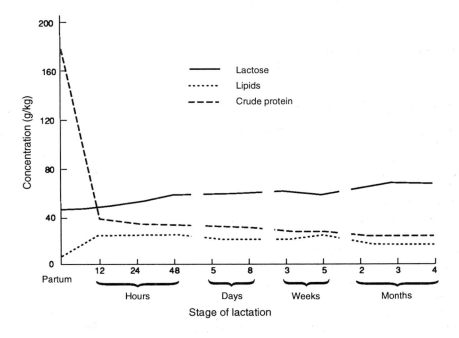

Graph 9.3 Changes in the concentrations of lactose, lipids and crude protein in milk during lactation (Ullrey, Struthers, Hendricks and Brent, 1966).

elevated at 500 mg/kg in colostrum but fall off rapidly in the first 12 hours and then continue to decline slowly throughout the rest of lactation. Phosphorous remains relatively steady at 400–500 mg/kg for the first 8 weeks of lactation and then concentration declines (Graphs 9.4 and 9.5).

It is evident that there is a trend for the concentration of all the components of milk to decline as lactation proceeds, this is nature's way of encouraging the foal to obtain its nourishment elsewhere by gradually reducing those available via its mother's milk (Smolders, Van der Veen and Van Polanen, 1990).

The average composition of milk during the main part of lactation in the mare is given in Table 9.2.

Table 9.2   The average composition of mare's milk during the main period of lactation.

Component	%
Water	89.0
Protein	2.7
Lactose	6.1
Fat	1.6
Ash (minerals, vitamins etc)	0.6
Total	100.0

Fat

The concentration of fats in mare's milk is relatively low in comparison with other species of mammal. Fat is present in milk in the form of globules of saturated fat, cholesterol and unsaturated fats as free fatty acids, phospholipids and triglycerides. The 8% concentration of triglycerides as a proportion of total fats in mare's milk is very much lower than the 79% in cows. These fat globules exist as an emulsion within the milk and contain a high concentration of short-chain fatty acids, less than 16 carbons in length.

Protein

Proteins during the main lactation are present in the form of 1.3% caseins and 1.2% whey. Caseins are unique to milk and have several functions. Under the influence of the stomach's acid pH they form a clot with the enzyme renin. This clot facilitates the digestion of proteins by the proteolytic enzymes of the digestive system. Caseins also contain essential amino acids and aid in the transport of minerals from the mare to the foal via milk. Caseins

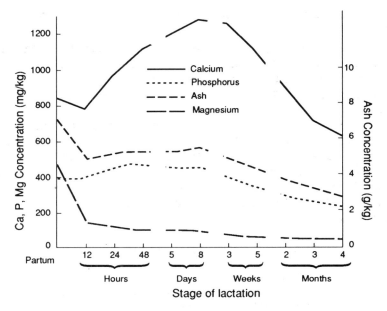

Graph 9.4 Changes in the concentrations of calcium, phosphorus, magnesium and ash in mare's milk during lactation (Ullrey et al., 1966).

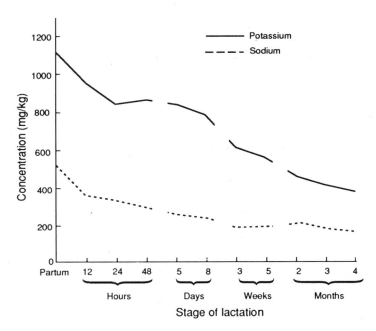

Graph 9.5 Changes in the concentrations of potassium and sodium in mare's milk during lactation (Ullrey et al., 1966).

associate with calcium, phosphate and magnesium ions to form micelles, and this allows a higher concentration of these minerals to be transported than is possible in a simple aqueous solution.

Two types of whey proteins are found in the mare's milk and unlike caseins do not precipitate in acid pH. The whey proteins are divided into those that are specific to milk and those that can be found in both milk and blood. In most mammals those specific to milk can be further subdivided into β lactalbumin and α lactalbumin. However, in the mare only α lactalbumin is found. It is a good source of amino acids and is rich in essential ones such as tryptophan. It is also one of the two component halves that make up the enzyme lactase synthetase, which is made up of A and B components: α lactalbumin, the B component, is the terminal enzyme in the synthesis of lactose, the major sugar component of mare's milk.

The second type of whey proteins found in mare's milk are ones also found in blood, serum albumin and serum globulin. Serum albumin is identical to blood serum albumin and is directly transferred unchanged from the blood through the lactating cell to the alveoli lumen. It is, therefore, only found in small concentrations unless there has been cell damage or haemorrhage within the mammary tissue.

Serum globulin, on the other hand, is the immunological fraction of milk and, therefore, its concentration is very high in colostrum. Antibodies attach themselves to these globulins and via these the foal attains its passive immunity to all the infections that the mare has raised antibodies against. Such passive immunity via colostrum is very important in mammals like the horse, which have a relatively thick placenta that limits the passage of antibodies and hence the attainment of passive immunity in utero. For further details on placental structure within the mare, see Chapter 5.

The digestive system of the foal is permeable to complete protein molecules for the first 48 hours of life, after which this ability to absorb large protein molecules is irreversibly changed. It is essential, therefore, that a new-born foal receives its colostrum well within 48 hours of birth, as after this time it cannot take advantage of the antibodies it carries and they will be broken down by proteolytic enzymes into their component amino acids and absorbed as such.

Lactose
Lactose is the energy component of milk. Unique to mammals, each lactose molecule consists of a molecule of galactose and one of glucose. In the intestine of the foal lactose is split into its two component parts, glucose and galactose, and galactose can then be

easily converted into glucose. Lactose is, therefore, in essence two molecules of glucose. The question then rises as to why lactose, not glucose, is present in milk, especially as it takes energy to convert glucose to lactose and vice versa.

The answer lies in the effect of glucose on the osmotic pressure of milk relative to blood. The osmotic pressure of the two must be the same, and the component of milk that has the largest effect on osmotic pressure is the small molecule of lactose. However, if glucose were present it would have an even greater effect on the difference in osmotic pressure. Additionally, one molecule of lactose gives rise to two molecules of glucose. That is, one molecule of lactose has twice the calorific value per molecule of glucose, and hence also per unit of osmotic pressure. It has also been suggested that lactose provides a more beneficial medium for intestinal activity and regulates bacterial flora and pH, so aiding the absorption of minerals.

Milk Synthesis

Milk is synthesised in the epithelial or lactating cells lining each alveolus. The precursors of, and components for, milk are obtained from small molecules in the blood system supplying the udder. The components cross the basal membrane of the lactating cells. There is little information on how they pass across this membrane, but as the molecules are small it seems likely that it is by diffusion. The protein, fat and lactose components of milk are then built up within the lactating cells and pass across the cell membrane to the lumen of the alveolus.

Each of the major components of milk will now be discussed in turn.

Protein

Proteins are made from amino acids within the lactating cells. The total amount of nitrogen that crosses the basal membrane is equal to the total amount within the milk; however, the amino acids do change. Non-essential amino acids are synthesised within the cell and are built up into proteins along the messenger ribonucleic acid (mRNA) within the ribosomes of the rough endoplasmic reticulum (RER). These proteins are then secreted into milk. Essential amino acids are passed unchanged across the basal membrane of the lactating cell and are incorporated into proteins, along with the non-essential amino acids synthesised within the cell.

Lactose

Blood glucose is the primary precursor of lactose. However, glycerol, acetate and amino acids are also thought to contribute. The amount of glucose absorbed by the gland is much more than is needed solely for conversion to lactose. The difference is used as energy to drive general cell metabolism. The conversion of glucose to lactose involves five enzymes, the fifth enzyme being lactose synthetase, which is made up of two components, A and B. As discussed previously, component B is the major milk protein α lactalbumin. The biochemical pathways involved in the conversion of glucose to lactose are summarised in Figure 9.5.

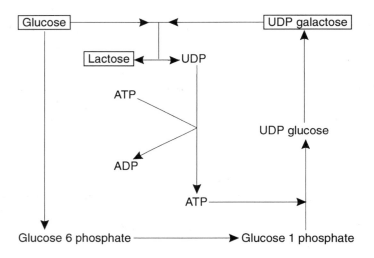

Figure 9.5   A summary of the conversion of glucose to lactose within the lactating cell.

Fat

The fat globules within milk are made up of esterified glycerol and free fatty acids, which aggregate to form a fat droplet emulsion within milk. There is much variation in the length of free fatty acids making up the fat globules in the milk from females of varying species. The horse tends to have a higher concentration of short-chain fatty acids (less than 16 carbon atoms in length).

The fatty acids are derived mainly from three sources: glucose, triglycerides and free fatty acids. Glucose carbon is a significant precursor of free fatty acids in the non-ruminant, for example the horse. Glucose is absorbed across the basal membrane and converted to acetyl CoA and on to malonyl CoA within the cytosol of the cell. Malonyl CoA is then built up, using a multienzyme complex, to free fatty acids, which tend to be of shorter chain, rarely greater than 16 carbon atoms in length. An alternative way by which the lactating cell obtains free fatty acids is via blood triglycerides, which are broken down into glycerol plus free fatty acids within the cell. The free fatty acids obtained from triglycerides are longer-chain fatty acids, typically 16 to 18 carbons in length. Triglycerides, therefore, are not a very important source of fatty acids in the mare. The triglycerides are either broken down into fatty acids and glycerol in the blood, similar to the way proteins are broken down into amino acids and then absorbed into the cell, or they are absorbed directly.

Glycerol that combines with the free fatty acids is derived again by three different methods: from the breakdown of triglycerides within the cell, by the absorption of free glycerol in the blood or finally from the breakdown of glucose within the cell.

The free fatty acids and glycerol within the cell combine by esterification within the endoplasmic reticulum. These molecules then aggregate together to form the fat droplets within milk.

Milk Secretion

All the components of the milk produced by the lactating cells have to pass across the apical membrane of the lactating cell into the alveolar lumen (Figure 9.6). The different components of milk pass by different mechanisms.

Fat

The fat droplet size increases as the free fatty acids and glycerol combine by esterification and the resulting molecules aggregate

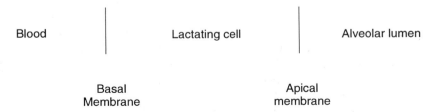

Figure 9.6   The route of passage for all milk components from the mare's blood supply on the left through the lactating cell to the lumen of the alveolus.

into increasingly larger droplets as they migrate towards the apical membrane. In the vicinity of this membrane strong London–Van der Waals forces attract these fat droplets and envelop them in the membrane, forming a bulge in the apical membrane surrounding the droplet (Figure 9.7).

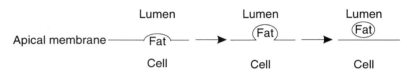

Figure 9.7   *The secretion of milk fat from the lactating cell (below) into the alveolar lumen (above).*

The droplet and surrounding plasmalemma move away from the apical membrane into the lumen forming a narrow bridge. This bulge then pinches off as the bridge gets narrower and releases the fat droplet plus surrounding plasmalemma into the alveolar lumen. The process is termed pinocytosis. Occasionally, part of the cell cytoplasm, sometimes including cell organelles, is enclosed in the bulge of apical membrane along with the fat droplets, and then gets secreted into the alveolar lumen along with the milk fat. The formation of these structures, which are termed signets, occurs more in the lower order of mammals, but they are evident occasionally in mare's milk.

Protein

Proteins are built up from their amino acid building blocks along the RER within the cell and then pass on to the Golgi apparatus. They accumulate as granules of proteins within the Golgi; this Golgi apparatus then migrates towards the apical membrane. The membrane of the Golgi apparatus fuses with the apical membrane and this releases the proteins into the alveolar lumen by reverse pinocytosis (Figure 9.8). By this reverse pinocytosis, plasmalemma lost during the secretion of milk fat is replaced during the secretion of milk proteins.

Lactose

The secretion of molecules of lactose, unlike that of milk fat and protein, is not visible using electron microscopy, and so the method of secretion is less clear. As discussed earlier in this chapter one of the mare's milk proteins is α lactalbumin, which is the B

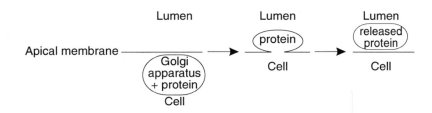

Figure 9.8 The secretion of milk protein from the lactating cell (below) into the alveolar lumen (above) by reverse pinocytosis.

component of the fifth and last enzyme involved in lactose synthesis. It, therefore, seems likely that the lactose secretion is closely linked to milk protein secretion. The A protein of the enzyme lactase synthetase is known to be closely associated with the membrane of the Golgi apparatus, but the B component (α lactalbumin) is synthesised, as are all other milk proteins, on the RER and then passed on to the Golgi apparatus. Whilst the B component is in the Golgi apparatus, it becomes associated with the A component already there, and together they form active lactase synthetase. This enzyme catalyses the conversion of UDP galactose and glucose to lactose. The lactose is then secreted along with the milk proteins by reverse pinocytosis (Figure 9.8).

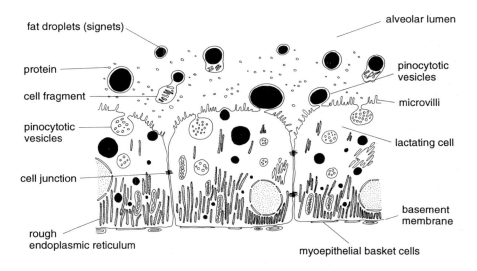

Figure 9.9 Diagrammatic representation of a mammary secretory cell illustrating the build-up of protein and fat for release into the alveoli.

Sodium and Potassium

Milk has a relatively high concentration of potassium in comparison with sodium, and this is similar to the relative concentrations within the cytoplasm of the cell. Sodium and potassium in milk are derived from the intercellular fluid. The components within the lactating cell are derived from the blood system, but the potassium:sodium ratio in blood is the reverse of that within the cell and milk. Therefore, there must be an active transfer system across the basal membrane. This is via a sodium pump, which pumps sodium away from the cell cytoplasm and into the blood system and pumps potassium the opposite way towards the cell. This maintains the high potassium:sodium ratio within the cell. As the potassium:sodium ratio in the milk is the same as that in the cell, then sodium and potassium are thought to pass across the apical membrane by simple diffusion.

Calcium, Magnesium and Phosphorus

The concentrations of calcium, magnesium and phosphorus are higher in milk than in the cell cytoplasm. Therefore, their passage must be via an active transport system. The exact mechanism is unclear but all three ions are known to be closely associated with the milk protein casein. It is, therefore, assumed that this association occurs within the Golgi where the casein proteins are built up. These ions are then passed into the milk, along with their associated proteins, via reverse pinocytosis.

Iron

Iron is also secreted in association with proteins by reverse pinocytosis, as it is specifically bound to a minor milk protein lactoferrin.

Water

Water passes to the alveolar lumen from the cell cytoplasm by osmotic pressure. Fat and protein molecules in milk are in the form of large droplets, and so their effect on osmotic pressure is minimal. However, lactose and free ions are much smaller and it is these that effect osmotic pressure and hence drive water diffusion from the cell into milk.

Colostrum

In colostrum, as discussed previously, there is a high concentration of proteins consisting of immunoglobulins and their associated antibodies. These immunoglobulins are combined into large corpuscles termed Bodies of Donne. The mechanism by which these are secreted is unclear. It is possible that engorgement of the lactating cell in late pregnancy results in the breakage of some of the junctions within the cell membranes, especially the basal cell membranes of the lactating cells. This allows the serum proteins to pass into milk unchanged. There is also evidence suggesting an active transport system for these serum proteins but the exact mechanism is as yet unclear.

Conclusion

Our present knowledge regarding equine lactation anatomy and physiology is still significantly lacking but by extrapolation from other species a reasonable picture has been developed. However, these assumptions cannot be confirmed or denied until research specifically on equine lactation is carried out in more detail.

References and Further Reading

Doreau, M. and Boulet, S. (1989) Recent knowledge on mare's milk production. A review. Livestock Prod.Sci., 22, 213.

Forsyth, I.A., Rossdale, P.D. and Thomas, C.R. (1975) Studies on milk composition and lactogenic hormones in the mare. J.Reprod.Fert.Suppl., 23, 631.

Jennes, R. and Sloane, R.G. (1970) Dairy Science Abstr 32, Review Article 158.

Leadon, D.P. (1984) Mammary secretions in normal, spontaneous and induced parturition in the mare. Equine Vet.J., 16, 259.

Mepham, B. (1976) The secretion of milk. Studies in Biology No. 60. Edward Arnold.

Ousey, J.C. (1984) Preliminary studies of mammary secretions in the mare to assess foetal readiness for birth. Equine Vet.J., 16, 263.

Peaker, M.; Rossdale, P.D. Forsyth, I.A. and Falk, M. (1979) Changes in mammary development and the composition of secretion during late pregnancy in the mare. J.Reprod.Fert.Suppl., 27, 555.

Smolders, E.A.A., Van der Veen, N.G. and Van Polanen, A. (1990) Composition of horse milk during the suckling period. Livestock Prod.Sci., 25, 163.

Tyznik, W.J. (1972) Nutrition and disease. In: Equine Medicine and Surgery. 2nd Ed. Ed. E.J. Catcott and J.R. Smithcors, p.239. American Veterinary publications, Illinois.

Ullrey, D.E., Struthers, R.D., Hendricks, D.G. and Brent, B.E. (1966) Composition of mare's milk. J.Anim.Sci., 25, 217.

CHAPTER 10

Control of Lactation in the Mare

THE control of lactation in most mammals, including the mare, is via both nervous and hormonal pathways. Unfortunately, information specific to the mare is very limited, though it is assumed that the control of lactation is very similar to that in other mammals. The information discussed in this chapter is gleaned from the limited experiments carried out on horses and extrapolation from other mammals where appropriate.

Lactation can be divided into three stages as far as its control is concerned: these are lactogenesis, galactopoiesis and milk ejection.

Lactogenesis

Lactogenesis refers to the initial milk secretion that occurs in late pregnancy prior to parturition, resulting in a build-up of colostrum within the mammary glands. In the mare lactogenesis is evident well before parturition, as demonstrated by the presence of lactose, proteins and fat within the mammary tissue well before lactation commences.

Control of lactogenesis is hormonal. High progesterone concentrations during pregnancy drive alveoli development, along with expansion of the mammary duct systems. In most mammals these high progesterone concentrations inhibit milk secretion, but progesterone levels in the mare were previously reported to be elevated right up until parturition. However, as discussed in Chapter 8, it is now thought that these seemingly high progesterone levels are in fact elevated 5α pregnane concentrations, which take up progesterone binding sites. Hence as far as the mare's system is concerned, the effect is the same as a fall in progesterone concentrations, and hence in the mare, as in other animals, progesterone inhibition of milk production is removed in late pregnancy. In addition, elevated prolactin, growth hormone and cortisol actively drive milk production. Lactogenesis, therefore, increases in late pregnancy and immediately prior to parturition (Forsyth, Rossdale and Thomas, 1975).

In other mammals, for example ruminants and humans, a placental lactogen has been identified and found to have an additional effect of driving lactogenesis. No such placental lactogen has been identified in equids.

Galactopoiesis

Galactopoiesis is the term given to the maintenance of milk production. Little information on it in the horse is available. It is assumed that control is similar to that in the sheep and the cow, where it is controlled by hormones, elevated prolactin, growth hormone and cortisol concentrations acting as driving agents. Galactopoiesis is driven by and mimics the foal's demand for milk, and this, in turn, dictates and governs the shape of the lactation curve and the quantity of milk produced.

Milk Ejection

Milk ejection, also termed milk letdown, is different from the other stages of lactation in that its control is via both nerves and hormones. A nervous reflex acts as the stimulus or afferent pathway of this reflex, and hormones form the efferent path. Nerve receptors within the teats are stimulated by the action of suckling, and the nervous afferent pathway is activated, resulting firstly in a localised effect that causes the contraction of the myometrial cells within the mammary gland. Secondly, this afferent nervous pathway sends a message via the central nervous system to the paraventricular nucleus within the hypothalamus of the mare's brain. The hypothalamus then activates the posterior pituitary, which in response produces the hormone oxytocin. The efferent pathway of the milk ejection reflex is formed by the oxytocin, which passes into the systemic blood system and hence to the mammary gland. The effectors that react to oxytocin are the myoepithelial basket cells surrounding each alveolus and the small ducts, causing them to contract and forcing milk out of the alveoli, along the ducts, to the gland cistern and on to the teat cistern ready to be removed by the suckling action of the foal.

In addition to the above control mechanisms, the central nervous system has an overriding effect on the control of lactation. For example stress, especially as a result of fear or shock, reduces the effectiveness of the milk ejection reflex by increasing the levels of

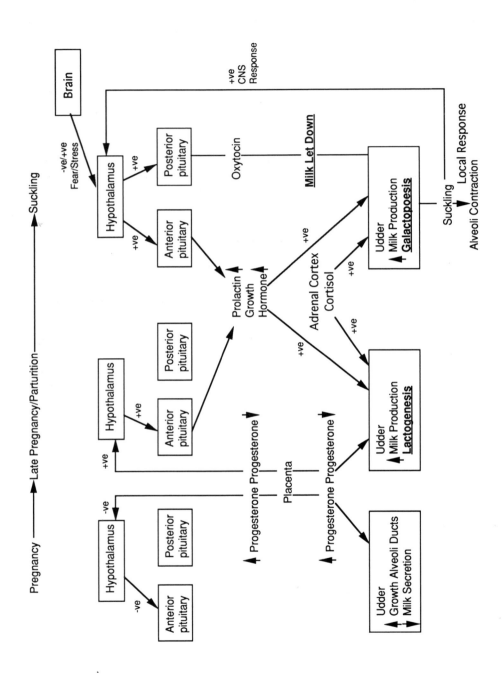

Figure 10.1   Schematic representation of the control of lactation in the mare.

adrenaline within the system. Adrenaline causes vasoconstriction, reducing the level of oxytocin arriving at the alveoli and hence the effectiveness of the reflex (Figure 10.1).

Conclusion

It is evident that there is much left to be learnt about the control of lactation in the mare. Because of the lack of direct commercial value, little research has been carried out into equine milk production, and as such many extrapolations and assumptions from the cow have been applied to the horse.

References and Further Reading

Doreau, M. and Boulet, S. (1989) Recent knowledge on mare's milk production. A review. Livestock Prod.Sci., 22, 213.

Forsyth, I.A. Rossdale, P.D. and Thomas, C.R. (1975) Studies on milk composition and lactogenic hormones in the mare. J.Reprod.Fert.Suppl., 23, 631.

Mepham, B. (1976) The secretion of milk. Studies in Biology, No. 60. Edward Arnold.

Tyznik, W.J. (1972) Nutrition and Disease. In: Equine Medicine and Surgery, 2nd. ed. Ed. E.J.Catcott and J.R Smithcors, p. 239. Illinois, American Veterinary Publications.

Breeding and Stud Management

CHAPTER 11

Selection of the Mare and Stallion for Breeding

THE choice of both mare and stallion for breeding can be a very complicated and time-consuming process. Often not enough importance is placed on this selection, the result being an oversupply of mediocre or poor stock on the market, and unnecessary difficulties with mares at covering, during pregnancy, at foaling and during lactation.

One of the most obvious selection criteria is that of performance or ability. If you are looking to produce a quality show horse, you will obviously be selecting for conformation and show success on both the mare's and the stallion's side and in their breeding. If you are looking to produce a racehorse or competition horse, you will obviously be interested primarily in the proven performance of the dam and sire and select accordingly. Finally if you are looking for a pleasure horse you will be interested primarily in temperament and possibly hardiness.

It is beyond the scope of this book to discuss in detail the wide range of performance criteria that a breeder may look for in order to achieve that specific ideal horse. These performance criteria vary considerably with the individual breeder and type of horse required. Many books and articles have been written on the subject but ultimately it is a personal decision by the individual breeder.

Another large area of selection criteria that should be investigated, regardless of the type of horse you intend to breed, is that of reproductive competence and the ability to produce a healthy offspring with minimal danger to the life and wellbeing of the mother. Today's horse, unlike the case in farm livestock, has been selected primarily for performance ability, often at the expense of reproductive competence. As a result there are potential reproductive problems that the breeder should be aware of in selecting both the mare and stallion when aiming to produce a healthy foal at minimum financial cost and danger to the mare.

The following sections will concentrate solely upon the criteria and techniques that can be used in the selection for reproductive competence of both mare and stallion and will assume that the selection criteria for performance and conformation have been already met.

The Mare

The selection criteria for reproductive competence in the mare can be divided as follows:

History – general and breeding
Temperament
Age
General conformation and condition
External conformation of the reproductive tract
Internal conformation of the reproductive tract
Infections
Blood sampling

History

The history of the mare includes both her general past and any specific breeding history. Many mares these days, especially those of any value, come with full documentation of their history. Such records are invaluable in assessing the mare's ability to produce a foal, as well as in easing management. If there are no records available it is a good idea to contact her previous owners and find out as much as possible about her past. Details on past breeding should be obtained, and examples of questions that should be addressed include:

Does she cycle regularly?
When does she first come into season?
How long does her season normally last?
Does she show oestrus well?
What are her characteristic signs?

This information will indicate, among other things, whether she will be easy to detect in heat and cover. Mares that do not show well tend to be hard to cover. This leads to frustration due to missed heats and wasted journeys to the stallion, all increasing costs. In addition, mares that are habitually hard or even dangerous to cover may be refused by some studs, reducing the pool of potential stallions or necessitating the use of artificial insemination.

Other questions will include whether the mare has bred before. If so, has she had problems holding to the stallion? This will necessitate return journeys to the stud or a prolonged period of time away. Has she had problems during any of her pregnancies? Has she ever reabsorbed or aborted? Does she or her family show incidence of twins? If a mare has a history of habitual reabsorption or abortion or the need for holding injections (artificial progesterone supplementation) post service, then she is really not a good bet for a brood mare. Such problems may indicate an inherent inability to carry a pregnancy to term, due to hormonal imbalance, uterine incompetence and/or genetic abnormality.

Repeated incidences of twins is a problem in that the likelihood of one or both of the foetuses aborting is high. If only one of the twins aborts the remaining one is often born weaker and smaller than would normally be expected. Spontaneous abortion of twins is not due to an inherent inability to carry a foal to term and such mares are perfectly capable of producing a foal, providing a single pregnancy is present. They may, however, need to be checked routinely for twin pregnancies and aborted, necessitating a return to service.

It is important to find out whether there are any incidences of problem foaling. It is rare for abnormal foetal positions to occur repeatedly, but past difficulties during parturition may have caused internal lacerations or damage, and it is particularly important to know if the cervix has been affected. Limited damage to the uterus, vagina and vulva will repair quite effectively, though they may leave areas of weakness. Damage to the cervix may cause it to become incompetent, allowing bacteria to enter the higher reproductive tract and cause infection and reduce conception rates. Conversely, adhesions may hinder the passage of sperm at covering. A damaged cervix may also be less able to dilate during parturition and necessitate a caesarean delivery.

Previous post foaling problems may also show up, including rejection or even habitual attacking of her foals. This could be a sign of a general temperament fault or may not show up in subsequent pregnancies. Records may show that the mare is not a good mother, producing an illthrifty foal that does not do well. It is, therefore, helpful to look at foal birth weight and subsequent development and growth rates. Such effects may be due to poor milk yield, which in turn may reflect faults in nutritional management rather than a specific mare problem, but if a mare has consistently produced foals that did not do well, it may be worth investigating further.

Records should also show any incidences of infections and the treatments given. Minor infections may show no long-term effects. However, infections of the reproductive tract are responsible to a large extent for the relatively high infertility rates in horses. Previous mastitis infections should also show up, and such mares should be examined to make sure that the udder has not suffered permanent damage.

Detailed mare records not only prove invaluable as an aid to selection; they also provide useful information for the stud to which you intend to send your mare. This is especially important if she is to stay at stud for a prolonged period of time and heat detection is to be carried out by them.

In addition to specific breeding records, information on the mare's general history is also very useful. This should indicate her vaccination status and also show up incidences such as accidents, especially if these involved injury to the pelvic area, abdominal muscles or internal injuries. All these may predispose the mare to having a problem pregnancy and may prevent natural parturition. Damage to limbs or muscles may also cause problems in late pregnancy when the mare is carrying the weight of a full-term foetus.

Past illnesses should also be shown up. Mares with a history of respiratory or circulatory disorders would possibly not stand up to the strain of pregnancy. Mares with disorders that are potentially inherited may be able to carry a pregnancy to term, but it is debatable whether they should be bred and hence perpetuate a heritable problem within the population.

Temperament

The temperament of the mare is obviously important for ease of management, especially at birth and when dealing with a young foal, when many mares tend to be antisocial. In addition, temperament is an inherited characteristic and hence that of the parents has a long-term effect on the temperament and character of the foal. The foal also picks up and copies many behavioural characteristics from its mother during its early life in the period leading up to weaning.

A mare with a quiet and gentle disposition is much easier to handle and manage. Such mares tend to show oestrus more readily and are, therefore, easier to cover. Anxious and highly strung mares with debatable temperaments may show signs of aggression towards a teaser or stallion even though in season; stress may often cause oestrus signs in such mares to be masked by those of aggression. Such aggressive mares are obviously more difficult and dangerous to cover. Many need to be twitched and/or hobbled, which again raises stress levels and is not conducive to optimum fertilisation rates or easy management. Such mares can, of course, be covered using artificial insemination but you still have the problem of detecting oestrus. Highly strung and nervous mares suffer from higher abortion rates due to the higher stress levels which in turn elevate cortisol levels that can disrupt the normal reproductive endocrine control mechanisms.

Age

The age of a mare at her first breeding has an effect on the ease of her pregnancy, and there are potential long-term effects especially in mares under five or over 12 years of age at first mating (Day, 1939). As discussed in Chapter 1 mares reach puberty at 18–24 months of age on average. It is, therefore, theoretically possible to breed a mare at 18 months to produce a foal by 2.5 years of age. However, evidence, especially from research into sheep, demonstrates that embryo mortality rates are significantly higher in maiden pubertal animals, and return rates in young maiden mares do tend to be higher than in older multiparous mares. Returns of such mares also tend to be longer than the normal 21 days, indicating that embryo mortality is the problem rather than fertilisation failure.

Young mares may not only be hard to get in foal but they may also suffer detrimental long-term effects if nutrition is inadequate. A horse does not attain its mature body size until on average five years old, so for a mare under this age, the demands of pregnancy are additional to those of growth. This should be reflected in her feed, to ensure that she has enough nutrients to satisfy the demands for maintenance, pregnancy and growth. To breed a mare early in life is a decision not to be taken lightly. She must be well grown for her age and in good but not fat physical condition. You must also have the wherewithal to feed her additional good-quality food, especially in late pregnancy, and to provide proper accommodation in order to minimise her body's demand for maintenance.

At the other end of the spectrum, maiden mares over the age of 12 may also find it difficult to carry a pregnancy and bring up a foal. There is no evidence to suggest that embryo mortality rates are any higher in old mares, but the mares tend to have been exposed to infections that affect their fertility and they can be difficult to cover. The mere fact that a mare has been around longer means that she has had a greater chance of exposure to reproductive tract infections, which, if severe, may have affected her ability to conceive. Older mares have spent the majority of their lives as barren mares without the attentions of the stallion; they therefore tend to reject or be antagonistic towards a stallion.

Older maiden mares may also have problems associated with how their past lives have been spent. Many such mares are ex-performance horses and as such they have been trained and kept in top athletic condition, which often results in a disruption of reproductive activity. Prime athletic condition in mares, as in other female animals, can result in a variety of reproductive malfunctions, for example, delayed oestrus, prolonged dioestrus, complete reproductive failure etc. These mares need a considerable readjustment period before they are physically and psychologically capable of bearing and rearing a foal. A prolonged rest period of at least six months usually allows the mare's system to settle down into a non-athletic state, whereas some mares take as long as 18 months to readjust to their new way of life.

If such a mare has been treated with drugs, illegally or legally, in association with her athletic career, her system will need time to rid itself of such effects, which again can disrupt reproductive function. The top athletic condition of a performance horse builds up muscles not necessarily advantageous to an easy pregnancy. Such muscle tone, especially in the abdominal and pelvic areas, can cause complications in pregnancy and parturition, and time should be given to allow relaxation of any such muscle tone.

Ex-performance mares are also more likely to have suffered accidents and injuries in the past. If these were severe they can cause the mare difficulties in late pregnancy when the weight she has to carry is considerably increased. Leg injuries especially may cause complications. Fractures of the pelvis and sacral area may also cause complications at parturition, precluding natural birth and necessitating a caesarean delivery. The preparation of ex-performance horses for breeding is also discussed in Chapter 12.

The ideal age to breed a maiden mare is at five to six years of age, when she will have reached her mature size and, therefore, not have the additional demand for growth on her system. She will also not be old enough to have become set in her ways and will be less likely to have developed aggressive tendencies towards the stallion.

Obviously, breeding for the first time at the age of five to six will not suit all individual systems of breeding. Most mares will not have proved their worth and so the decision to breed her may not yet have been made. One increasingly popular way of dealing with mares of this sort is the use of embryo transfer, which will be discussed in Chapter 21. Fertilised ova can be collected from performance mares that have been mated and subsequently transferred into recipient, non-performance brood mares. This system allows performance mares to breed but not at the expense of their performance careers.

Once a mare has borne one foal, she is much more able to cope with the next pregnancy and is much more likely to breed successfully as an older mare well into her teens.

If a mare has been a brood mare all her life, then age is less important. She may naturally be barren during the occasional year, suggested to be nature's way of allowing recovery and regeneration. More care and attention is needed with older age. It is not normally advisable to breed with a mare over 20 years old though this does largely depend on the breed type and condition of the individual mare. Mares of the native type tend to breed more successfully in later life than the hotblood/warmblood type animals, though of course there are exceptions to every rule.

General Conformation and Condition

A mare's general conformation is of importance, not only to ensure that her offspring are well conformed, but also to ease her pregnancy. She needs to have a strong back and legs to enable her to carry the considerable extra weight of the foetus during late pregnancy. She should have correct pelvic conformation with a pelvic opening adequate for a safe delivery. Ideally she should also possess good heart and lung room across the chest and have plenty of room in the barrel. Rather fine tucked-up mares tend in general to be poorer breeders or have more problems during pregnancy.

A mare's general body condition also has very important implications on reproductive activity. The body condition of horses can be classified on a scale of 0–5, 0 being emaciated and 5 obese. The ideal condition score for a mare at mating is 3. Such mares have a good covering of flesh, and the ribs may be felt with some pressure, as can the vertebrae of the back bone. Mares in condition score 3 have been shown to have the highest fertilisation rate and subsequent reproductive success (Plate 11.1).

Overthin mares suffer from ovulation and fertilisation failure, and extreme emaciation results in the complete suspension of all reproductive activity, nature's way of ensuring that the mare's system does not have to cope with the additional strain and demands of pregnancy while it is in such a poor condition. Thin mares may also show prolonged anoestrus or very long/delayed oestrus cycles.

At the other end of the spectrum, overfat mares suffer from similar problems of ovulation failure (Plate 11.2). It has also been suggested that overfat condition results in excess fat deposition on the reproductive tract, limiting its ability to move and expand with a developing pregnancy. Fat may also be deposited on the ovaries and around the fallopian tubes, interfering with the process of

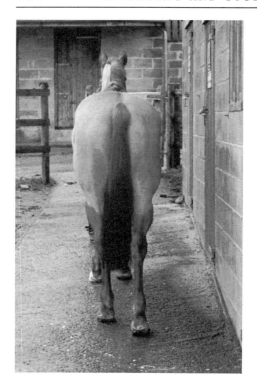

Plate 11.1

Mare in condition score 3, the ideal condition for covering.

Plate 11.2

Mare in overfat condition.

ovulation and the collection and funnelling of ova down the fallopian tubes.

It has been demonstrated that an increasing plane of nutrition 5 to 6 weeks prior to mating, with an accompanied gradual improvement in body condition up a score of 3, results in the best ovulation rates and reproductive success (Van Niekerk and Van Heerden, 1972). This process is termed flushing and is a common practice in sheep and cattle management. It is discussed in further detail in the following Chapter 12.

Reproductive Tract External Examination

Poor external conformation of the mare's reproductive tract can have severe implications on reproductive performance. The mare's reproductive system naturally has a series of three seals providing the internal system with protection from bacterial invasion. These are the outermost external vulval seal, the inner vestibular or vagina seal and the innermost cervical seal as illustrated in Figure 11.1.

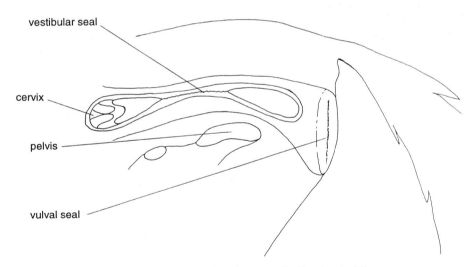

Figure 11.1   The three seals protecting the reproductive tract of the mare.

The outer vulval seal is formed by the perineal area of the mare, the anus, vulva and surrounding skin. If the conformation of the mare's perineal area is incorrect, the vulval seal becomes incompetent and allows bacteria and airborne pathogens to challenge the vagina and upper reproductive system. Such an incompetent vulval seal is illustrated in Figure 11.2 and Plate 11.3, which show how a sinking of the anus causes the vulva to tip and lie in a position not directly in line below the anus but to protrude. This non-vertical positioning of the anus and vulva will not only result in an incompetent vulval seal but will also allow contamination of the vulval lips by faecal matter, which is invariably sucked into the vagina along with air and airborne pathogens. This sucking in of air into the vagina, especially as the mare moves, is termed pneumovagina or windsucking (from the vulva). Vertical alignment of the anus and vulva allows faeces to roll away from the vulval opening. The greater the angle of the line from the anus to the vulva, the higher the incidence of faecal contamination of the vagina. Pneumovagina is a common cause of reproductive tract infection and hence infertility, especially when it causes ballooning of the vagina. Such a condition is prevalent in Thoroughbred mares, mares in poor condition and mares with reduced vulval tone due to injury, damage, age, oestrus or foaling (Pascoe, 1979).

Mares with such poor conformation can be helped by a Caslick's operation (Caslick, 1937). This involves cutting the vulval lips along their length and suturing them together to form a natural seal, leaving a free open area at the lower end for the release of urine (Plate 11.4). Such an operation successfully prevents pneumovagina and hence bacterial contamination. However, such mares are difficult to breed from, and unless artificial insemination is used, the vulval lips have to be opened at mating and resutured after, with a repeat process at foaling.

The second seal protecting the reproductive tract is the vaginal seal formed by the apposition of the floor and the roof of the vagina plus the hymen if still present. If the floor of the pelvis is low this allows the floor of the vagina to drop and the vaginal seal to gape. This poor conformation of the pelvic floor often accompanies an incompetent vulval seal due to the resultant tipping of the vulva. In such cases neither the vulva nor vagina offers any protective seal and therefore, because many of these mares suffer from pneumovagina, the cervix is bombarded with airborne pathogens and faecal contamination. Such mares run a very high risk of being infertile.

The height of the pelvic floor may be assessed by the use of a sterile probe inserted into the vagina and resting on its floor (Figure 11.3). In a normal conformation at least 80% of the vulva should lie below the pelvic floor in order for the vaginal seal to be fully competent. In a mare with a low pelvic floor, contamination of the reproductive tract can be prevented if perineal conformation and hence the vulval seal are competent.

The final seal preventing infection reaching the upper reproductive tract is the cervix. This final seal is highly challenged if both the

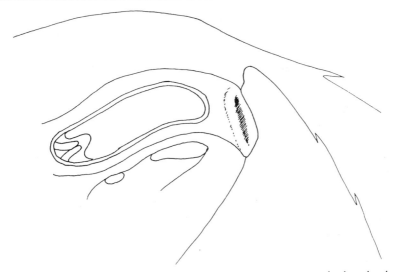

Figure 11.2   The conformation of a mare with an incompetent vaginal and vulval seal.

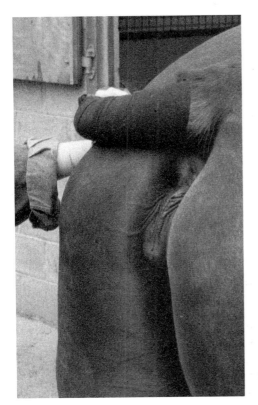

Plate 11.3

A mare with incorrect perineal conformation leading to an incompetent vulval seal.

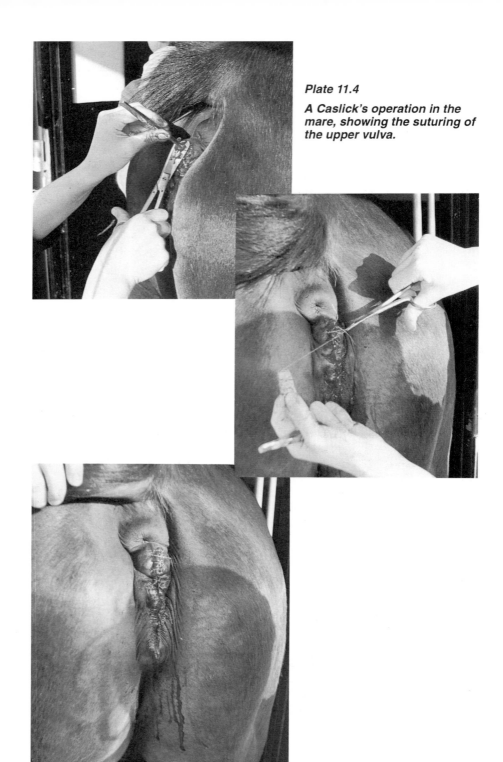

Plate 11.4

A Caslick's operation in the mare, showing the suturing of the upper vulva.

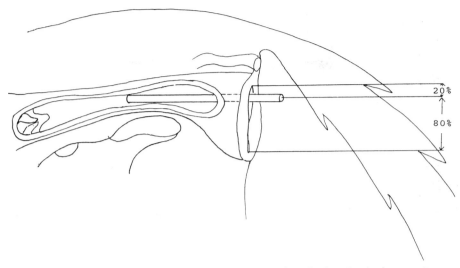

Figure 11.3   Diagram to illustrate the assessment of vaginal and vulval competence and hence the likelihood of pneumovagina.

vulval and vaginal seals are incompetent. If, in addition, the cervical seal has been damaged, usually at foaling, the mare is invariably infertile due to infection. Even if the cervix is competent, the bacterial challenge in some environments may be too great for it to cope with.

The external perineal conformation and hence vulval seal competence can be assessed by eye. The competence of the vaginal seal and cervix can be accurately assessed only by speculum. Speculum examination is discussed in detail in the following section.

Reproductive Tract Internal Examination

Examination of the internal reproductive tract of the mare is a skilled veterinary surgeon's job, necessitating the use of several examination techniques. Information given by these assessments can be indispensable in assessing the reproductive potential of a mare. The use of some of the following techniques may be limited by financial implications as many are expensive to perform. The value of the mare and any potential offspring must be borne in mind when assessing which technique to use and how far to take the assessment process.

Vulva and Vagina
The internal vaginal conformation is assessed by means of a

speculum (Figure 11.4). The most commonly used is a Caslick's speculum (Plate 11.5).

The speculum consists of an expandable hollow metal tube with a light source attached. The sterilised and well-lubricated speculum is inserted into the mare's vagina in the closed position; once in place, it is expanded slowly, opening up the vagina. The light source then illuminates the mucous membranes lining the vagina and the cervix, allowing examination in detail. Disposable speculums of plastic or coated cardboard are also available but these need an independent light source and hence involve the use of two hands, making techniques such as swabbing very difficult. The use of a vaginal speculum must be accompanied by several precautions as when it is in position it opens up both the vulval and vaginal seals, allowing bacterial contamination and challenge of the cervix. Hence, the procedure should be carried out under conditions as sterile as possible in a dust-free environment. The

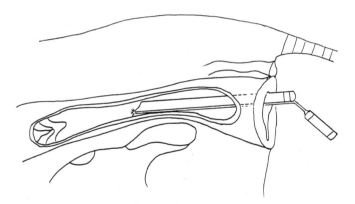

Figure 11.4   Vaginal speculum examination of the mare's vagina. The independent light source allows the inside of the vagina to be illuminated and hence easily viewed.

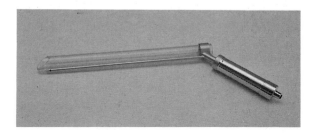

Plate 11.5

A vaginal speculum.

mare's tail should be bandaged prior to speculum insertion and the
perineal area washed thoroughly (Plate 11.6).

A more sophisticated instrument that can be used for viewing the
mare's vagina and cervix is the endoscope. The older hollow
endoscope is rigid and, therefore, has limitations to its use; the
newer flexible fibre optic endoscope is more versatile. As in humans,
the endoscope can be used to view any internal structures, either by
viewing through the body cavity or via the oral/nasal cavity to the
digestive/respiratory system, or alternatively via the rectum into the
large intestine or finally via the vagina to view the reproductive
system right up to the fallopian tubes.

The endoscope consists of a series of flexible carbon fibre filaments
with a light source and camera attached at the outer end. The
flexible carbon fibre rod is passed through the vagina and up into the
inner reproductive tract. One set of the carbon fibre filaments allows
the passage of light from the light source down the endoscope,

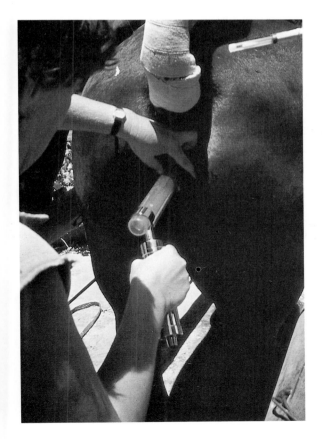

Plate 11.6

*Use of a vaginal
speculum in the mare.*

illuminating the internal reproductive tract. The other set allows the transmission of the image back to the camera and on to a television monitor.

Use of the endoscope is expensive and is usually confined to examination of the upper, less accessible parts of the reproductive tract (Plate 11.7).

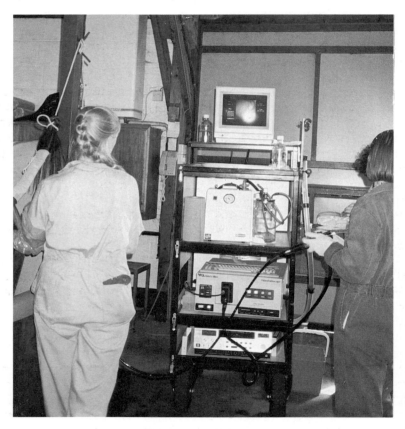

Plate 11.7 Endoscopic examination of the reproductive tract of the mare. The attendant on the right controls the optic fibre camera, while the attendant on the left guides the endoscope within the mare's reproductive tract. The image produced is viewed on the TV monitor. (Thoroughbred Breeders Equine Fertility Unit, Newmarket)

In the healthy mare both speculum and endoscope should show the mucous membranes lining the vulva and vagina as a healthy pink colour. Any mucus present should be clear and not cloudy or yellow-white in colour. An infected vagina appears red, inflamed and possibly covered in a cloudy mucus. Infections of the vagina are not uncommon and will be discussed in the following section on infections and also in Chapter 19.

The vulva and vagina should be free from bruising, irritation, scars and tears. They are particularly susceptible to damage and injury at parturition and so they should be carefully checked for scarring and resultant adhesions.

The vagina is also susceptible to adhesions, usually caused by damage at foaling or possibly mating. Adhesions are caused by scar tissue and may cause the mare to be unserviceable or cause her pain at mating and hence lead to rejection of the stallion. Artificial insemination is a possible alternative but she may still have problems at foaling.

Neoplasms of the vulva and vagina are relatively rare, though if present, they may cause problems at covering, with possible rejection of the stallion. Three kinds of neoplasms can be seen in horses. Firstly there are melanomas, malignant growths of the melanin-containing pigment cells, particularly prevalent in grey mares and not necessarily confined to the vulva and vagina. Secondly are carcinomas or malignant neoplasms of the epithelial cells. Finally there are papillomas or benign neoplasms. If a mare does show evidence of neoplasms it is debatable whether she should be mated, as there is evidence to suggest that the tendency to develop some neoplasms is heritable.

Finally pooling of urine within the floor of the vagina may be seen and the resultant stagnant urine will obviously increase the chance of infection. Urine pooling is more evident in older multiparous mares in which vaginal tone has been lost. It may also result from past vaginal damage or injury.

Cervix

The cervix can be considered as the connection between the outer reproductive tract and the inner more delicate and less resistant tract. As such, its competence as a seal is very important and this can be assessed by means of a speculum as described above for vaginal examination. The cervix varies greatly with the stage of the oestrous cycle and pregnancy and as such can aid in the diagnosis of pregnancy and the stage within the oestrous cycle. During oestrus the cervix is fairly relaxed and its tone flaccid. This is to ease the passage of sperm through the upper tract to the waiting ova. The cervical tone increases and the seal becomes tighter during dioestrus and to greater extent during anoestrus. At this stage the cervix changes from pink to white in colour and as oestrus ends also changes from wet/oedematous in appearance to dry. During pregnancy the cervix is again tightly sealed and white in colour with a mucous plug acting to enhance the effectiveness of the seal.

The cervix can also be examined via rectal palpation, and it should correspond with the state of the ovaries and the rest of the reproductive tract for the stage within the cycle. If not, infections or abnormalities should be suspected. Infection of the cervical mucosa, or cervicitis, will be discussed in the following section dealing with infections. Cervicitis is characterised by a red/purple swollen cervix often protruding into the vagina and covered with mucus, clearly seen at speculum examination. In severe cases cysts may be evident and may result in adhesions and long-term permanent damage preventing correct closure of the cervical seal and resultant infection. Adhesions as a result of cervical damage may close it completely, preventing the passage of sperm to the upper tract and also the draining of fluid from the uterus and beyond, resulting in serious uterine infections. Minor cervical damage and/or adhesions can be helped by surgery.

Uterus
Examination of the mare's uterus is a more complicated procedure as the uterus lies within the inner and less accessible reproductive tract. An appreciation of its gross appearance and structure can be obtained by rectal palpation and/or ultrasonic scanning. The inside colour and condition of the mucous membranes of the uterus may be inspected by means of an endoscope and further examined by uterine biopsy.

Rectal palpation is a commonly used and relatively efficient way of identifying structural abnormalities in the upper reproductive tract of the mare (Figure 11.5 and Plate 11.8). The procedure involves restraining the mare in stocks and inserting a well-lubricated gloved hand and arm through the anus and into the rectum of the mare. The wall of the rectum is fairly thin and so the reproductive tract that lies immediately below it can be palpated through the rectal wall. The procedure, however, must only be carried out by an experienced operator, as rupturing the thin rectal wall is relatively easy. Rectal palpation can be used to assess the tone, size and texture of the uterus, uterine horns, fallopian tubes and ovaries, and as such is a useful aid in the diagnosis of the stage of oestrus cycle, pregnancy and abnormalities.

Ultrasonic scanning is a newer and more expensive method of assessment but gives a visual impression of the upper reproductive tract rather than the tactile interpretation obtained with rectal palpation. The ultrasonic scanner is based on the Doppler principle: when ultrasound hits an object it is reflected at a different frequency, depending on the density and tone of the object. In the case of ultrasonic scanners used in the assessment of a mare's reproductive status, the transducer or ultrasonic emitting and

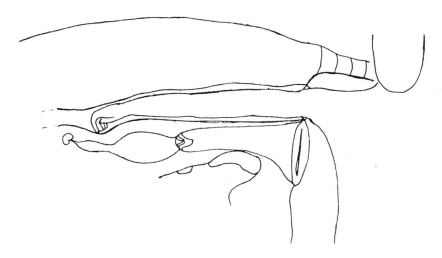

Figure 11.5 Rectal palpation in the mare allows the practitioner to feel the reproductive tract through the relatively thin wall of the rectum.

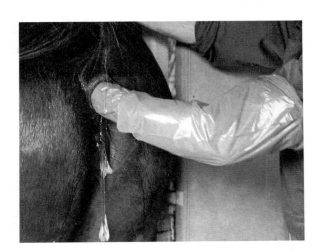

Plate 11.8

Rectal palpation in the mare.

receiving device can either be placed on the mare's flank or inserted into the rectum and directed towards the reproductive tract (Plate 11.9).

Placing the transducer on the flank is more common when the ultrasonic scanner is being used to detect pregnancy, as there is some evidence to suggest that rectal palpation in early pregnancy may be associated with a higher incidence of abortion. The emitted sound waves are absorbed by fluid and the more flaccid structures, whereas the firmer, more dense structures reflect back the sound waves. The transducer picks up these returning ultrasonic waves and transduces them into a visual picture on a television monitor. On the picture produced the fluid areas that absorb the waves are shown up as black in colour and the dense structures that reflect the waves show up as white, areas between the two showing up as varying degrees of grey (Plate 11.10).

With aid of this scanner the size and texture of the uterus, uterine horns, fallopian tubes and ovaries, along with any abnormalities, can be detected (Ginther, 1984a,b; McKinnon, 1987a,b).

By means of rectal palpation and ultrasonic scanning the healthy uterus should appear either enlarged and flaccid with little tone if the mare is in oestrus or turgid with plenty of tone if she is in dioestrus. If the person palpating works their way gradually from either uterine horn down and across the uterine body, running the horns through the hands and between the fingers and the thumb, then local thickenings, cysts or fibrotic areas can be detected. The diameter of each horn can be estimated and compared. Both horns should match. Mares that have had a history of infections may show ventral outpushings of the uterus, especially at the junction between the uterine body and the uterine horns. These mares may well have difficulty in carrying another pregnancy to term, as they often lose the foetus at the time of implantation. Incomplete involution of the uterus post partum can also be detected. Normally the uterus will return to its pre pregnancy state within four weeks of birth. In older mares, or those that have suffered weakening or rupture of the uterine wall, involution may take a lot longer, and in severe cases it will never be complete. This leaves a flaccid area of uterus with inadequate tone compared to the remainder. Again, such mares may well have difficulty carrying a pregnancy to term. Haemorrhages within the broad ligaments, which hold up the reproductive tract within the body cavity, may be picked up as local hard or firm swellings up to 40mm in diameter; these are often due to excessive strain, accidents or damage. Such mares may be incapable of carrying a pregnancy to term, if the broad ligaments are unable to support the weight of a full-term foetus.

An endoscope may be used in the mare to examine the internal uterine surface. It is passed through the vagina and guided on into the uterus via rectal palpation. Prior to the procedure all the usual precautions should be taken to minimise any chances of bacterial contamination. Once the endoscope is in place sterile air or oxygen is passed into the uterus and hence eases viewing of the mucous membranes and uterine epithelium (Plate 11.7).

The normal healthy uterine epithelium is pale pink in colour and

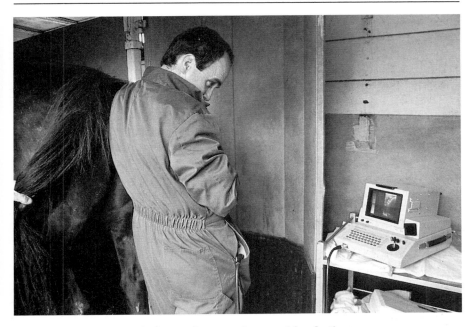

Plate 11.9   The use of ultrasonic scanning machine in the mare.

Plate 11.10

The image produced by an ultrasonic scanner. Fluid areas show up as black and solid structures as white.

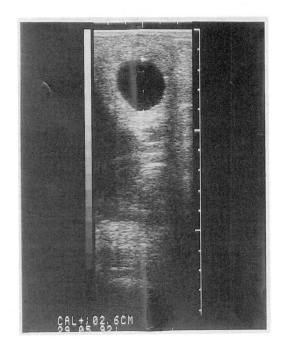

any mucus present is clear. Any signs of redness within the uterus, especially if this is also associated with cloudy or creamy mucus, is indicative of uterine infection or endometritis. Evidence of cysts, blood clots, thickening of the endometrium or scar tissue may indicate possible future problems in the mare's reproductive life.

The use of the endoscope is as yet limited due to the cost of the instrument and its delicate nature. Endoscopes at present are normally restricted to research use but are available if individual breeders wish to have their mares examined to the full extent. If there are any areas of concern within the mare's uterus, picked up either by rectal palpation or endoscopy, then a uterine biopsy may be performed to aid further investigation.

Uterine biopsies involve the removal of a small section of uterine lining by means of a small pair of forceps. These pieces of tissue are then examined cytologically to identify and investigate any suspected abnormalities (Kenney, 1978). Uterine biopsies can be taken with an uterine endoscope but the samples taken tend to be too small to be of any real diagnostic use. The best uterine biopsies are taken using uterine biopsy forceps (Figure 11.6 and Plate 11.11).

For this procedure the mare should be restrained as for rectal palpation. The perineum should be thoroughly washed and all instruments sterilised. The mare's tail should also be bandaged to further prevent contamination. The biopsy forceps are then passed into the vagina of the mare and guided up through the cervix by a well-lubricated gloved hand. The index finger guides the forceps through the cervix and into the uterus. The gloved hand is then removed and inserted into the rectum to guide the forceps to the section of the uterus that is to be biopsied. Samples taken from the mid uterine horn have been shown to be representative of the uterine endometrium as a whole (Bergman and Kenney, 1975), so they are suitable for routine biopsies where no specific abnormality is being investigated. In biopsies to further investigate suspected abnormalities, samples should be taken from each area in question and another sample from a seemingly normal area of endometrium. Once the area to be biopsied has been identified, the forceps are guided there with aid of the hand inserted in the rectum. The jaws of the forceps are opened and the section of endometrium to be sampled is pushed between the jaws by this hand. The jaws are then shut and the forceps drawn back slowly until the pressure of the endometrium is felt. A sharp tug will then release the sample which can be withdrawn through the vagina. The sample is immediately removed and fixed by fixation fluid in readiness for future histological examinations. Biopsies can be taken at any time of the oestrous cycle but standard biopsies are normally taken in mid dioestrus in order to standardise results.

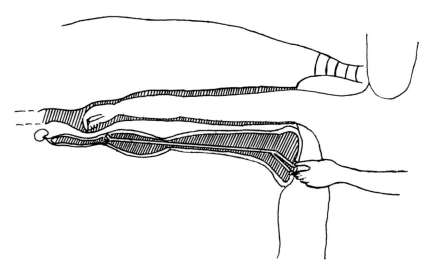

Figure 11.6   The use of uterine biopsy forceps in the mare. The arm inserted via
the rectum is used to ease the wall of the uterus into the jaws of the
biopsy forceps.

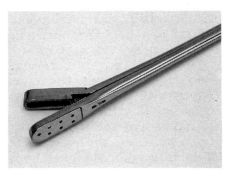

Plate 11.11

Uterine biopsy forceps.

Such samples can allow identification of abnormalities plus
indicate evidence of uterine degeneration. The examiner looks for
changes in cell structure, especially in luminal cells which may be
diagnostic of inflammation, fibrosis or necrosis. The procedure is
reported not to be disadvantageous to the mare's reproductive
cycle, though some people report a delay in the next oestrous
period (Kenney, 1977).

These four methods discussed above for the assessment of a
mare's uterine competence are not always used as standard
selection criteria, but may be used to elucidate any indications of

abnormalities picked up by the simpler routine procedure of rectal palpation. The choice ultimately lies with the owner and may be partly determined by the value of the mare and that of any prospective offspring.

Fallopian Tubes

The competence of the fallopian tubes is more difficult to assess. As mentioned previously, uniformity in size and shape can be assessed via rectal palpation. The fallopian tubes should feel wiry and uniform in consistency when rolled between the palpater's fingers. Severe abnormalities may be detected this way, especially adhesions connecting the infundibulum to the uterus or ovaries and scar tissue. Salpingitis (inflammation of the fallopian tubes) is hard to detect, though it has been reported to be relatively common in mares. Any such inflammation may hinder the passage of ova along the fallopian tube and even result in complete blockage of the oviduct. Ovarian tumours may also be detected but their presence is rare.

Assessment of fallopian tube blockage is very difficult. The starch grain test is relatively successful but complicated to carry out and, therefore, of limited practical use. A starch grain solution is injected through the mare's back at the sub luminal fossa, using a long needle. The ovary is manipulated, via rectal palpation, to lie immediately beneath the needle, and when it is in place, the starch grains are injected, approximately 5 ml in total, onto the surface of the ovary. Washings are collected from the anterior vagina and cervix 24 hours later. The presence of starch grains in the flushings indicates that the fallopian tube associated with the ovary on to which the starch grains were injected is patent or open.

A more invasive form of investigation is the use of laparotomy under general anaesthesia. Such cases involve the exteriorisation of the uterine horns and fallopian tubes through an incision either in the flank or ventrally. The fallopian tubes and ovaries can then be examined in detail. Such techniques are expensive and highly invasive; they also run the risks associated with general anaesthesia and are therefore rarely used.

Laparoscopy involving the insertion into the body cavity via the abdominal wall of a rigid endoscope with a light source is a further possibility, allowing the upper tract to be viewed in situ within the abdominal cavity. This technique is less invasive but requires an anaesthetic and does not allow such detailed examination.

Ovaries

Ovarian activity can be assessed most simply by rectal palpation and more extensively by ultrasonic scanning. The appearance of the mare's ovaries varies considerably with season and reproductive activity. These changes are detailed and discussed in Chapter 1. Rectal palpation has traditionally been used to assess the stage of the mare's oestrous cycle and therefore help in the timing of mating. It can be used in this context of selection criteria for a mare to ensure that the mare is reproductively active and that follicles and corpora lutea are produced regularly. It also can be used to ensure that the reproductive stage shown within the ovaries is synchronised with developmental changes in the remainder of the mare's tract. Such techniques can also indicate the presence of adhesions and neoplasms as well as cystic follicles, ova fossa cysts and other ovarian abnormalities which may disrupt the mare's reproductive activity. Further detail on the assessment of the reproductive stage of the mare by ovarian examination is given in Chapter 13.

Laparoscopy has been used experimentally to elucidate ovarian problems. Its use in practical stud farm management is as yet limited due to the need for a general anaesthetic (Plate 11.12).

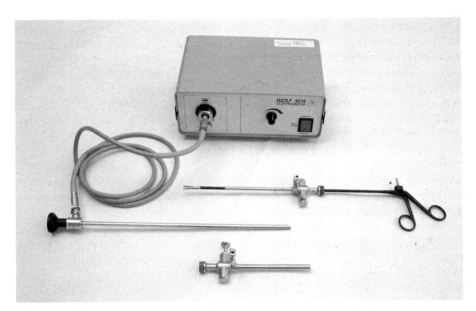

Plate 11.12   A laparoscope used to view the internal structures of the mare's reproductive tract via the abdominal cavity. The trocars pictured in the foreground are inserted through puncture wounds in the mare's abdomen. Through the larger is passed the laparoscope for viewing the internal organs, and through the other trocar is passed the manipulating forceps for moving internal organs.

Infections

Mares are notorious for being susceptible to reproductive tract infections. Not only can these cause temporary and possibly permanent infertility but they can be transferred relatively easily to the stallion and on to other mares. One of the standard procedures used to detect such infections is swabbing. Swabs may be taken from the uterus, cervix and/or clitoral area and assessed for pathogenic organisms, for example *Klebsiella aerogenes*, *Pseudomonas aeruginosa* and *Haemophilus equigenitalis* that are sexually transmissible. Such swabs can also give an indication of endometritis or cervicitis if organisms such as *Streptococcus zooepidemicus*, *E.coli* or *Staphlococcus* are present.

There are several methods of collecting samples via swabs but basically the clean sterile swab should be inserted in through the vagina which is held open by a vaginal speculum. For uterine swabs the process should only be carried out during oestrus when the cervix is relaxed and moist, easing the swab's passage, and when the sample obtained is most likely to contain pathogenic organisms truly reflecting the endometrium sampled. The swabs may be guided in through the cervix using a speculum or guided via a gloved, lubricated arm, as for uterine biopsies. Once the swab is in the lumen of the uterus, it is rotated against the endometrium to absorb uterine secretions and bacteria.

One of the problems with this technique is the possible contamination of the swab by either airborne pathogens or bacteria present in the cervix and/or vagina. In order to minimise this the mare should be prepared and washed as for biopsy and, in addition, a guarded swab may be used. Such swabs are contained within two sterile tubes. The first tube is passed through the cervix and the second tube is telescoped into the uterus. The swab can then be pushed out through the second tube far enough to reach the endometrium. A reversal of this process will reduce the amount of contamination on retraction of the swab (Figure 11.7 and Plate 11.13).

Once the sample has been taken the swab can be applied to a growth medium and incubated and subsequently examined for specific bacterial colonies. Incubation under specific temperature, humidity, atmospheric pressure and oxygen content can aid the identification of specific pathogenic types. Alternatively, swabs can be smeared on to glass slides for cytological examinations.

Cervical swabs are easier to obtain and may be collected at any time of the cycle. The samples may be taken by a similar process to that described above. Obviously the swab is not passed through the cervix but absorbs the secretions of the cervix itself. Clitoral swabs can also be easily taken usually from the clitoral sinuses around

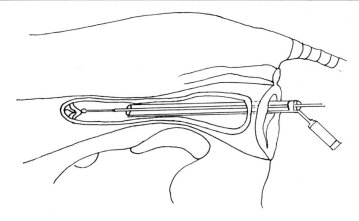

Figure 11.7 Cervical swabbing in the mare.

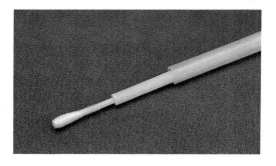

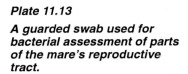

Plate 11.13

A guarded swab used for bacterial assessment of parts of the mare's reproductive tract.

the clitoral fossa. Gentle squeezing of the clitoris may produce smegma secretions for swabbing; care, however, should be taken as some mares might object. The clitoral fossa harbours the contagious equine metritis (CEM) bacterium that is notorious for causing infertility. The urethral fossa may also be swabbed, especially if a urine tract infection is suspected.

Swabbing is a very good, efficient and quick method of assessing bacterial contamination of the mare's reproductive tract, but its results must be treated with some caution. The presence of bacteria within the vagina and cervix does not necessarily indicate uterine infection, especially if the cervical seal is fully competent. The natural flora of many mares contains pathogenic organisms. The actual process of taking a sample also increases the chance of reproductive tract contamination by airborne bacteria. The technique must be carried out under as clean and sterile conditions as possible. In addition, swabbing is best carried out during oestrus, as at this time, due to high oestrogen levels, the body's ability to cope with vaginal pathogens is enhanced by increased leucocyte production.

Providing the possible drawbacks are borne in mind, swabbing is a simple procedure to carry out in the routine selection of mares for breeding. Indeed as a result of the Thoroughbred Breeders Association Code of Practice 1978 it is advised that all mares should be swabbed prior to arrival at the stud and again at the oestrus of service to ensure the absence of CEM. Only mares with a negative certificate will be accepted at the stud and only those with a second negative will be covered.

Blood Sampling

If there are reasons to believe that a mare may have problems in carrying a foal to term, but no anatomical abnormalities have been detected using the previously discussed techniques, then the problem may lie in hormonal inadequacies. The endocrine profiles of mares can be determined by sequential blood sampling and using the accurate radioimmunoassay. Any deviations from the norm can be picked up and the specific area of failure, ie follicle development, ovulation signal, oestrus behaviour, can be identified. In the light of these results appropriate hormonal therapy may be possible in an attempt to compensate for natural deficiencies, or it can be decided that such a mare is not worth the risk or cost of such therapy. Details on the normal endocrine profiles for mares are given in Chapter 3.

Blood samples can also be used to indicate the general health status of a mare, bringing to light specific deficiencies possibly related to diet, or high leucocyte counts indicative of latent infections. More details on what information can be gathered from blood sampling is given in the following section on the selection of the stallion.

Detailed above are all the possible techniques that can be used to aid in the selection of a mare for breeding. Obviously the extent to which you take the selection procedures depends on the finances available, the value of the mare, the potential value of any progeny, the extent of any abnormalities picked up and finally on personal choice.

The Stallion

The selection criteria for reproductive competence in the stallion can be divided up into the following areas:

History – general and breeding
Temperament and libido
Age
General conformation
Reproductive tract examination
Semen analysis
Blood sampling
Infections
General stud management

These will be discussed in turn.

History

Records of a stallion's past are invaluable in aiding selection and as in the mare the stallion's history can be divided into his breeding and his general history. Records for stallions do not tend to be as detailed or as readily available as for mares. However, records of his past breeding performance, if available, will answer such questions as:

When does his season normally start and end?
How many mares is he used to covering in a season?
What are his return rates like?

The answers to these questions will indicate his reproductive ability. Stallions with short seasons will be less able to cover as many mares and may suffer from low libido. The number of mares he has served per season in the past and the return rates, along with semen analysis will give you an indication of what work load he will be capable of doing. If his return rates are high, especially if a significant increase is seen with an increase in work load, this may indicate the natural limit of the number of mares he is able to cover. The routine of covering depends on the individual stud and what the stallion works best under. Examples are two covers per day for six days with a day's rest, or two covers per day for eight days followed by two days' rest or variations on this theme. Most stallions need a rest day but should be able to cover mares at the rough frequency of the system given above. If there are indications that a stallion is not capable of such work loads and requires more rest days than indicated to maintain his fertility rates, then his selection should be queried, especially if you are looking for a stallion to purchase. Return rates are a good guide to a stallion's ability but it must be remembered that the fertility of a stallion is only as good as the fertility of the mares he is covering, so if a stallion has had a season of poor mares, his high return rates will not be a true representation of his own ability.

Records should also indicate the stallion's temperament and any specific characteristics that he might have. It is to be hoped that his bad habits, especially those that might prove dangerous, will also be indicated. To be forewarned is to be forearmed and might lead you to reject an unsatisfactory stallion.

Any previous semen analysis should be detailed in his records. Many Thoroughbreds have a semen analysis carried out routinely at the beginning of each season. This, along with a blood sample taken normally at the same time, allows any potential problems to be picked up in time for action to be taken before the season starts in earnest. Some native studs, especially those running less valuable stock, tend to restrict semen analysis to prior to purchase or if a problem is indicated. Any past infections of the reproductive tract should also be detailed in his records, along with any treatment given and its success. The long-term effects of infection should be shown up in the stallion's work load and return rates for that season and any subsequent seasons.

The stallion's general history should indicate his vaccination and worming status along with incidence of injuries and details of any past accidents. Damage to his hind quarters or limbs may restrict his ability to mount a mare. Artificial insemination may be an alternative, but even so he will need the occasional mount for the collection of samples, and the number of mounts per mare fertilised will be significantly reduced. Such stallions are thus not really a good bet for purchase. Again, as with the mare, if there are any suggestions that any damage or weaknesses may be heritable it is not a good idea to select that stallion. Injuries to a stallion's genitalia, usually as a result of a kick from a mare, will cause degenerative tissue and scar tissue within the testes, which will reduce the sperm-producing cells and hence fertility rates of the stallion. Severe damage resulting in the removal of a testicle should also be noted in a stallion's records to reassure potential purchasers that he is not a rig. Such stallions are capable of fertilising a mare but the work load may have to be reduced and this should be borne in mind. Severe injuries to a stallion during mating often have psychological effects drastically reducing his libido, possibly to such an extent that he is unwilling to mate.

Past illnesses should also show up on records. As with mares, illnesses associated with the respiratory or circulatory systems may indicate that the stallion will not be capable of working a full season, limiting the numbers of nominations to him that are available or that can be sold. Again, if there is a possibility that such weaknesses could be heritable, the stallion should be avoided. Any illnesses resulting in a fever can disrupt the developing spermatogonia and hence spermatogenesis due to the elevated temperature. This may result in temporary infertility though this may not be evident for several months as the spermatogenic cycle takes approximately 60 days. Systemic infections, for example strangles or flu, can cause inflammation within the testis and if this results in a significant amount of tissue degeneration, permanent sub-fertility or even infertility may result.

The sorts of stallion records that should be available to a potential purchaser are illustrated in Figure 11.8.

Date of Mares Arrival	Mare/Owner	Service Date	Return	Return	Preg. Det. Date	Preg Result	Live foal D.O.B.	Comments
16/1/91	Trefaes Fancy Mr T. James	20/3/91	-	-	20/4/91	✓	Live 20/2/92	Foal at foot D.O.B. 12/3/91
18/1/91	Katie Jane M. Davies	15/3/91	7/4/91	-	30/4/91	✓	Abortion 10/12/91	Maiden mare
30/3/91	Penpontpren Mina Mrs M. Morel	30/3/91	21/4/91	-	15/5/91	✓	Live 15/3/92	Barren Mare
1/2/91	Lluest Megan Lluest Stud	30/4/91	21/5/91	11/6/91	5/7/91	✗	-	Foal at foot D.O.B. 30/3/91 delayed uterine involution
12/2/91	Rose Mr. A. Jones	18/3/91 10/4/91	-	-	8/4/91 3/5/91	✓ Twins ✓	- 30/3/92	Foal at foot D.O.B.12/3/91 Twins detected 8/4/91 Aborted & served again

Figure 11.8   The type of records that should be kept for each stallion during his breeding season, indicating the mares he has been put to and the result, along with any noteworthy comments.

Temperament and Libido

The temperament of the stallion is very important, for the ease of management and as it is a heritable trait. A stallion of a quiet and kind disposition is a great asset and will be much easier and safer to handle (Plate 11.14). A stallion that is rough to his mares will not only run the risk of inflicting permanent damage to them but may also be hurt himself if they retaliate. A rough stallion will prove unpopular and it may be difficult to get him enough mares to make his use economical. Some protection in the form of neck guards can be given to mares that are mated to stallions that tend to bite during covering, but no protection can be given against stallions that are downright vicious, and they should be avoided at all costs.

Bad behaviour in many stallions is a direct result of the conditions and management under which they are kept. Therefore, especially in the case of a stallion that seems to have developed bad habits later in life or after a change of owner or management, the conditions under which he is kept should be looked into before he

Plate 11.14   *A well-behaved stallion is an asset to any stud, easing his management and reducing the danger to his handlers.*

is completely ruled out for covering a mare, though as a potential purchase he is not a good choice as such habits are difficult to break. Bad behaviour tends to perpetuate itself, as, due to the potential danger, such stallions are kept in and confined for longer periods of time and hence away from other companions. Their boredom is, therefore, exasperated and their bad habits develop further. Other habits to look out for are the usual signs of boredom, for example, weaving, crib biting and windsucking, as he may pass these habits on to other horses within the same yard.

Some stallions also develop the habit of self-masturbation, rubbing his extended penis against the underside of his belly even to the extent of ejaculation. Such a habit is not only potentially embarrassing for owners but it may lead to the loss of precious semen and hence reduce the work load that can be expected of that stallion. A ring may be fitted to the penis to prevent this habit but avoidance is the best. Some stallions also indulge in self-mutilation, especially after mating, biting themselves in areas where the smell of the mare lingers. Thorough washing post mating can reduce the incidence but adds to management time and expense.

When selecting a stallion to purchase it is best, therefore, to avoid stallions that show evidence of either of these traits, though, as there is no significant evidence that such behaviour is heritable, there is no reason why the stallion should not be chosen to cover an individual mare.

A stallion's reproductive temperament and willingness to cover is termed his libido, which partly determines his reproductive potential. Libido is governed, like all other sexual activity, by season, as discussed in Chapter 4. Hence, those stallions with longer seasons tend to show a higher libido and, therefore, willingness to mate early on in the season, extending the time in which they can be worked. Ideally, if selecting a stallion to purchase, he should be seen teasing and covering a mare. A stallion with a low libido will need to mount a mare several times before ejaculation, taking up to 30 minutes to cover a mare, or he may fail completely; he may also show initial interest very reluctantly. The number of mounts per ejaculation and the time between actual intromission and ejaculation are good indications of libido. The number of mounts per ejaculation should be as near to one as possible and the time between intromission and ejaculation a matter of seconds (5–10 seconds on average).

Age

The age of the stallion is less important than that of the mare as far as reproductive ability is concerned. The significance of age in the selection of the stallion depends on what that stallion is required for, ie for a single mating to a selected mare or as a potential purchase for long-term future use. If you are selecting him for the service of a single mare, then as far as you are concerned, he will be required to perform a couple of times and that is it; his age is of limited importance providing he is capable of doing the job. However, if you are looking to select a stallion for purchase and, therefore, long-term future use, you want to ensure that he is young and fit enough to give you plenty of covering seasons ahead but old enough to have proved his worth.

As far as a lower age limit is concerned, most colts have reached puberty at 18 to 24 months of age, as discussed in Chapter 2. Puberty is classified as the age at which an individual is capable of producing the gametes ready for fertilisation and subsequent embryo development. A colt can, therefore, in theory be used as soon as he reaches puberty, but care must be taken to introduce him to the job gradually and not to overwork him too soon or give him awkward mares that may affect his as yet delicate ego and reproductive confidence. Further details on early stallion management are given in Chapter 19. The purchase and use of such young stallions is risky, as they have no proven record of performance.

As far as an upper age limit is concerned, this really depends on the stallion's general health and condition. If he has no problems with lameness, stamina, wind, injury etc, he may well be capable of working well into his teens and even twenties, though in the later years his work load may well have to be reduced.

As discussed in the case of mare selection, if an older ex-performance horse is being considered, it must be borne in mind that he will require a prolonged period of time to adjust physically and psychologically to his new role in life. Details of the problems associated with performance horses as stallions are given in Chapter 12.

General Conformation and Condition

A stallion's general conformation is of importance, not only as it will be passed on to his offspring, but also to ensure that he is capable of withstanding a full breeding season. A stallion of poor conformation in the limbs, especially hind quarters, will also be weak in this area and may, therefore, be unable to withstand the heavy work load of a full season, limiting his economic viability.

Good general condition and physical fitness are very important for the breeding stallion. The condition of a stallion, like that of the mare, can be classified on a scale of 0 to 5, 0 being emaciated and 5 overfat. The optimum body condition for a stallion in work is 3, that is, he is well muscled up and in fit working condition (Plate 11.15). Particular note should be made of his physical ability to cover mares. He should be free of all signs of lameness, especially in the hind limbs. His legs should be checked before and after exercise and a comparison made, to ensure that there is no sign of swelling, a possible indication of weakness that he would not hold up to the strain of a heavy breeding season. He should be free of all conditions such as arthritis, spinal or limb injury, wobbler syndrome or laminitis causing pain that might either put him off mating or make it impossible. A stallion's feet should also be in excellent condition and regularly trimmed to ensure they stay that way. Adequate heart room in a broad chest is also very desirable, and if doubt is placed on the stallion's cardiovascular system, an electrocardiography may be conducted. The nutritional demands during the breeding season are similar to those of a performance horse, the work load of the two being approximately equivalent. Further details on stallion management are given in Chapter 19, which is devoted entirely to this subject.

Stallions in a condition score of less than 3 tend to have a low libido, and are physically unable to stand a heavy work load. If the stallion's condition is very poor, spermatogenesis will also suffer and, therefore, sperm quality will fall off and in parallel his fertilisation rates. At the other extreme overfat stallions also tend to have a low libido. They tend to be lazy and may be incapable of mounting a mare, and in addition, the extra weight puts additional strain on the mare at mating and may cause her damage.

There is no evidence to suggest that increasing body condition as the season approaches has any effect on spermatogenesis, unlike the effect seen on ovulation in the mare.

Reproductive Tract Examination

An external examination of the stallion's reproductive genitalia is an essential selection procedure, as his ability to perform is a function of the condition of his reproductive organs. He should have two testes that function normally so that testosterone is produced and spermatogenesis occurs. The two testes can be felt through the scrotum and palpated to ensure they are of a similar

Plate 11.15   A stallion in good, fit, well-muscled working condition ready for the breeding season.

size and consistency, move easily within their tunicas, and are not warm to the touch. Occasionally the left testis is slightly larger than the right, but the difference should be slight and should not be accompanied by an increase in heat. The surface of the testis should feel smooth with the occasional blood vessel being felt running under the skin. Any adhesions preventing the testes slipping up and down easily within their tunica are likely to be due to past injuries, that, in turn, may be an indication of scar tissue or fibrous tissue within the testes. This not only reduces the volume of spermatic tissue but may also interfere in the sperm production within the remaining tissue. Excessive fat within the scrotum may be a sign of poor conditioning and will also increase the insulation of the testes, with the danger of increasing testes temperature and, therefore, decreasing sperm production.

Cancerous growths within the testes are normally rare but should be checked for. The skin of the scrotum should be checked for dermatitis, an inflammatory reaction of the skin which can again affect sperm production because of an increase in testicular temperature. The position of the epididymides should also be felt. Their normal position in the non-retracted relaxed testis is on the cranial (abdominal) side of the scrotum. Positioning elsewhere may be indicative of testes torsion or twist.

The penis and prepuce of the stallion should also be examined for injury, squamous cell carcinoma, summer sores, sarcoids and general infections. Examination can be carried out at washing prior to semen collection and should be a routine selection procedure.

The spermatic chord or vas deferens leaving the testis, plus its blood and nerve supply, passes up into the body of the stallion via the inguinal ring, which should be free from adhesions and hernias. The vas deferens then passes up to the penis and out and is joined along its length by the accessory sex glands that produce seminal plasma. These glands are of obvious importance in ensuring the longevity of sperm survival. However, as they lie within the body cavity they cannot be felt by external palpation, so rectal palpation similar to that carried out in the mare is used to feel the testis, vas deferens and accessory glands through the rectum wall.

The vas deferens should be felt entering the body cavity at the inguinal ring and should both, one either side, be smooth and of uniform diameter along their entire lengths. Alongside the vas deferens as they enter the body cavity lies the spermatic artery. The pulse of the spermatic artery should also be checked. Very low pressure, or a drop between successive examinations, may be indicative of a haemorrhage, blood clot or tumour, or the release of body fluids into a localised infection site. The accessory sex glands can be palpated individually and their texture, size and shape compared to the norm for that breed and size of animal. The paired glands, the seminal vesicles, should also be checked for symmetry. The function of the accessory glands can be assessed at semen evaluation as will be discussed in the following section.

Semen Evaluation

The evaluation of stallion's semen is a routine selection procedure. If a stallion is to cover mares throughout the breeding season with consistent success, his semen has to meet various minimum parameters. In many studs all stallions routinely have their semen evaluated at the beginning of each season. The quality of his semen has a direct effect on the stallion's ability to consistently and successfully cover a number of mares throughout the season. Ideally the collection of semen for evaluation would be carried out daily for a period of seven days and hence daily production of semen could be calculated, but this is impractical in many situations. In most systems, semen is collected from the stallion or he is used, and then he is rested for four to seven days. Two samples of semen are then collected for evaluation with an interval of at least one hour between. Collection of semen is normally by means of an artificial vagina (AV). Details of the collection and evaluation procedure are given in Chapter 20. The normal parameters for semen are given in Table 11.1.

Table 11.1   Normal acceptable parameters for stallion semen.

Colour	Milky white
Volume	30–300 ml
Concentration	$30–600 \times 10^6$ / ml
Live/dead ratio	6.5 : 3.5
Morphology	>65% morphologically normal
Motility	minimum 40% actively motile
Longevity at room temperature	45–50% alive after 3 hrs 10% alive after 8 hrs
pH	6.9–7.8
White blood cells	<1500 ml
Red blood cells	<500 ml

Blood Sampling

Blood sampling of stallions can be used to assess their general health and can indicate low grade infection, blood loss, cancer, nutritional deficiencies or parasite burdens. Low red cell counts, ie below 10×10^6/ml, indicate that the stallion is anaemic. Further tests can then ascertain whether this is due to iron or copper deficiencies or parasitism. Packed cell volume is a quick and easy assessment of red cell:fluid balance. Normal levels are 40–50%. The colour of the supernatant in packed cell volume tests is also indicative of any problems. Normally it is straw-like in colour. Discolouration can indicate problems; for example, red/pink supernatant indicates a breakdown of red cells and release of haemoglobin. Haemoglobin levels themselves are a well-known indicator of anaemia: levels <12–17 g/l indicate problems.

A high white blood cell count that is above 12,000–14,000/ml indicates the presence of disease, especially in the form of an infection or cancer. A low total protein index, that is the protein content of serum after clotting, of <15 g/l indicates blood loss, starvation or liver, kidney or gastrointestinal disease. Fibrinogen levels may also be indicative of abnormalities, high levels of >10 g/l suggesting inflammatory, neoplastic or traumatic disease (Table 11.2).

Table 11.2   Stallion blood parameters indicative of disease.

Red blood cell count	<10 x 10^6/ml
Packed cell volume	<40–50 %
White blood cell count	>12,000–14,000/ml
Total protein index	<15 g/litre
Haemoglobin	<12–17 g/litre
Fibrinogen	>10 g/litre

Any stallion showing these characteristics should not be considered for use until the problem has been isolated and treatment, if possible, has begun.

Infections

Like the mare, the stallion is susceptible to sexually transmitted diseases and as such all stallions should be tested for infections prior to purchase, either to eliminate them or to allow treatment to commence immediately prior to their use.

As with the mare, infections are identified by the use of swabs, which should be taken from the urethra, the urethral fossa and the prepuce of the stallion's penis. Swabs should be taken from the erect penis, erection being encouraged by a mare on heat or by the use of tranquillisers. Three different swabs must be used and it is best to take the urethral fossa sample last, as this one, unlike the others, can cause discomfort to the stallion and hence objection, often in the form of kicking. The stallion's penis has a natural microflora of bacteria and fungi and these should be distinguished from venereal disease bacteria, eg *Haemophilus equigenitalis*, *Klebsiella aeruginosa* and *Pseudomomas aeruginosa*.

Swabbing is routinely carried out in many studs on all stallions well before the season starts. This allows time, if infections are identified, for treatment to begin well before the beginning of the breeding season. Further details of infection of the stallion's reproductive tract and the effect on reproduction are given in Chapter 19.

Chromosomal Abnormalities

Chromosomal abnormalities are well documented in the mare but less so in the stallion (Long, 1988). Such configurations as 63XO:64XY mosaic, 64XY 13 quarter/deletion are associated with inability to impregnate mares or very low fertility rates, despite apparently normal genitalia (Halnan and Watson, 1982; Halnan, Watson & Pryde, 1982).

General Stud Management

If your selection of a stallion is not for purchase but rather for use on one of your mares, you will be interested in the management at the stud at which he stands. There are several things that will concern most owners selecting a stud to send their mare to, and these will be discussed in turn.

The system of breeding used is of prime importance. Is the stud geared up to have visiting mares to stay or are you expected to bring the mare for the day, having detected that she is in heat yourself, and take her away the same day after covering? Some studs allow the mares to stay a few nights but only have limited facilities and may well expect mares to live out. This obviously has a bearing on the distance it is possible to travel. You should also look at the method of covering, varying from pasture breeding to intensive in hand breeding. The various methods are discussed in detail in Chapter 13. Some studs will expect the mare to be taken home as soon as she has been covered, whereas others will allow her to stay for re-covering if she has not taken the first time and will only allow her home after a positive pregnancy detection (PD) at scanning and/or rectal palpation, usually 3–4 weeks post mating.

Many of the smaller native pony type studs do not have the facilities to foal down visiting mares. Mares have to be brought to the stud very soon after foaling, which can be traumatic and dangerous for the foal and really precludes using a stud that is too far away. Larger studs, especially Thoroughbred studs, tend to have the mares brought in to foal at the stud. They are normally brought 4–6 weeks prior to foaling: this allows the mares to be covered on their foal heat without the danger of travelling a young foal, and it also allows the foal to be born in the place where it is to spend the first few weeks of its life. As such, it will gain invaluable antibodies in utero to that specific environment, avoiding the possible challenge to its immune system of a strange environment so early in life.

The system that you choose is ultimately a personal choice depending on your priorities and the finances available. The more intensive systems tend to be associated with the Thoroughbred industry, where expense is of less significance but hygiene and safety to the horses are of paramount importance. In this system the mares are taken to the stud to foal down and are subsequently covered and left until they have a positive PD. In such systems the service fees are high and the costs of keep and veterinary inspection are great, but the offspring are worth considerable amounts of money and the risks are lower. At the other end of the spectrum native studs will serve a mare that arrives in their yard and within half an hour she can be on her way back home. Stallion fees are low, costs are low but the offspring is not that valuable and the risks are higher.

The daily management at the stud should also be investigated and matched as closely as possible to the mare's normal routine. If not, her routine at home should be slowly altered to that at the potential stud to minimise the changes the mare has to adjust to. All animals on the stud should be wormed regularly and vaccinated, and documented proof is usually required to prove that visiting mares are also adequately protected. At the more intensive studs, mares standing to valuable stallions will also require CEM negative certificates.

A good impression of the standard of management of a yard can be gained by a general visit. The yard, whatever system in use, should be clean and tidy, all the mares and stallions should be in good condition, the pasture good and the animals contented. If the mare is to foal there, the foaling facilities should be clean, safe and roomy with a good system for 24 hour monitoring by skilled staff. The facilities of the yard and the equipment and expertise available will reflect the type of stallion and his nomination fee.

Further details on the management systems and principles for both the mare and stallion at breeding are given in Chapters 12–19. When examining potential studs it is as well to bear in mind that the ideal is not normally achieved. It is unrealistic to expect a yard standing the cheaper and normally more native type stallions, with stud fees in the order of £50, to have the facilities found in a Thoroughbred stud standing stallions with nomination fees up to £50,000.

References and Further Reading

Amann, R.P. and Graham, J.K. (1933) Spermatozoal Function. In: Equine Reproduction. Ed. A.O. McKinnon and J.L. Voss. Lea and Febiger, Philadelphia, p. 715.

Bergman, R.V. and Kenney, R.M. (1975) Representation of an uterine biopsy in the mare. Proc. 21st Ann.Conv.Am.Ass.Equine Practnrs, pp. 355–362

Bowling, A.T. and Hughes, J.P. (1993) Cytogenetic Abnormalities. In: Equine Reproduction. Ed. A.O. McKinnon and J.L. Voss. Lea and Febiger, Philadelphia, p. 258.

Brook, D. (1993) Uterine Cytology. In: Equine Reproduction. Ed. A.O. McKinnon and J.L. Voss, Lea and Febiger, Philadelphia, London, p. 246.

Caslick, E.A. (1937) The vulva and vulvo-vaginal orifice and its relation to genital tracts of the thoroughbred mare. Cornell Vet., 27, 178.

Day, F.T. (1939) Some observations on the causes of infertility in horse breeding. Vet.Rec., 51, 581.

Doig, P.A. and Waelchi, R.O. (1993) Endometrial Biopsy. In Equine Reproduction. Ed. A.O. McKinnon and J.L. Voss, Lea and Febiger, Philadelphia, London, p. 225.

Ginther, O.J. and Pierson, R.A. (1983) Ultrasonic evaluation of the reproductive tract of the mare: Principles, equipment and techniques. J.Equine Vet.Sci. 3, 195.

Ginther, O.J. and Pierson, R.A. (1984a) Ultrasonic evaluation of the reproductive tract of the mare: Ovaries. J.Equine Vet.Sci. 4, 11.

Ginther, O.J. and Pierson, R.A. (1984b) Ultrasonic anatomy of equine ovanes. Theriogenology, 21, 47.

Halnan, C.R.E. and Watson, J.I. (1982) Detection of G and C band karyotyping of genome anomolies in horses of different breeds. J.Reprod. Fert.Suppl., 32, 626.

Halnan, C.R.E., Watson, J.I. and Pryde, L.C. (1982) Prediction of reproductive performance in horses by karyotype. J.Reprod.Fert.Suppl., 32, 627.

Kenney, R.M. (1977) Clinical aspects of endometrium biopsy in fertility evaluation of the mare. Proc. 23rd Ann.Conv.Am.Ass.Equine Practnrs., pp.355–362.

Kenney, R.M. (1978) Endometrial biopsy technique and classification according to interpretation. Proc.Soc. Theriogenology, pp. 14–25.

Kenney, R.M., Kent, M.G., Garcia, M.C. and Hurtgen, J.P. (1991) The use of DNA index and karyotype analysis as adjuncts to the stimation of fertility in stallions. J.Reprod.Fert.Suppl., 44, 69.

LeBlanc, M.M. (1993) Vaginal Examination. In: Equine Reproduction. Ed A.O. McKinnon and J.L. Voss. Lea and Febiger, Philadelphia, London, p. 221.

LeBlanc, M.M. (1993) Endoscopy. In: Equine Reproduction. Ed. A.O. McKinnon and J.L. Voss. Febiger, Philadelphia, London, p. 255.

Long, S. (1988) Chromosomal anomolies and infertility in the mare. Equine Vet.J., 20, 89.

McKinnon, A.O. (1987a) Ultrasound evaluation of the mare's reproductive tract. Part 1 Compend.Cont.Ed.Pract.Vet. 9, 336.

McKinnon, A.O. (1987b) Ultrasound evaluation of the mare's reproductive tract. Part 11 Compend.Cont.Ed.Pract. Vet. 9, 472.

Nett, T.M. (1993) Reproductive Endocrine Function Testing in Stallions. In Equine Reproduction. Ed A.O. McKinnon and J.L. Voss. Lea and Febiger, Philadelphia, London, p. 821.

Parker, W.A. (1971) Sequential changes of the ovulatory follicle in the estrous mare as determined by rectal palpation. Proc.Ann.Mtg.Vet., 1971.

Pascoe, R.P. (1979) Observation on the length and angle of declination of the vulva and its relation to fertility of the mare. J.Reprod.Fert.Suppl., 27, 299.

Pickett, B.W. (1993) Reproductive Evaluation of the Stallion. In Equine Reproduction. Ed A.O. McKinnon and J.L. Voss. Lea and Febiger, Philadelphia, London, p. 755.

Pierson, R.A. and Ginther, O.J. (1985) Ultrasonic evaluation of the corpus luteum of the mare. Theriogenology 23, 795.

Ricketts, S.W., Young, A. and Medici, E.B. (1993) Uterine and Clitoral Cultures. In Equine Reproduction. Ed A.O. McKinnon and J.L. Voss. Lea and Febiger, Philadelphia, London, p. 234.

Shideler, R.K. (1993) History. In: Equine Reproduction. Ed A.O. McKinnon and J.L. Voss. Lea and Febiger, Philadelphia, London, p. 196.

Shideler, R.K. (1993) External Examination. In: Equine Reproduction. Ed A.O. McKinnon and J.L. Voss. Lea and Febiger, Philadelphia, London, p. 199.

Shideler, R.K. (1993) Rectal Palpation. In: Equine Reproduction. Ed A.O. McKinnon and J.L. Voss. Lea and Febiger, Philadelphia, London, p. 204.

Van Neikerk, C.H. and Van Heerden, J.S. (1972) Nutritional and ovarian activity of mares early in the breeding season. J.S.Afr.Vet.Ass., 43, 351.

Varner, D.D. and Schumacher, J. (1991) In: Equine Medicine and Surgery, 4th. Ed. Ed Colahan, P.T., Mayhew, I.G., Merritt, A.M. and Moore, J.N. Vol.2, Chapter 9.

CHAPTER 12

Preparation of the Mare and Stallion for Covering

IT is essential that preparation of the mare and stallion for covering starts in plenty of time prior to mating. It is of no use making a last-minute decision that you wish to put a mare in foal or stand a stallion and then wondering why she does not hold or your return rates are high. A preparation time of at least 6 months is required in order to maximise the chance of conception. This chapter will concentrate mainly on this period, with some reference to any earlier preparation that may be required.

Preparation of the Barren Mare

If a mare is destined to be a brood mare, she must be brought up with this aim in mind with close monitoring of her general condition and growth. Horses normally reach puberty at 1.5–3 years of age, depending on the breed and the nutritional status during early life. The large coldblooded heavy horse types and more native pony types tend to be late maturers and hence tend to reach puberty later than the finer warm- or hotblooded types. If you are considering breeding a mare at or near puberty before she has reached her full size, then her stage of development and her general condition are of utmost importance. Breeding her before she has attained her mature body size gives her system the additional burden of pregnancy as well as her own continued growth and development. Providing she is well grown and in good body condition, she should be able to cope with such additional demands, but this must be borne in mind when considering her management throughout her pregnancy, especially her nutritional management.

Today, many horses, both mares and stallions, come into breeding relatively late in life, having had a successful performance career, as a result of which they have been chosen for breeding. In such cases, both mares and stallions are older and more set in

Plate 12.1   Many successful stallions are currently or have been performance horses. As such they need careful management in order to perform well at both jobs.

their ways and have been managed all their lives as athletes, not breeding stock (Plate 12.1).

Performance horses must be allowed plenty of time to unwind and be let down both physically and psychologically. This should start at the latest during the autumn prior to the planned mating. Work loads should be slowly reduced to a maintenance level allowing a 6 month wind down period. This is usually adequate for most horses, though much variation between individuals is seen, and some may take as long as 18 months to readjust. The effect of intensive training to realise maximum athletic potential has detrimental effects on reproductive performance in all mammals. This can be clearly demonstrated in women athletes that regularly do not ovulate during training. Mares in peak athletic condition will characteristically demonstrate abnormal oestrus cycles, often showing delayed oestrus, silent heats and oestrus behaviour not accompanied by ovulation. However, given a long enough winding-down period, a mare should start showing regular oestrous cycles and be successfully mated.

Careful attention also has to be given to the mare's nutrition and, related to that, her exercise. Barren or maiden mares should be in good, not fat, body condition at covering. The condition of mares and stallions can be rated on a scale of 0 to 5, 0 being emaciated and 5 grossly fat. The body condition score to aim for at mating is 3, that is the mare will be lean and fit. In addition, it has been shown that better conception rates are obtained in mares that have been on a rising plane of nutrition prior to covering, especially in the last 4–6 weeks. This system of nutritional management is termed flushing and is practised regularly in the agricultural industry, especially with breeding ewes. This flushing of barren mares has also been reported to advance the start of the breeding season by as much as 30 days, providing the mare is not overfat or too thin (Plate 12.2). Poor condition and excess condition, especially in the categories 0 and 5, severely affect reproductive performance, delaying the onset of the breeding season and in severe cases suspending all oestrous cycles. The best regime is to get the mare in condition score 2–3 in the autumn prior to covering. As the season approaches, gradually replace some of the mare's roughage intake with concentrate feeds in the last 4–6 weeks, so increasing the energy intake. It is this increase in energy intake that has the desirable effect on conception rates.

If the mare is young, it may also be pertinent to supplement her diet during the last 4–6 weeks with protein, calcium, phosphorous and vitamin A, as the requirements for these are higher in young maiden mares than mature mares.

Plate 12.2   Mares being flushed on lush pasture prior to mating in order to encourage optimum fertilisation rates.

Exercise is also important in the barren mare, helping to maintain body condition and prevent obesity. Barren mares are often ridden regularly during the preparation period, especially if they are stabled in. If the mare concerned is an ex-performance horse it must be borne in mind that this period of preparation is designed to allow her to unwind (Plate 12.3). All mares should also be turned out daily and in an ideal world barren mares, especially those not broken, would live out except in the most inclement weather, thus getting ad lib exercise (Plate 12.4).

A period of 6 months' preparation also allows a mare not bred before to be tested for infection or for any damage due to old infections to be identified. Time is, therefore, available for any active infections to be diagnosed and treated and a full recovery to be achieved, and any damage due to old infections can be investigated and either helped in its repair or the mare discarded from the breeding scheme, saving the cost of a stud fee. If she is of sufficient genetic merit, the possibility of embryo transfer can be investigated and arrangements made if she is to enter such a scheme.

All barren mares, especially maidens, should be introduced to the handling systems, buildings and surroundings associated with breeding in the autumn. Old timers need to be reintroduced only a few weeks before the season starts. Any changes in diet must be introduced gradually and also the introduction of new companions should be done early enough to allow a settling down period.

Preparation of the Pregnant Mare

Much more care has to taken in the preparation of the pregnant mare for re-covering the following spring, and the presence of a pregnancy limits to a large extent how she can be managed. One advantage with a pregnant mare is that she has seen it all before, at least in the previous year, and no physiological adjustment time or winding down period is required. However, a careful eye should be kept on her condition in order to prevent obesity. Once a pregnant mare has got overfat it is very difficult to do anything about it, especially in late pregnancy, without endangering the foetus. Prevention is therefore much better than cure and it is essential that the mare is in a fit condition prior to going to the stallion initially and that condition 3 is maintained throughout pregnancy and into her next covering. Flushing of pregnant mares, as discussed above for barren mares, is not advised and has no effect on conception rates to the foal heat. If there is a reasonable period of time between foaling and re-covering this can be used to try to

Plate 12.3   Gentle riding provides ideal exercise for barren mares, in order to keep them fit for covering and any resultant pregnancy.

Plate 12.4   If mares are not broken to ride, they must be turned out to get enough exercise to keep fit.

adjust body condition, but care must be taken as alteration in nutrition will affect milk yield and hence the foal at foot.

Exercise is an important aid in maintaining a body condition of 3 in the pregnant mare. Pregnant mares can be safely ridden well into the sixth month of pregnancy; some can be ridden further on into pregnancy but this depends on the individual. By the sixth month all strenuous work must be excluded. Mares that are not broken and those in late pregnancy should be turned out every day to help maintain fitness, blood circulation and prevent boredom. In an ideal world mares would live out with plenty of opportunity to exercise: such systems are popular in temperate latitudes without the risk of adverse weather conditions. Exercise not only helps to prevent obesity but also maintains fitness in the mare and her muscle tone, both of which will be needed at and after parturition. Further information on this is given in Chapter 14.

Preparation of the Stallion

If a colt is destined to become a working stallion he must be brought up from his early days with this in mind, especially as far as discipline is concerned. Many stallions become hard to handle and in some cases downright dangerous because discipline and respect for authority have not been established in early life.

Most of the problems encountered in stallions previously used as performance horses are behavioural abnormalities. They will have had several years during which they have been actively discouraged from displaying any sexual behaviour. As a result they may well be severely inhibited at their first sight of an oestrus mare, anticipating punishment. They will often find it hard to revert to natural stallion behaviour and need varying amounts of time to unwind and adjust to their new job in life. Many stallions take a few seasons to completely adjust and some never really do achieve complete adjustment. Stallions' libido may also be affected, and such stallions may, as a result, always prove to be slow to react to an oestrus mare and show clumsy mounting behaviour.

As well as their physiological and psychological adjustment, attention should be paid to a stallion's nutrition and exercise during this preparation period. They must be fit, not fat. A heavy covering season will take a lot out of a stallion and he will often lose condition over the season. This loss in condition is minimised if the stallion is fit and in a body condition score of 3 when the season commences (Plate 12.5). Both excess and low body weights reduce a stallion's libido and hence ability to perform. Exercise helps to keep a stallion in good condition preventing him from becoming

Plate 12.5 A stallion should be in a fit, not fat, condition, that is condition score 3 at the beginning of the breeding season.

overfat and maintaining his muscle tone and stamina. Stallions often have a tendency to be overfat as they are regularly kept individually in stables or paddocks away from each other and from mares. In natural conditions they, of course, would be free to exercise as and when they wanted.

If stallions are badly behaved it is tempting to keep them confined nearly all the time with only limited turnout. This just perpetuates the problem and accentuates any misbehaviour due to boredom. Some stallions can be safely ridden, providing an excellent form of exercise as well as good discipline (Plate 12.6). In less intensive studs, usually where stock is less valuable, some quiet stallions are turned out in July at the end of the season with either their mothers, an old mare or other quiet mares. They can then be brought back into riding work and even hunted over the winter. This system allows a rest period after the season, followed by a fitness regime prior to the start of the next season. It also provides them with another purpose in life, which often greatly helps in their discipline.

At least four weeks prior to his first mare the stallion, if kept away from the stud in the non-breeding season or if he is new to a stud, should be brought in. He should then be introduced or

Plate 12.6 *Riding provides an excellent form of exercise as well as good discipline.*

reintroduced to the yard, handling systems, buildings, surroundings and especially the covering area in plenty of time to allow familiarisation prior to the first covering. Any changes in diet should be introduced slowly, before the season starts, along with any new companions to allow a settling down period.

General Aspects of Preparation

Many performance horses have been on various drug regimes either legal or illegal during their performance lifetime. Plenty of time must be allowed for the body to rid itself of any of these drugs, especially corticosteroids. Corticosteroids are used as anti-inflammatory drugs for the treatment of injuries and can have serious detrimental effects on reproductive performance in both stallions and mares and can impair the body's ability to fight infection which is of particular significance in the mare. The use of anabolic steroids to boost muscle development and hence performance also has severe detrimental effects on reproductive performance and should be cleared out from the system well in advance of the breeding season.

Many breed societies recommend, and in some cases require, that all mares and stallions should be swabbed prior to covering. Infections of the reproductive tract are a major cause of infertility in the mare. Many studs, especially in the Thoroughbred industry, require all mares being covered by their stallions to hold a current CEM (contagious equine metritis) certificate prior to arriving at the stud and also to be swabbed again for CEM at the oestrus at which they are going to be covered. Stallions may also be swabbed to test for venereal diseases prior to each season and also during the season if infection is suspected.

The above discussion details the general management principles for the mare and stallion during the preparatory 6 months before covering.

As the majority of equine matings are today influenced by man, we have, therefore, attempted to control the timing of covering to our advantage. There are several methods by which the timing of mating can be manipulated, firstly, by altering the beginning of the season and, secondly, by manipulating the timing of the mare's oestrus within that season. This manipulation of the oestrous cycle in the mare can also be considered as part of her preparation for covering.

Manipulation of the Oestrous Cycle in the Mare

Advancing the Breeding Season

As discussed in Chapter 3, the horse is classified as a long-day breeder with a breeding season extending from March to November in the northern hemisphere and August to March in the southern hemisphere. In the Thoroughbred industry, and increasingly so in other breed societies, the birth of all foals is registered as the first of January, regardless of its actual birth date. It is, therefore, essential, in order to achieve maximum advantage in the racing industry, to ensure that the mare foals as soon after 1 January as possible. This means that, as the mare has an 11 month gestation, you are looking to cover mares at the beginning of February. Hence the arbitrary covering season in Thoroughbreds in the northern hemisphere runs from 15 February to 1 July and in the southern hemisphere from 15 September to 31 December. Other breed societies especially ones associated with show horses and ponies, have similar arbitrary breeding seasons. Even in societies that do not stick to set breeding seasons, it is an advantage to have foals born as early as possible in the year in order to maximise their size during the showing season and enhance their chances of success.

The existence of an arbitrary breeding season that does not correspond with the natural season and the desire that foals be born as early in the year as possible are major limiting factors in improving the naturally poor fertilisation rates (40–50% to a single service) characteristic of mares, especially in the Thoroughbred industry. Manipulation of the mare's breeding cycle is, therefore, required to advance the timing of oestrus in order to accommodate these artificial limits. There are several means by which the breeding season in the mare may be advanced, including the use of light and hormone therapy or a combination of the two.

Light Treatment
Only about 10% of mares voluntarily show oestrus and ovulation in the non-breeding season. Artificial manipulation of light will significantly increase this percentage. The idea of light treatment is to mimic increasing day length prior to its natural increase and so fool the mare into thinking that spring has arrived. Light is regarded as the most important environmental factor controlling reproduction, and its manipulation gives relatively encouraging results.

Light treatment of mares to advance the breeding season was first pioneered by Burkhardt (1947). Today's regimes are based on work by Kooistra and Ginther (1975). Mares may be introduced to a 16 hour light : 8 hour dark regime either suddenly or after a gradual build-up. Light can be delivered from a 200 watt incandescent source or equivalent per 12 ft × 12 ft loose box. A slightly lower watt light source may be used if the stables are lined with reflective material. Light treatment can be started any time from October to mid December. Within 4 weeks the first sign of its effect, coat loss, will be evident, and ovarian activity normally occurs as early as 6–8 weeks after treatment commences. The season may be advanced by anything up to 3 months, but there is considerable individual variation (Kooistra and Loy, 1968; Kooistra and Ginther, 1975; Oxender, Noden, and Hafs, 1977).

The effects of light manipulation are enhanced by increases in the ambient temperature and increased nutritional levels. As stated, this treatment gives rather variable results, and it is very expensive, especially with the additional use of extra feed and heating (Allen, 1978). There have, therefore, been various investigations into viable cheaper alternatives in the form of hormone treatment.

Exogenous Hormonal Treatment

The use of exogenous hormones seemed to be a potentially viable alternative, as such treatment works well in anoestrous ewes and is regularly used in such animals to induce breeding in the non-breeding season. Unfortunately, the mare's ovary seems to be relatively insensitive in comparison with other farm livestock. Treatment of the anoestrous mare with pregnant mare serum gonadotrophin (PMSG), human chorionic gonadotrophin (hCG) and follicle stimulating hormone (FSH) by single or multiple injection results in little if any response in the majority of cases. However, the natural crude equine pituitary extract, if injected daily over a period of two weeks, does induce oestrus out of season (Douglas, Nuti and Ginther, 1974; Lapin and Ginther, 1977). Such a long period of treatment is necessary because in the mare the ovary requires a prolonged period of FSH priming in order to develop any follicles to such a stage that they can react to an ovulating agent (see Chapter 3). Short-term treatments with PMSG, hCG etc will not develop follicles to an adequate stage to allow them to react to any ovulation stimulus.

The use of gonadotrophin releasing hormone (GnRH) has also been investigated. GnRH is naturally responsible for inducing FSH and LH production by the pituitary gland (see Chapter 3). Both GnRH infusion and injection over a period of time have been reported, in work carried out in the United States, to give success rates of up to 83% (Johnson, 1987; Hyland et al., 1987; Hyland and Jeffcote, 1988; Palmer and Quelier, 1988).

Further development into the use of GnRH has been investigated in New Zealand. In the regime of hormone therapy an attempt was made to mimic the natural concentrations of FSH, LH and progesterone in the normal oestrous cycle. A single injection of GnRH is followed by daily injections of progesterone for 8 days. This regime is repeated twice more to give three artificial cycles. This treatment of GnRH and progesterone, though not effective in mares in deep anoestrus, is reasonably successful in the transitional stage from anoestrus to the breeding season, that is late December–January in the northern hemisphere and late May–June in the southern hemisphere (Evans and Irvine, 1976, 1977, 1979).

Oestrogen treatment can also be used to advance the breeding season. However, it has to be administered during the luteal phase of a normal cycle for a period of 10–20 days towards the end of the breeding season. This use of oestrogen has a delayed effect, as it causes the mare to go into immediate anoestrus which lasts for two to three months and is followed by regular oestrous cycles at which conception can occur. The timing of oestrogen treatment can, therefore, be geared to ensure that the mare returns to oestrus in advance of the natural breeding season (Nishikawa, Sugie and Harada, 1952; Nishikawa, 1959).

Progesterone has also been used to induce oestrus and ovulation in anoestrous mares, but seems only to be effective in late anoestrus. Such mares will react to progesterone administration via injection or orally for 10–12 days. Ovulation accompanied by oestrus will occur within 7–10 days after the cessation of treatment. A further alternative in late anoestrous mares is the use of prostaglandin F2α (PGF2α) in a series of two or three injections at 48 hour intervals. Ovulation was reported in 73% of mares treated this way (Allen, 1980; Jochle et al., 1987).

Combination Treatments
A combination of light treatment and hormone therapy gives encouraging results. It is known that it is the drop in progesterone prior to ovulation that allows LH and FSH levels to rise, maturing and ovulating follicles. However, without this progesterone decline the reaction in concentration of LH and FSH is limited and, therefore, ovulation does not occur or occurs in the absence of oestrus. The resulting corpus luteum of any such ovulations is often incompetent, and only after the second ovulation of the season are acceptable conception rates achieved. As the system settles down at the beginning of the season the mare may show delayed return to oestrus or abnormal cycles. If, however, progesterone is administered and then withdrawn, it mimics the natural decline in progesterone and induces the normal rises in LH and FSH required for ovulation and results in a competent corpus luteum (Freedman, Farcia and Ginther, 1979; Allen 1980).

As discussed previously, progesterone alone is of use only if the mare is in the transitional stage between anoestrus and the breeding season. If, however, this transitional stage is brought forward by the use of light treatment, progesterone may then bring such mares into oestrus. Light treatment should be as described earlier: 16 hours light and 8 hours dark introduced during November/December in the northern hemisphere or March/April in the southern hemisphere. Progesterone can then be easily administered in the form of the synthetic progestagen allyl trenbolone or Regumate given orally for a period of 10–15 days (Palmer 1979; Squires, Stevens, McGlothin and Pickett, 1979; Heesman, Squires, Webel, Shideler and Pickett, 1980; Scheffrahn, Wiseman, Vincent, Harrison and Kesler, 1980).

There are several methods by which progesterone may be administered to a mare, the most popular at present being injections or oral administration of a synthetic progestagen like Regumate. The advantage of injection is that you can be certain each individual mare has received her allotted dose, though it is more expensive, especially when you consider the costs of the veterinary surgeon, who now by law in the United Kingdom should administer all injections to horses. Oral administration has the advantage that no vet is required, but some mares may refuse to take the feed in which it is mixed. If only part of the food is eaten it is impossible to know how much progesterone has been taken. Furthermore, all mares have to be fed individually to prevent one mare gorging herself on another's food and, therefore, being exposed to higher levels of exogenous progesterone.

Intervaginal sponges impregnated with synthetic progestagen, very much along the lines of the intervaginal sponges used in ewes, are being experimentally developed for use in the mare. These are inserted at the beginning of the desired time of progesterone supplementation and remain in situ for as long as progesterone supplementation is required, normally up to 15 days.

Some of the best and most consistent results have been obtained using the following regime: light treatment of 16 hours light:8 hours dark for 6–8 weeks followed by 10 daily injections of progesterone and oestradiol combined. PGF2α is then administered on day 10 to induce luteolysis. In studies up to 82% of mares ovulated within 9–16 days after final treatment and achieved a conception rate of 65%. Though such a treatment is relatively successful, its commercial feasibility is reduced by the costs involved.

Though there are very many ways of advancing the breeding season in the mare, most are too expensive in time and effort, as well as in the drugs used, to make them commercially viable, and they tend to give unreliable results. Nevertheless, many large studs, especially those involved in the racing industry, routinely use light treatment, occasionally supplemented by Regumate, with some success.

Synchronisation of Oestrus

In addition to advancing the breeding season there are other areas in which the oestrous cycle of the mare can be manipulated to ease mare management. In many countries of the world, eg South America and the United States, mares are run in large herds that roam over vast areas of land, and in such systems handling needs to be kept to a minimum. It would, therefore, be ideal if mares could all be treated in batches right through from conception to birth and foal rearing, as this would ease management considerably. In order for this to be successful a reliable and exact method of timing ovulation and oestrus is required. Such treatment would also alleviate the need for teasing, rectal palpation etc, and would be most useful in conjunction with artificial insemination eliminating the need for a stallion on site.

Synchronisation of mares already in the breeding season can be achieved by using either progesterone or PGF2α. Progesterone treatment, eg 100 mg/day, will supress oestrus in large mares and three to seven days after terminating the treatment mares will come back into oestrus with an associated fertile ovulation (Loy and Swann, 1966; Van Niekerk et al., 1973).

Such treatment can also be used to suppress reproductive activity completely in the case of performance mares, in which ovulation and oestrus would be a nuisance. Progesterone withdrawal can be timed to allow the mare to show oestrus at a less inconvenient time, or reproductive activity can be suppressed for the whole season as long as daily progesterone administration is continued. Oral administration is an alternative to injection, and it is hoped that intervaginal sponges will also soon be an alternative (Palmer, 1979; Squires, Stevens, McGlothin and Pickett, 1979). Work on the use of intervaginal sponges was first reported in France (Palmer, 1979). They were left in the vagina for 20 days and on withdrawal an injection of PGF2α was given. Ovulation and oestrus occurred 24 hours later, with a 71% fertilisation rate.

As explained, the administration of progesterone inhibits pituitary function, that is the secretion of LH and FSH, and its removal allows these hormones to rise. An alternative non-steroidal pituitary inhibitor, methallibure, may also be used with a similar effect to synchronise mares. Like Regumate, this is administered orally; however, it tends to have detrimental effects on appetite and so reduces food intake.

A further alternative to progesterone in the synchronisation of mares is PGF2α, which is a natural luteolysin and works well in mares as in other farm livestock. Lower dose rates, though, are required for the mare than for other livestock. Presumably this is due to the natural systemic effect of PGF2α in the mare as compared to the local effect apparent in sheep and cattle (Allen and Rowson, 1973). Several artificial analogues of PGF2α have been prepared for use in the mare, the most well known being Equimate. It is one of the most advanced hormone treatments available and provides a valuable reproductive management tool for the stud farm manager. Some mares do, however, show side effects of sweating, diarrhoea and other symptoms arising from smooth muscle activity, though research suggests that fewer of these side effects result from use of artificial analogues than from exogenous natural PGF2α (Cooper 1981).

PGF2α or its analogue, if injected between days 6 and 15 of the cycle, results in oestrus and fertile ovulation 2–5 days after injection. Insemination can then be carried out at a fixed time of 48 and 96 hours post injection (Bowen, Niang, Menard, Irvine and Moffat, 1978; Loy, Buell, Stevenson and Hamm, 1979). If the stage of the mare's cycle is known, a single injection of PGF2α is thus a very effective means of synchronisation. However, in many studs, especially those in the Americas that run large herds of mares, the stages of the various mares within their cycles are unknown. In such cases a second injection of PGF2α 12–14 days later, with the optional inclusion of hCG treatment, gives excellent synchronisation results. The following regime is an example:

DAY 0   PGF2α injection
DAY 6/7   hCG injection
DAY 14   PGF2α injection
DAY 21   hCG injection
DAY 22   inseminate

Success rates of 75–95% synchronisation have been reported for this system (Palmer and Jousett 1975). hCG is used as part of the treatment to enhance the release of LH. Two injections of PGF2α are used, as the corpus luteum of the mare is only susceptible to PGF2α during the period Day 5–16 of the cycle, that is when it is dominant within the system. Outside this time PGF2α has no effect. Mares taken at random from a herd will fall into 2 groups, say A and B for ease of description. Group A are mares in the period Day 0–5 of the cycle and group B are those in the period Day 5–21. If PGF2α is administered to all mares, those in group A will not respond, as their corpora lutea are not dominant, but those in group B will be advanced to Day 0, oestrus, either by induced luteolysis or because they were near that stage anyway. The second injection is administered 14 days later when group A will be between Day 14 and Day 18 and group B will be at Day 14. All mares, therefore, are capable of responding to the second injection of PGF2α and all will be accelerated on to oestrus, Day 0. Success rates of up to 95% synchronisation have been reported with two injections of PGF2α, especially if they are used in conjunction with hCG (Allen, Stewart, Cooper, Crowhurst, Simpson, McEnry, Greenwood, Rossdale and Ricketts, 1974; Palmer and Jousset 1975).

A combination of PGF2α and progesterone has also been used in the following regime:

DAY 0–10   progesterone
DAY 7   PGF2α injection
DAY 15/16   hCG injection

Again, synchronisation rates of up to 90% have been reported, especially with the additional use of hCG (Holten, Douglas and Ginther, 1977).

All these methods of synchronisation can be used to considerably ease the management of large herds of mares and are especially useful in combination with artificial insemination. They can also be used on a much smaller scale to reduce the need for rectal palpation, ultrasonic scanning or teasing to assess whether a mare is in heat. This is especially helpful in dealing with awkward mares that do not show oestrus well and may be hard to handle. These synchronisation techniques can also be used to induce ovulation in mares with some abnormal ovarian conditions. These cases will be discussed further in Chapter 19. There is, however, still no 100% reliable and simple-to-administer method of controlling oestrus in the mare, but research throughout the world is bringing this possibility nearer.

Manipulation of Breeding Activity in the Stallion

The breeding season in the stallion, as in the mare, is governed by photoperiod, and the stallion reacts to increasing day length in a similar manner to that seen in the mare. The breeding season of the stallion can, therefore, be advanced by the introduction of a 16 hour light: 8 hour dark regime in November/December. Continual stimulation, however, produces a refractoriness and a return to normal seasonal changes, despite the altered photoperiod (Argo, Cox and Gray, 1991). Manipulation of stallion reproduction is not as essential as manipulation in the mare, as it is the mare's reproductive cycle that is normally limiting to performance, not the stallion.

Conclusion

It is evident that the preparation of both the mare and stallion for covering needs considerable thought, especially if the animals concerned are maiden, with the added complication of an athletic career behind them. However, providing adequate time and forethought are put in, most horses that are anatomically and physiologically correct are capable of breeding regardless of their previous careers.

References and Further Reading

Allen, W.E. and Alexeev, M. (1980) Failure of an analogue of gonadotrophin releasing hormone (HOE-766) to stimulate follicular growth in anoestrous pony mares. Equine Vet.J. 12(1), 27–28.

Allen, W.R. (1978) Control of Oestrus and Ovulation in the Mare. In: Control of Ovulation. Crighton, Foxcroft, Haynes and Lamming, Eds., pp. 453–468. Butterworths, London.

Allen, W.R. (1980) Hormonal control of early pregnancy. Vet.Clin.No.Am. (Large Anim. Pract.) 2, 291.

Allen, W.R. and Rowson K.E.A. (1973) Control of the mares oestrous cycle by prostaglandins. J.Reprod.Fert. 33, 539–543.

Allen, W.R., Stewart, F., Cooper, N.J., Crowhurst, R.C., Simpson, D.J., McEnry, R.J., Greenwood, R.E.S., Rossdale, P.D, and Ricketts, S.W. (1974) Further studies on the use of synthetic prostaglangin analogues for inducing luteolysis in mares. Equine Vet.J., 6, 31–35.

Argo, C.M., Cox, J.E. & Gray, J.L. (1991) Effect of oral melatonin treatment on the seasonal physiology of pony stallions. J.Reprod.Fert.Suppl., 44, 115–125.

Bowen, J.M., Niang, P.S., Menard, L., Irvine, D.S, and Moffat, J.B. (1978) Pregnancy without estrus in the mare. J.Equine Med.Surg.2(5) 227–232.

Burkhardt, T. 1947. Transition from anoestrus in the mare and the effects of artificial lighting. J.Agric.Sci.Camb. 37, 64–68.

Cooper, M.J.(1981) Prostaglandins in veterinary practice. In Practice, 3(1), 30–34.

Douglas, R.H., Nuti, L, and Ginther, O.J. (1974) Induction of ovulation and multiple ovulation in seasonally anovulatory mares with equine pituitary factors. Theriogenology, 5, 133.

Evans, M.J, and Irvine, C.H.G. (1976) Measurements of Equine Follicle stimulating hormone and luteinizing hormone: response of anoestrous mares to gonadotrophin releasing hormone. Biol.Reprod., 15, 277–284.

Evans, M.J, and Irvine, C.H.G. (1977) Induction of follicular development, maturation and ovulation by gonadotrophin releasing hormone administration to acyclic mares. Biol.Reprod., 16, 452–462.

Evans, M.J, and Irvine, C.H.G. (1979) Induction of follicular development and ovulation in seasonally acyclic mares using gonadotrophin releasing hormones and progesterone. J.Reprod.Fert.Suppl., 27, 113–121.

First, N.L. (1973) Synchronisation of estrus and ovulation in mare with methallibure. J.Anim.Sci., 36, 1143–1148.

Freedman, L.J., Farcia, M.C. and Ginther, O.J. (1979) Influence of photoperiod and ovaries on season reproductive activity in mares. Biol.Reprod. 20, 567–574.

Heesman, C.P., Squires, E.L., Webel, S.K., Shideler, R.K. and Pickett, B.W. (1980) The effect of ovarian activity and allyl trenbolone on the oestrous cycle and fertility in mares. J.Anim.Sci., 51 Suppl.1, 284(Abs.).

Hyland, J.H. and Jeffcote, L.B. (1988) Control of transitional anoestrus in mares by infusion of gonadotrophin releasing hormone. Theriogenology, 29, 1383.

Hyland, J.H., Wright, P.J., Clarke, I.J., Carson, R.S., Langsford, D.A. and Jeffcote, L.B. (1987) Infusion of gonadotrophin-releasing hormone (GnRH) induces ovulation and fertile oestrus in mares during seasonal anoestrus. J.Reprod.Fert.Suppl., 35, 211.

Jochle, W., Irvine, C.H.G., Alexander, S.L. and Newby, T.J. (1987) Release of LH, FSH and GnRH into pituitary venous blood in mares treated with PGF analogue, luprostiol, during the transition period. J.Reprod.Fert.Suppl., 35, 261.

Johnson, A.L. (1987) Gonadotrophin releasing hormone treatment induces follicular growth and ovulation in seasonally anoestrus mares. Biol.Reprod., 36, 1199.

Holton, D.W., Douglas, R.H. and Ginther, O.J. (1977) Estrus, ovulation and conception following synchronisation with progesterone, prostaglangin F2α and human chorionic gonadotrophin in pony mares. J.Anim.Sci., 44(3), 431–437.

Kooistra, L.H. & Ginther, O.J. (1975) Effects of photoperiod on reproduction activity and hair in mares. Am J.Vet.Res. 36, 1413–1419.

Kooistra, L.H. and Loy, R.G. (1968) Effects of artificial lighting regimes on reproductive patterns in mares. Proc.Ann.Conv.Am.Assoc.Equine Practnrs., 159–169.

Lapin, D.J. and Ginther, O.J. (1977) Induction of ovulation and multiple ovulations in seasonally-anovulatory mares with equine pituitary extract. J.Anim.Sci., 44, 834.

Loy, R.G., Buell, J.R., Stevenson, W. and Hamm, D. (1979) Sources of variation in response intervals after prostaglandin treatment in mares with functional corpus lutea. J.Reprod.Fert.Suppl., 27, 229–235.

Loy, R.G. and Swann, S.M. (1966) Effects of exogenous progestagens on reproductive phenomena in mares. J.Anim.Sci., 25, 821.

Nishikawa, Y. (1959) Studies on reproduction in the horse. Jap.Racing Association, Tokyo.

Nishikawa, Y., Sugie, T. and Harada, N. (1952) Studies on the effect of daylength on the reproductive functions in horses. 1. Effect of daylength on function of ovaries. Bull.Nat.Inst.Agric.Sci., 3, 35.

Oxender, W.D., Noden, P.A. and Hafs, H.D. (1977) Estrus, Ovulation and serum progesterone, estradiol and LH concentrations in mares after increased photoperiod during winter. Am.J.Vet.Res. 38, 203–207.

Palmer, E. (1979) Reproductive management of mares without detection of oestrus. J.Reprod.Fert.Suppl., 27, 263–270.

Palmer, E. and Quelier, P. (1988) Uses of LHRH and analogues in the mare. Proc.11th.Int.Cong.Anim.Reprod. and A.I. pp. 338–346.

Palmer, B. and Jousset, B. (1975) Synchronisation of oestrus in mares with prostaglandin analogue and hCG.J.Reprod.Fert.Suppl., 23, 269–274.

Parker, W.G. (1971) Sequential changes of the ovulating follicle in the oestrous mare as determined by rectal palpation. Proc.Ann.Conf.Vet. College. Vet.Med.Bio.Med.Sci. Colorado State Univ., Fort Collins.

Scheffrahn, N.S., Wiseman, B.S., Vincent, D.L., Harrison, P.C. and Kesler, D.J.(1980) Ovulation control in pony mares during early spring using progestins, PGF2α, hCG and GnRH. J.Anim.Sci., 51 Suppl., 325(Abs.).

Squires, E.L., Stevens, W.B., McGlothin, D.E. and Pickett, V.W. (1979)

Effect of an oral progestagen on the oestrous cycle and fertility of mares. J.Anim.Sci., 49, 729–735.

Van Niekerk, C.H., Coughborough, R.I. and Doms, H.W.H. (1973) Progesterone treatment of mares with abnormal oestrus cycle early in the breeding season. J.S.Afr.Vet.Ass., 44, 37.

Van Nierkerk, C.H. and Van Heerden, J.S. (1973) Nutritional and ovarian activity of mares early in the breeding season. J.S.Afr.Vet.Ass., 43, 351.

CHAPTER 13

Mating Management

CHAPTERS 1–4 discuss in detail the anatomy and physiology of reproduction in the horse and attendant control mechanisms. In summary the mare is a seasonal, polyoestrus spontaneous ovulator, that is her reproductive activity is governed by season, only naturally being operational during the months of March to November in the northern hemisphere and August to March in the southern. Within this season she shows several oestrous cycles, coming into oestrus at regular intervals of 21 days. She ovulates spontaneously in synchrony with her sexually receptive oestrus periods and does not require an external stimulus to induce ovulation to occur.

The breeding cycle of the mare is diagrammatically represented in Figure 13.1.

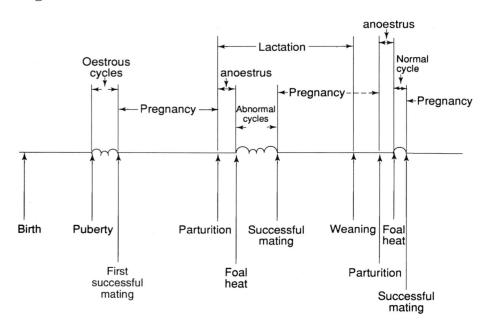

Figure 13.1   Diagrammatic illustration of the breeding cycle of the mare.

208

Puberty occurs at between 10 and 36 months of age, at which time oestrous cycles begin. The duration of gestation is 11 months, and the mare returns to oestrus within 4–14 days post parturition. This first oestrus is termed the foal heat. Lactation in the mare naturally lasts 10 months though in today's breeding system man normally dictates its length at nearer 6 months. Overriding this whole pattern is the seasonal control on reproduction.

Mating

The mare will only allow mating to occur when she is in her sexually receptive oestrus phase, which may also be referred to as season and/or heat. This period of sexual receptivity occurs regularly at 21 day intervals during the breeding season and lasts in the order of 2–10 days (average 5 days). There is a period of proestrus either side of true oestrus as she gradually becomes receptive and as she gradually goes out of oestrus. Oestrus is synchronised with ovulation and ensures that the mare is mated at the optimum time to ensure the successful meeting of ova and sperm and hence fertilisation.

Ovulation occurs normally about 24 hours before the end of oestrus so in an average mare it will be on day 4 of oestrus. There is, however, considerable variation between mares. Sperm are reported to survive for 24–48 hours within the mare's reproductive tract, but the exact time must remain in doubt as most experiments on longevity are carried out *in vitro*. The ova, on the other hand, are thought to survive in the order of only 8 hours, though longer times have been reported in *in vitro* work. The timing of mating is, therefore, very important.

Natural Mating

In the natural system stallions run with their mares and detect mares in oestrus at will, examining them daily for signs of sexual receptivity. This process is leisurely and unrushed and the signals used by the stallion to detect oestrus are smell and taste rather than sight. Natural courtship may occur over several days as the mare slowly progresses from dioestrus to proestrus and on into full oestrus, and takes place between a mare and stallion that are well known to each other. When she is fully in oestrus and receptive, the courtship, which can be quite awesome at times, culminates in mating. Courtship of a mare in full oestrus follows the following pattern. The stallion will normally start by fixing his eyes upon a mare that he suspects is in full oestrus. He will arch his back and neck and draw himself up to his full height, and he will become restless and will pace around pawing the ground and stamping his feet. He will show the typical facial grimace termed flehman or tasting of the air, often accompanied by a roar (Plate 13.1).

This whole process takes some time during which the mare, if she is truly receptive, will stand quietly and possibly nicker in response if interested. Once the mare's interested response has been ascertained by the stallion, he will have the confidence to approach her, normally from the head, working his way slowly down her neck; nickering and talking to her, he may nudge her slightly or even lightly bite her neck. All the time he is watching for her response and for any sign that she may object. If he feels confident he will then work his way down her flanks and on to her hind quarters and to her perineum. All the while he may stop and pause to reassure himself that she is still interested. If everything is still amicable, he will nudge her vulva and clitoris. If the mare is in full oestrus she will stand still, passive throughout the whole procedure, showing by curling her tail to one side, urinating, possibly a yellow-coloured urine, or she will take up the urinating stance. She will expose her clitoris by opening and closing the flap

Plate 13.1   The typical facial grimace, termed flehman, shown by a stallion when in the presence of an oestrus mare.

of skin covering it; termed winking, this can be clearly seen in mares with dark skin as a flash of pink mucous membrane at the bottom of the vulva.

If she is not in oestrus she will show hostility to the stallion, who will be unable to get closer to her than the initial advance. In nature the stallion will then turn away and forget about her for the time being, transferring his attentions elsewhere, and return to her later that day or the next (Klingel, 1982; Keiper and Houpt, 1984; Kolter and Zimmermann, 1988).

Once the stallion is sure that the mare is truly receptive he will mount her. His penis normally becomes erect as he approaches the mare and she shows interest. Mounting will only occur when his penis is fully erect. If the stallion is very sure of himself and the mare, he will often mount her directly and ejaculate immediately. A stallion with less confidence or one not so sure of the mare may nudge or push her forward slightly prior to mounting or half mount her a couple of times first, in order to test the water before he commits himself and runs the risk of being kicked. The number of mounts per ejaculate tends to decrease as spring and summer approaches, the number being higher at the beginning of the season as the stallion's reproductive activity, like the mare's, is governed by photoperiod. During the transition period into and out of season the stallion's libido tends to be lower, resulting in the need for more stimulation before successful mating can occur. Ejaculation follows a varying number of pelvic oscillations during which time the stallion may dance on his back feet. Successful ejaculation is signalled by rhythmic flagging of the tail followed by a terminal inactive phase when the stallion remains on the mare while penile erection subsides. This final terminal phase lasts in the order of 1–3 minutes and the stallion will not naturally dismount until his penile erection has completely subsided (Plate 13.2).

It has been demonstrated, by means of an open ended artificial vagina, that ejaculation begins as the pelvic thrusts cease and consists of 6–9 jets of semen, each a result of urethral contraction. The volume of each successive jet decreases and 70% of the total seminal volume is in the first three jets. The remaining jets are made up mainly of secretions of the bulbourethral glands making up the gel fraction of stallion semen (Kosiniak, 1975).

In this natural system a stallion will cover a mare in oestrus many times, up to 8–10 matings in 24 hours. Nature's system works extremely well as it has been perfected over thousands of years of evolution. It is a system, however, that is rarely used these days, but is seen occasionally in pony studs, or with horses run on large expanses of land with minimal managerial input. For such a system to work completely naturally the stallion will only mate the

Plate 13.2   *The final terminal stage post ejaculation, when the stallion remains inactive as his penile erection subsides.*

mares within his herd and no outside mares may be introduced solely for covering. The introduction of foreign or strange mares causes a disruption in the hierarchy and can result in jealousy and hostility from other mares, and uncertainty between the stallion and the introduced mares. The nearest system to this practised today is carried out in some pony studs with native type animals where the stallion is run out with groups of mares. Visiting mares are put in their own group and the stallion is moved around between the groups and allowed to cover the mares at will when they come into oestrus.

Such a system has many disadvantages from the breeder's point of view, when he is attempting to maximise the financial return from his stallion. This natural system limits the number of mares he can cover in a season as each mare gets covered 8–10 times, whereas if mating is timed correctly, in theory only a single service is required. As the financial return from stallions increases, so does their value. Man then feels he has to interfere to protect his investment and maximise the number of mares covered per season and minimise the risk of injury to both himself and the stallion from awkward mares. This interference by man is the start of many of the problems associated with stud management today.

Mating in Hand

Today, we control the whole life of the horse to such an extent that there is very little similarity between the natural life of the horse and the life we impose on him for our own ends.

Today, we segregate fillies and colts early on in their lives, often as soon as weaning, ie at six months of age. Naturally they would run as a herd and the colts would get put in their place and learn respect at a young age, with little chance of serious damage. They also get used to having mares around keeping them in order and only accepting their attentions if they want to. In the intensive systems of today this introduction of the young stallion is removed and, in addition, we expect stallions to cover mares that are unknown to them. They are, therefore, in danger of being kicked and of fighting because the development of a hierarchy has been bypassed. An additional consideration is the vastly increased value of stallions, and, therefore, the need to minimise the number of mounts per successful service and optimise his total use and number of offspring.

To overcome the potential danger to stallions and to optimise their use we tend today to mate in hand. Man, therefore, has total control over the number of mares covered by each stallion. However, one of the major drawbacks of this system is that there is the need for man to interpret the mare's oestrus signals before the stallion is allowed near her. As discussed previously the stallion uses the senses of smell and taste to detect mares in oestrus; man, however, has to rely on the sense of sight only. This obviously leads to inaccuracies and can prove very unreliable. Hence we use a teaser stallion.

Teasing

The teaser is often purely kept to detect mares in oestrus, if he is injured or damaged by an objecting mare, then, as his value is low, it is not the end of the world. As stated a teaser may be kept solely for teasing, in which case he must be allowed to mate the occasional mare; otherwise his libido will suffer and hence his effectiveness at his job. Alternatively, a pony stallion is used in some large horse studs. He has the libido to act as a teaser but is too small to be in any danger of successfully covering the mares. In many native pony breeding studs the stallion to be used acts as the teaser as well. This allows the reproductive state of the mare to be confirmed, but does not give the stallion the protection of using a separate teaser. However, such stallions tend to be less valuable. Native mares show more readily, and the mare may well be teased over the stable door initially, giving some protection.

The principle behind teasing is that as soon as the mare is thought, by visual signs interpreted by man, to be in oestrus, she is brought in contact with an entire male horse under varying conditions in an attempt to confirm the visual diagnosis. The objective is to cause the mare to display oestrus activity and, therefore, confirm man's detection of oestrus. Once she is confirmed as being in true oestrus then she can be mated. This system allows the stallion to mate the mare only at the most optimum time, towards the end of oestrus, maximising the chance of conception and minimising the risk of damage to the stallion by a dioestrus mare.

There are problems associated with this system, mainly because the courtship that naturally takes place over a prolonged period is concentrated into a short space of time and forces the attentions of the stallion upon the mare. Some mares object to such forced attentions without full courtship even if they are in full oestrus, and such objection may mask the signs of oestrus. Mares with foals at foot often object, occasionally violently, to the removal of the foal prior to teasing and mating, a practice normally carried out to protect the foal. In the natural system the foal would still be in close vicinity to the mare, but seems to know instinctively to keep its distance at such times. Teasing some mares before feeding or turnout can give erroneous results, and environmental conditions of extreme heat, cold, rain or wind may mask signs of oestrus. Some mares need a longer time of teasing before they can be coaxed into showing and in a busy stud working to a tight schedule there is little time for extended teasing and hence she may never seem to be ready to cover. Again some mares will only show under certain circumstances, ie only in the covering yard, or when the perineum is being washed or the tail bandaged prior to service or when a twitch is applied. This is where the mare's records are invaluable to identify any idiosyncrasy. In summary, teasing is a valuable technique but it can give variable results and can never be as accurate as the stallion himself.

The signs to be looked for in a mare being teased have already been detailed under the discussion on the natural system of mating. In summary, she will be docile; accept the attentions of the stallion; will show, ie take up the urination stance and wink her clitoris; and she will show a general lack of hostility towards, and signs of acceptance of, the stallion (Plate 13.3).

Plate 13.3 A mare showing to the stallion by lifting her tail to one side and displaying a passive, quiet reaction to his attentions.

Trying Board

There are several methods of teasing depending on the stud, the value of the stock and the facilities available. One of the most common methods is leading the mare and teaser in hand up to a trying board. They are introduced one on either side of the board and their muzzles are allowed to meet. The board provides protection for both and is usually high enough to allow just the horses' head and necks to reach over. It is solid in construction, possibly made of wood and ideally twice the length of the horses. Its top is usually covered by curved rubber or the equivalent, so that if the stallion and mare should attempt to attack each other over it, there will be minimum injury (Figure 13.2 and Plates 13.4a–d).

The teaser is allowed to stretch his muzzle along the mare's neck,

Plates 13.4a–d   Mare being teased over a trying or teasing board. The mare and stallion should be introduced initially nose to nose and the stallion allowed to work his way down towards the mare's vulva (a, b, c). If she is in oestrus she will show little or no objection. A mare not in oestrus will usually object violently (d).

a

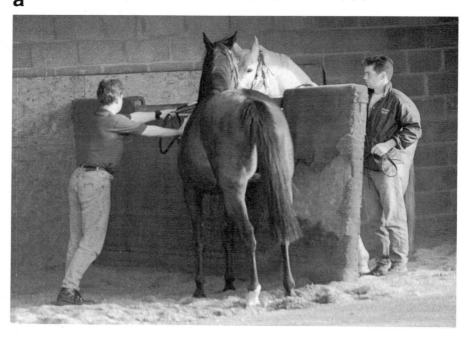

b

c

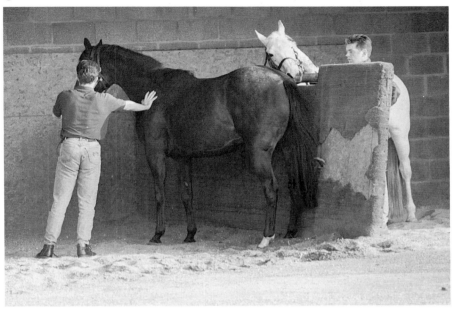

d

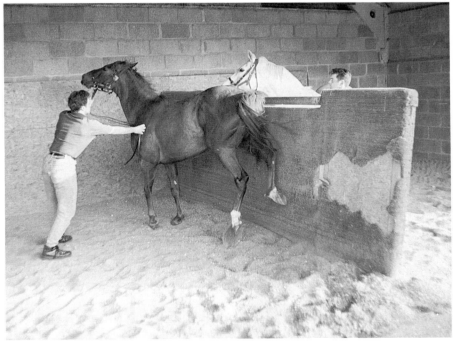

possibly gently nipping her; this mimics the stallion's natural first approach. The attitude of the mare to this attention is closely watched, signs of hostility—laid back ears, squealing, biting and kicking out—being indications that she is still in dioestrus and not ready. In contrast leaning towards the male and raising of the tail, in addition to the other signs of oestrus, indicate that she is ready to be covered. If the mare is interested then her flank can be turned towards the trying board and the teaser allowed to work his way further down her body. It is, however, very important that direct contact with the mare's genital area is avoided to prevent the transfer of disease to other mares teased. After a few minutes of such attention most mares will show definite signs if in oestrus, although some mares may take longer.

A stable door can be used as an alternative to a purpose-built trying board, with the teaser in the stable and the mare introduced to him over the door. This is a popular practice in the smaller more native type studs, but can be dangerous unless the teaser is well known and is not likely to try to climb out to get at the mare.

Teasing over the Paddock Rail

An alternative and quite popular system is used in America and South Africa, as well as in Britain and Europe, especially in large studs running large numbers of mares. The teaser is led by hand to the paddock rail and mares, which are normally in relatively small groups in individual paddocks, can be teased daily across the fence (Plate 13.5). A permanent trying board is built into the paddock rails or a movable trying board is placed there. The mares are free within the paddock and their reaction to the teaser is noted. Most mares will come up to the teaser by the fence and show definite signs of oestrus, others may show hostility. Some may seem uninterested or show no distinct signs of interest. Such mares should then be brought up to the trying board by hand, as previously described, to be tested individually. This is a good method for use with mares that are turned out as it greatly reduces time and labour. It is best to avoid teasing mares by this method immediately after turnout or just before feeding as this can give erroneous results.

This system has potential problems when mares have foals at foot. Conflicting reports have indicated that there is considerable danger to the foal, or that foals naturally keep out of the way as they do in nature and that mares, especially in small groups, are very careful to avoid damage to their foals.

Walking the teaser past the paddock fence on a regular basis is similar to the above. This is of particular use if the teaser can be

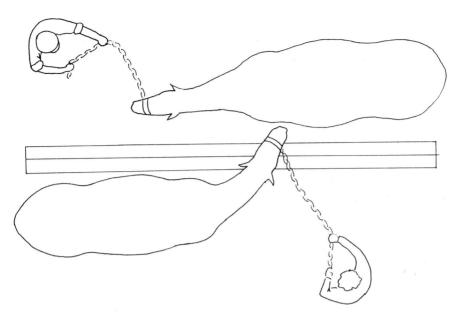

Figure 13.2 *Trying over a teasing board allows the mare and stallion to be introduced to each other in relative safety. This method also gives some protection to the handlers.*

Plate 13.5 **Teasing over the paddock rail is an alternative method. Mares may be let loose in the field to approach the stallion at will or as in this case can be caught and teased individually.**

ridden or requires regular exercise in hand. His daily route can then be organised to pass via the paddocks containing mares likely to be in oestrus. As the teaser passes, the general reaction of the mares is watched and those that show interest are then caught up and taken to the covering yard.

Not all mares react to these two forms of teasing. Those that are less demonstrative in oestrus or shy may well be missed. It is essential, therefore, that in such systems a detailed record of the mare's normal behaviour in oestrus is available. The great advantage of these systems is the reduction in the labour and time required to detect mares ready for covering, and it is a good method for use with turned out mares.

Leading Teaser through the Paddock

Other less popular methods include leading the teaser in hand through the paddock. This can be extremely dangerous and must only be practised when both mares and teasers are well known to the handler and to each other. Even so, stallions in the presence of oestrus mares can be extremely unpredictable.

Teasing in Chutes and Crates

In the southern hemisphere, including South Africa and South America, mares tend to run in large herds and are handled less frequently. In such enterprises the mares may be run into chutes or crates and held individually for a short period while being teased from outside (Figure 13.3). This system also reduces labour and enables large numbers of mares to be teased in as a short a period of time as possible and with limited handling. However, the mares need to be accustomed to this system or erroneous results may be obtained.

Teasing Pen

A further alternative system is to confine the teaser in a railed or boarded area in the corner of a paddock or paddocks. This system is particularly popular in Australia and South America. The area confining the teaser normally has high boards with a grill or hole through which the teaser can put his head (Figure 13.4). An alternative is the use of a small pony or miniature breed stallion confined by a stout fence. If he escapes it is not too disastrous as his size limits his ability to cover the mares, though it is not actually impossible. These are again good systems for teasing a large number of mares, but they require frequent observations for

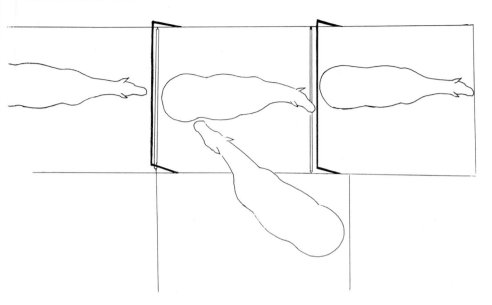

Figure 13.3 *Teasing of mares using chutes is popular in South America. The stallion is restrained in a single pen, and the mares are run through the chute and stopped momentarily by the stallion to gauge their response.*

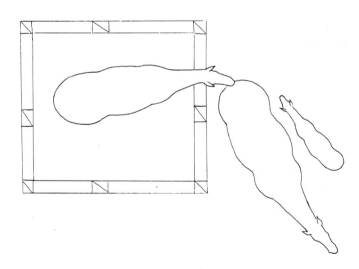

Figure 13.4 *Use of a teasing cage confines the stallion or teaser and allows any mares in oestrus loose in the field to approach at will.*

signs of oestrus in the mares and it may also be difficult to pick up shy mares. It must also be remembered that the teaser can only be confined in the railed area for relatively short periods of time.

Along the same lines as the teasing pen is a central teasing pen surrounded by a series of individual pens in which mares can be held (Figure 13.5).

A further variation on this theme is the use of two adjacent boxes divided by a grill, one for the stallion and another for the mare. Yet another variation involves the teaser being confined in a stable in the corner of a yard with the mare free within the yard to show to the stallion at will. Such an arrangement is seen at the National Stud in Newmarket (Plate 13.6).

These three systems are good for use with difficult mares as they can be left alone to show in their own time and can be observed from a discreet distance.

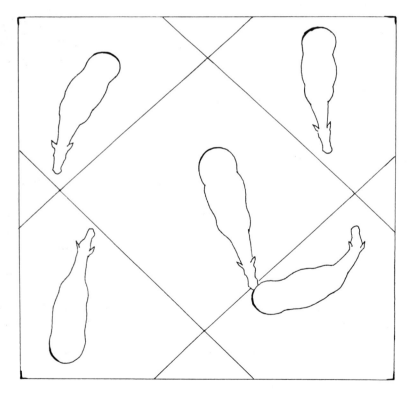

Figure 13.5   *The use of a central teasing pen to try mares in oestrus allows mares to be teased over a prolonged period of time and with minimal interference from man.*

Plate 13.6 *Use of a stable in the corner of a covering yard (open window in the background) for the confinement of a teaser, allowing the mare free within the yard (in the foreground) to show at leisure. (The National Stud, Newmarket)*

Vasectomised Stallions

Vasectomised stallions can be used to run out with mares. This can be especially useful with maiden or difficult mares; those mounted can then be covered by the intended entire stallion. This system has the obvious danger of running unaccustomed mares and stallion out together and is therefore of limited use, especially with valuable stock. However, it is an extremely reliable method of detecting oestrus. The other major disadvantage is that this system allows intromission to occur and, therefore, enables the ready transfer of venereal disease organisms from stallion to mare and vice versa. This can be averted by surgical retroversion of the penis, causing it to extend caudally between the hind legs on erection and, therefore, making intromission impossible (Belonje, 1965).

Hermaphrodite Horses

Hermaphrodite horses are very useful as teasers but are very rare and, therefore, not really a viable alternative.

Conclusions on Teasing

In any system of teasing, direct contact between the male's muzzle
or penis and the female genitalia must be avoided at all costs.
Direct contact risks the transfer of disease to successive mares via
the stallion.

Not all mares show under the above systems and some may require
additional stimulation such as gentle rubbing of the perineum.

If the mare has a foal at foot it may well become agitated with all
the unaccustomed attention to its mother and disturb her. To
reduce this the foal may be penned within reach or sight of its
mother while she is teased. However, this can cause more problems
with some mares and foals, and in such cases it may be best to
remove the foal completely from sight and sound of the mother.
Careful observation of difficult mares may be the key as they may
not show until they are returned to their loose box, or may only
show to the foal or other mares.

In all teasing it must be remembered that a mare's reaction is
very variable and is not 100% reliable. Diagnosis is often, therefore,
confirmed by veterinary examination.

Veterinary Examination

Routine veterinary examination to confirm a mare's reproductive
state is used in many studs, especially those running valuable
stallions with the aim of optimising their use. The technique is
used to back up teasing and confirm diagnosis and hence optimise
the timing of covering. There are three types of veterinary examination that may be used in this context: rectal palpation, vaginal
examination and the more recently developed ultrasonic ovarian
assessment. They can be used to confirm a mare's sexual state,
correlate coitus to 24–48 hours prior to ovulation and diagnose venereal infections. For all techniques the mare should be
restrained in stocks (Plate 13.7).

Rectal Palpation

Rectal palpation is carried out as described in Chapter 11. It is
used to feel the ovaries through the rectum wall (Plate 13.8).
Details of the measurements of the reproductive tract can be
correlated to parts of the hand and then translated into exact
measurements. The practitioner is looking to assess follicle size,
consistency and position in order to estimate the time of ovulation.
Most follicles need to be 3 cm before ovulation, though reports of

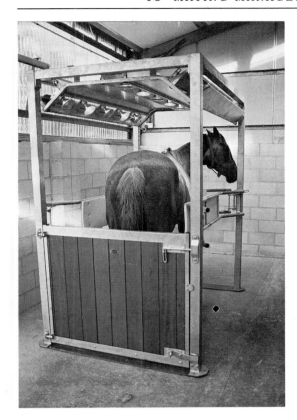

follicles up to 10 cm have been produced (Greenhof and Kenney, 1975). Follicles that are about to ovulate tend to be thin walled and the contents under reduced pressure and feeling like soft fluctuating swellings on the ovary surface. In contrast, atretic or regressing follicles feel tense and are thin walled. In practice if a follicle greater than 3 cm in diameter is detected, the mare may be expected to ovulate within 3 days. If in addition it feels soft then it may be expected to ovulate within 24 hours.

Greenhof and Kenney have carried out a lot of work (1975) correlating follicular characteristics and ovulation. A recent corpora lutea is characterised by a soft friable mass occupying the cavity of the ruptured follicle. Twenty-four hours post ovulation the corpora lutea feels firm and a pit may be felt. Later on it feels similar to a developing follicle and the two can quite easily be confused. If the mare's ovaries are inactive because of hormonal inadequacies or anoestrus, the ovaries are 2–5 cm in length and feel firm over their entire surface, with no evidence of follicles or corpora lutea.

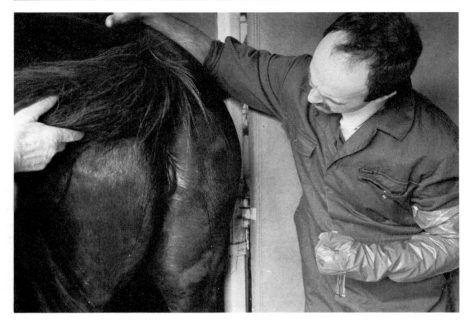

Plate 13.8   Rectal palpation in the mare, used to assess ovarian activity.

The uterus may also be felt by means of rectal palpation and used as an aid in determining the reproductive state of the mare. The tone, size and thickness of the uterus are assessed. Early pregnancy is characterised by a discrete swelling of the uterus at the uterine horn/uterine body junction. If the mare is in dioestrus then, due to elevated progesterone levels, the uterus feels less flaccid and has more tone than is evident during oestrus.

Rectal palpation is a popular and regular procedure in many Thoroughbred studs in the United Kingdom. During the 1970s work by Voss et al. (1973) and Voss and Pickett (1975) indicated that over an experimental period of three years palpated mares had lower conception rates to artificial insemination, using undiluted semen, than unpalpated controls. Subsequent work casts doubt on the general validity of these reports and indicates that such effects are only seen in mares not accustomed to the procedure of rectal palpation, the effect being mainly due to stress of handling rather than the technique itself.

Vaginal Examination

Vaginal examination is a less popular technique but nevertheless can be a useful tool in confirming a mare's sexual state. The mare is prepared as for speculum examination, described in Chapter 11. The speculum allows illumination and hence observation of the mucosal membranes of the vagina and cervix. During oestrus the vagina appears red/pink in colour with a fluid secretion high in chlorine ions, peaking just post ovulation. The cervix at this time appears relaxed and its lining oedematous. Shortly after oestrus the cervix begins to contract and becomes paler in appearance with a thinner secretion. During dioestrus the cervix is closed, pale in colour and dry, and it may protrude some distance into the vagina. The vaginal secretion at this time is sticky and the vaginal walls tend to adhere to each other, making speculum examination more difficult. If the mare is already pregnant then the cervix will be tightly closed and white to pink in colour with a central sticky mucous plug. The vaginal secretion is thick, thickening even more as pregnancy progresses. The anoestrus cervix, on the other hand, is completely relaxed, pale in colour and dry.

Ultrasonic Ovarian Assessment

In more recent years ultrasonic assessment of the mare's reproductive tract, which was initially used for pregnancy detection, has been extended to use in determining the optimum time for covering mares. The ultrasonic detector is used in a similar way to that for pregnancy detection and can be used to assess ovarian function and, therefore, appropriateness of covering. Developing follicles can be seen and their size and, hence, nearness to ovulation can be estimated. Corpora lutea can also be detected and assessed to a certain extent for viability (Ginther and Pierson, 1984 a and b; Pierson and Ginther, 1985 a and b).

Preparation for Covering

Once it has been determined, by whatever method used, that the mare is in oestrus and ready for covering, then attention must be turned to the preparation of the mare and the stallion for mating. This depends entirely upon the system used for mating and varies from the strict codes of practice in the Thoroughbred industry to practically no preparation in the case of native pony studs. We will consider in greatest detail the most cautionary preparation as practised in the Thoroughbred industry. Other studs dispense with some, if not all, of the preparation techniques that will be detailed below.

The Mare

The mare is prepared with all eventualities in mind. She is bridled and restrained, her tail bandaged and the perineal area washed thoroughly. When washing the mare, gloves should be used and there should be a different swab of cotton wool for each swipe. Each swipe should be from the buttocks towards the perineum and the cotton wool used discarded immediately to prevent contamination of the washing solution. If soap is used, all traces must be thoroughly rinsed away, as soap and disinfectants can act as spermicides. They may also upset the natural microflora of the genital tract, which in turn may reduce fertilisation rates due to microfloral imbalance within the vagina and cervix. A hose is a good alternative to washing and is quite popular in the United States. Once the mare has been washed, felt covering boots may be fitted to her back feet, which should have had the shoes removed prior to arrival at the stud (Plate 13.9).

Once in the covering yard she may also have a nose or ear twitch applied, depending on her temperament and past behaviour at covering. Some studs twitch as standard, believing that prevention is better than cure as far as damage to expensive stallions is concerned. If she has the reputation of being particularly bad-tempered she may need one of her fore legs held up in a carpal flexion or hind leg hobbles fitted to prevent her lunging forward and objecting when the stallion mounts. Hobbles are particularly popular in the United States and also in Australia, but they are not commonly used in the UK. It is questionable that a mare requiring such drastic restraint is truly in oestrus, as conception rates may be adversely affected, due to both the incorrect timing of service and the stress of such treatment. Other articles used occasionally at covering are blinkers, hood or blindfold, especially of use in highly strung maiden or difficult mares. A mare being covered by a stallion that tends to bite his mares can be protected by use of a shoulder or whither pad, with or without a biting roll (Plate 13.10).

On very rare occasions tranquillisers may be administered 15–30 minutes prior to covering; this can be of use in particularly nervous or vicious mares. If such extreme restraint is necessary, then the use of such a mare for breeding really should be questioned, as there is a reasonable risk that these antisocial temperament characteristics will be passed on to her offspring and so perpetuated.

As discussed in Chapter 12, Thoroughbred studs also require all mares to have a second bacterial swab taken for CEM at the oestrus of service. A negative result must be obtained before service will be allowed. Swabs are taken from the uterus, urethral opening and clitoral area in an attempt to isolate Haemophilus equigenitalis,

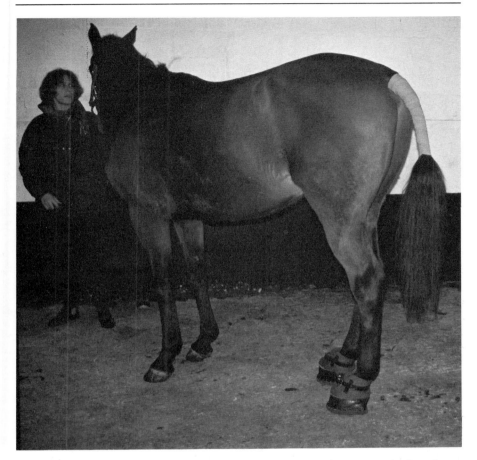

Plate 13.9 Mare ready for covering, tail bandaged, perineum washed and wearing covering boots.

Klebsiella aeroginosa and *Pseudomonas aeroginosa* (Plate 13.11). This is carried out to protect the stallion from infection and hence passage on to other mares at covering and also to avoid a service that is likely to fail.

The Stallion

A stallion's behaviour during the covering season can be very unpredictable and, therefore, dangerous. In order to reduce the potential danger stallions can be racked up, ie chained to the back wall of their stables as a regular routine during the day. When a mare is to be brought to them they can be racked up prior to her

Plate 13.10   From the left: breeding roll, withers pad, covering boots, and neck guard with biting roll. (The National Stud, Newmarket)

Plate 13.11

Swabbing a mare to check for venereal disease infections prior to covering.

arrival. This prevents them becoming overexcited by seeing the mare arrive and makes catching an enthusiastic stallion much safer. Most stallions have a specific bridle for exercise and for breeding and they get to know what the job in hand is by the bridle produced. For covering in intensive in hand systems a stallion is normally tacked up with his covering bridle and held on a long rein. A pole, which may be used as an alternative to a rein, gives a degree more control to unpredictable stallions.

When he is tacked up he is led from his box to a washing down area where his penis is washed along with the rest of his genital area, his belly and inside his hind legs with warm, clean, soapy water. Antiseptic, once very popular, may be used with care. If so, the area must be thoroughly rinsed as antiseptic and soap act as contraceptive agents. The stallion is then led to join the mare waiting in the covering area.

The Mare and Stallion

Such preparation of the mare and stallion represents the one extreme of taking full precautions to avoid all possible transfer of bacterial infection to the stallion and hence on to successive mares. In the Thoroughbred industry where the value of stock is high, such precautions are economically justifiable. In other studs, however, with progressively lower turnovers and less valuable stock, preparation becomes less cautionary and less labour intensive. It is interesting to note that several of these procedures are being introduced into the less intensive studs. For example, several breed societies now recommend swabbing for CEM and some restrict the use of the yearly premium stallions to mares with negative CEM certificates as one of the conditions of the award.

The other extreme to that practised in the Thoroughbred industry is that seen in many native studs where no preparation of the mare or stallion is practised except the possible bridling of the pair for restraint and removal of the mare's hind shoes. Such extensive systems run the high risk of disease transfer. One of the saving graces is that native type stock seem to suffer less from genital infections than the typical Thoroughbred, though of course they are not completely immune.

Covering

The style and management of the actual covering procedure varies widely between the two extremes of intensive in hand covering and the extensive pasture breeding practice. In hand covering is the norm within the Thoroughbred industry in most countries, and with the increasing value of stock, in hand breeding is becoming more widely used.

In such a system the stallion to be used is initially introduced to the mare for covering over a trying board, protecting him during the initial contact. The mare and stallion are given plenty of time to allow her to show. When she is ready she is led away from the trying board to the area used for covering. At this stage when she is in position for covering she may be further restrained by means of a twitch or hobbles if required. The stallion is then led into the covering area and allowed to approach the mare at an angle several feet from her nearside to avoid startling her and causing her to kick out.

At this stage for each horse there should be at least one handler wearing stout footware and a hard hat. The stallion handler should also carry a stick to reprimand him if he gets too boisterous, and in many traditional systems there is also an assistant to help the stallion gain intromission if necessary or to hold the mare's tail. A fourth handler may also be present to hold the mare steady and prevent her tottering too far forward. No more than four handlers should be required, and it is essential that all know their exact job and that an emergency procedure has been worked out beforehand.

Once the stallion's penis is fully drawn or erect he may be allowed to mount. A period of teasing over the trying board is required by some stallions; otherwise they need to spend a prolonged period of time with the mare in the covering area, a potentially more dangerous situation, prior to full erection. The stallion must be fully drawn before mounting is allowed. There are two blood pressures within the stallion's penis at erection: turgid and intromission pressure. Attainment of intromission pressure is essential to avoid the stallion hurting himself on entry into the mare (Plate 13.12).

As the stallion is allowed to mount the mare all handlers must stand on one side of the two animals, usually on the left, to reduce the chance of being kicked. If problems do occur, both the stallion and the mare can be pulled towards their handlers, turning their hindquarters away from anyone likely to be kicked.

As the stallion mounts the mare she will probably totter forward, which is to be allowed as long as she does not move too far (Plate 13.13). However, if she is allowed to walk forward too far, she will put a strain on the stallion, who will rest more of his weight on her back and cause possible strain or injury. Tottering too far forward can be prevented by mating in front of a protective barrier and this arrangement can also be used to protect handlers from mares that tend to strike out with their forelegs at mating.

Plate 13.12   *A stallion must be fully erect before being allowed to mount a mare.*

Plate 13.13   *The mare may totter forward at covering. She may be allowed to do so to a limited extent, but walking too far forward can put undue strain on her back.*

Traditionally, as mentioned, an assistant stallion handler was present at covering with the job of pulling the mare's tail to one side and guiding the penis into the vagina. However, this is no longer considered good practice as any unexpected external stimulus may put the stallion off by disrupting the normal sequence of events leading up to ejaculation.

Successful ejaculation is signalled by the rhythmic flagging of the stallion's tail (Plate 13.14). This is a pretty obvious signal in most stallions, but if in doubt contractions of the urethra at ejaculation can be felt by a hand placed along the underside of the penis. However, such external interference may, as discussed previously, disrupt the natural sequence of events.

After ejaculation the stallion should be allowed to relax on the mare where he will appear to go limp (Plate 13.15). He should then be allowed to dismount at will, and this may be several minutes after ejaculation. During this period of relaxation the rose of the penis, that increases considerably in size at ejaculation, returns to its normal size, allowing the penis to be withdrawn. Early withdrawal can cause considerable damage to the stallion and possibly the mare, especially if she is a maiden. Immediately he has dismounted, the mare should be turned towards her handler on her left to avoid hitting the stallion should she lash out. The stallion should similarly be turned towards his handler after the mare has been walked forward, again to prevent him hitting the mare or handlers if he kicks out in his excitement.

After covering, the stallion should be allowed to wind down. His penis may be washed while still erect, usually with warm soapy water; at the same time his genitalia can be examined for abrasions. Many practitioners advocate walking the mare around slowly after covering to prevent her straining and so losing any of the semen deposited.

If the stallion and mare are of similar size this system works well. Occasionally their sizes may vary and if so, certain precautions should be taken. A very large stallion put on to a small mare may cause her problems in late pregnancy because of the large size of the foetus near term. The reverse situation poses no such problems. In order to assist in the mating of horses of unequal size, some covering yards have a dip in the floor or use a breeding platform to even up the horses' heights (Figure 13.6). In America hydraulic breeding platforms are becoming popular; these can be adjusted to the exact height required.

A breeding roll may also be needed if a large stallion is used to cover a small mare, especially if he is well endowed. The breeding roll is placed between the mare's perineum and the belly of the stallion, preventing him from plunging too far into the mare and

Plate 13.14 *The flagging of the stallion's tail is normally a reliable sign of successful ejaculation.*

Plate 13.15 *The quiescent period post ejaculation allows time for the rose penis to return to its pre-ejaculation size and allow safe dismounting of the stallion.*

Figure 13.6   *A horseshoe-shaped breeding platform may be used to increase the height of a small stallion for easing the covering of a tall mare.*

causing her damage. The breeding roll is usually a padded leather roll 15–20 cm in diameter and in the order of 50 cm long, and to maintain conditions as sterile as possible, it can be kept covered with a disposable examination glove.

Covering Yard

The covering yard can be any area that is quiet with a good surface and away from the melee of the general yard, and a paddock or open yard (Plate 13.16) can be quite easily used. Many studs practising in hand breeding have specially designed covered areas that provide a clean, dry and safe area for use at mating (Plate 13.17). If large enough, this covering yard can double up as an exercise area (Plate 13.15).

A specially designed area should be at least 20 m by 12 m roofed and have two sets of wide doors to allow horses to be brought in and taken out in different directions if needed. The floor is very important indeed. It should be dust-free, and suitable surfaces include clay, chalk, peat moss, woodchip. Nowadays, especially in the United States, rubber matting is becoming more popular (Plate 13.18). It is very important that the surface be non-slip. Materials like concrete are not advisable but are occasionally used.

Plate 13.16   Covering in a field or paddock is common practice in smaller studs.

Variations

As mentioned previously, there are many variations on the above theme, many being more cost effective and used on the smaller native studs. The alternatives are the use of a small open paddock or yard or railed area with a non-slip surface, possibly of earth or grass. In hand mating is often practised in such yards. Many stallions, especially in the Welsh cob industry, successfully cover mares in hand in a convenient field near the main yard. This worked very well historically when stallions were walked around from farm to farm covering mares in any convenient paddock or area. If covering is to occur in a large field, the stallion used must be well behaved and controlled. If he escapes in such a large area it may take a long time to catch him and in the meantime he may have wreaked havoc with mares in adjacent fields.

The other extreme to in hand breeding is pasture breeding, where the stallion is allowed to run free with his mares to cover them at will. This system is the nearest to the natural system but can only be used in a static herd of mares with very limited introduction of visiting mares. Such pasture breeding is popular in South Africa

Plate 13.17   A specifically designed indoor covering yard. (The National Stud, Newmarket)

Plate 13.18

Rubber matting is also a suitable non-slip surface in a covering yard.

and South America, where mares are run in large herds over wide expanses of land with stallions running freely with them.

In such open systems any new or foreign mare to be covered has to be introduced before the breeding season commences in order to allow her to settle down and find her place within the hierarchy of the herd before mating begins (Bristol, 1982; Ginther, 1983; Ginther, Scraba and Nuti, 1983).

A halfway house between in hand breeding and pasture breeding is to allow the stallion to be loose in the field and introduce the mare to him individually bridled and restrained. The stallion is then allowed to mount her at will. This system allows visiting mares to be successfully covered in a system that is as near to the natural system as possible. Stallions used in this system must be well behaved and their characteristics known well by the mare handler so any signs of trouble can be identified and avoiding action taken.

As mentioned previously, a major problem when covering mares is the danger to any foals at foot (Plate 13.19). The majority of mares come to stud with their foals and are covered while the foals are still with them, often at a very young age. If mating is to occur in a large open area the foal may be left loose and will normally keep well away from the proceedings, but within sight of his mother, reducing her anxiety and stress. If mating is to occur in a small enclosed covering yard, or if the mare is difficult, the foal may be removed for its own safety and put in a loose box within sight of the mare and with a handler to watch it. Some mares will become distracted by the antics of the foal in the loose box. In such cases more success may be achieved by removing the foal altogether out of sight and earshot. The system used depends entirely on the character of the mare and foal and will not always be the same with the mare in successive years due to the differences in foal characteristics.

Plate 13.19

Mares with foals at foot may prove difficult to tease and cover successfully.

Conclusion

It is evident that a vast array of management practices are used in covering mares. The system ultimately chosen is up to the individual stud and largely depends upon financial considerations.

There is an increasing move towards the intensive systems of breeding horses, as this provides more protection for both handlers and horses, but often at the expense of reproductive performance.

References and Further Reading

Belonje, C.W.A. (1965) Operation of retroversion of penis in the stallion. J.S.Afr.Vet.Ass., 27, 53.

Bristol, F. (1982). Breeding behaviour of a stallion at pasture with 20 mares in synchronised oestrus. J.Reprod.Fert.Suppl., 32, 71.

Ginther, O.J. (1983) Sexual behaviour following introduction of a stallion into a group of mares. Theriogenology, 19, 877.

Ginther, O.J. and Pierson, R.A. (1984a) Ultrasonic evaluation of the reproductive tract of the mare: Ovaries. J.Equine Vet.Sci., 4, 11.

Ginther, O.J. and Pierson, R.A. (1984b) Ultrasonic anatomy of equine ovaries. Theriogenology, 21. 471.

Ginther, O.J., Scraba, S.T. and Nuti, R.C. (1983) Pregnancy rates and sexual behaviour under pasture breeding conditions in mares. Theriogenology, 20, 333.

Greenhof, G.R. and Kenney, R.M. (1975) Evaluation of reproductive staus of non-pregnant mares. J.Am.Vet.Med.Ass., 167, 449.

Kosiniak, K. (1975). Characteristics of successive jets of ejaculated semen of stallions. J.Reprod.Fert.Suppl., 23, 59.

Keiper, R. and Houpt, K. (1984) Reproduction in feral horses: an eight year study. Am.J.Vet.Res. 45 (9), 91.

Klingel, H. (1982) Social organisation of feral horses. J.Reprod.Fert.Suppl., 32, 89.

Kolter, L. and Zimmermann, W. (1988) Social behaviour of Przewalski (Equus przewalskii) in the Cologne zoo and its consequences for management and housing. Appl.Anim.Behav.Sci., 21, 117.

Pierson, R.A. and Ginther, O.J. (1985a) Ultrasonic evaluation of the corpus luteum of the mare. Theriogenology, 23, 795.

Pierson, R.A. and Ginther, O.J. (1985b) Ultrasonic evaluation of the pre-ovulatory follicle in the mare. Theriogenology, 24, 359.

Voss, J.L., Pickett, B.W. and Back, D.G. (1973) The effect of rectal palpation on fertility rates in the mare. Proc. 19th Ann.Conv.Am.Ass.Equine Practnrs., 33:

Voss, J.L. and Pickett, B.W. (1975) The effect of rectal palpation on the fertility of cyclic mares. J.Reprod.Fert.Suppl., 23.

CHAPTER 14

Management of the Pregnant Mare

BEFORE the management of the pregnant mare is considered it is essential to ascertain that the mare is indeed pregnant.

Pregnancy Detection

Pregnancy detection is done for several reasons. Firstly, it is carried out to determine as soon as possible whether the mare is pregnant, so that she can be re-covered by the stallion on her next oestrus if not. Secondly, it is necessary for sale or for insurance purposes, being particularly relevant if the mare is in the last two-thirds of her gestation. Thirdly, if the agreement at covering was no foal, no fee, it establishes whether a fee is due. 1 October is the customary date on which the buyer of a stallion nomination has to pay in the northern hemisphere; the corresponding date in the southern hemisphere is 1 April.

The mnemonic device AEIOU describes well the major aspects that should be looked for in a pregnancy test:

A accurate
E early
I inexpensive
O once only
U uncomplicated

Several methods of detection are available but none as yet really meets all the above criteria, though further research brings these targets ever closer.

Manual

Manual methods of pregnancy detection are the oldest and still the most common in use. They are also the cheapest. A major advantage is that they give immediate results, though they can only be accurately used after day 20 post coitum. Examination is carried out via rectal palpation or cervical/vaginal examination.

Rectal Palpation

Rectal palpation allows the uterus to be felt through the rectum wall. Initial work on this technique was carried out by Day (1940). This, plus subsequent work, suggests that with experience an accurate diagnosis can be made 20–30 days post coitum. Detection of the equine pregnancy is reported possible prior to Day 18, although occasionally with awkward mares it is not possible until Day 50. The accuracy of this technique is not 100%, but is still very good. By Day 60 the accuracy of detection is very good indeed, as at this stage the age of the foetus can be estimated to within 1 week and the presence of twins can also be detected.

Prior to Day 60 of gestation the examiner should proceed as follows. The mare should be prepared under as near sterile conditions as possible as described in detail in Chapter 11. After inserting a gloved hand into the rectum, the examiner should pick up the ovary through the rectum wall and work down the fallopian tube to the uterine horn and onto the uterine body, then proceed back up the other side of the uterus to the other ovary. As the equine pregnancy invariably implants at the junction of the uterine horn and body, a pregnancy can be detected as a discrete swelling either side of the midline at this junction. In addition, the uterus feels turgid, not flaccid as in the non-pregnant state.

Post Day 60, in the case of twins, the foetal sacks merge and the diagnosis of twins is less accurate. In addition, the foetus seems to move towards the uterine body, and due to its size palpation of the margins of the foetal sack is difficult. The whole uterus and the uterine horns become progressively less turgid and more distended and no discrete swellings can be detected. As a result there is an increased chance of inaccuracies in pregnancy detection at these later stages. However, as pregnancy progresses further it becomes possible to feel the head of the foetus through the uterine wall.

Post Day 200 accuracy of detection returns to near 100%, as at this stage you can readily feel the foetus through the now much thinner uterine wall. Pregnancy may also be obvious by the mare's external appearance (Figure 14.1).

Pregnancy detection by this method of rectal palpation is most commonly conducted between Day 20 and 35 of gestation and is considered accurate and safe. Van Niekerk (1965), however, reported that rectal palpation between Day 25 and 31 is best avoided as this is the stage at which the equine placentation changes from a choriovitelline form of the initial chorionic girdle

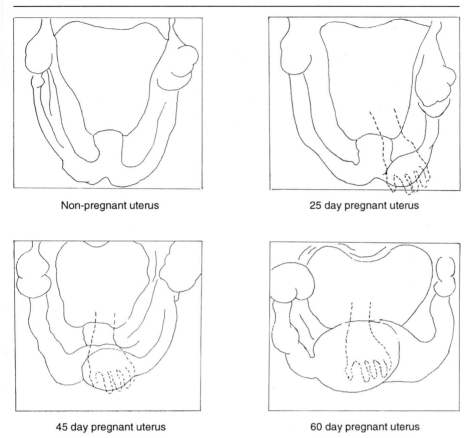

Non-pregnant uterus

25 day pregnant uterus

45 day pregnant uterus

60 day pregnant uterus

Figure 14.1   The expected size of the foetal sack at various stages of pregnancy when undertaking rectal palpation.

attachment to the more permanent chorioallantoic placentation. Further details of equine placentation are given in Chapter 5.

The major disadvantage of rectal palpation is that it cannot be used early enough to detect mares not holding and thus allow them to be returned to the stallion for service within one oestrus cycle. However, this technique is quick, simple and gives immediate results.

Cervical/Vaginal Examination
Examination of the mare's cervix and vagina via speculum can be used as an aid to pregnancy detection, but it is not accurate enough to be used alone. The pregnant cervix is white/light pink in

colour with a sticky mucus. The vagina is also very sticky with an opaque mucus, making the insertion of a speculum difficult. The major problem with this technique is the significant variation in cervical/vaginal characteristics between mares, though in conjunction with rectal palpation it can be a useful aid to diagnosis in awkward mares.

Abortion Risk

There are conflicting reports associating manual manipulation of the mare's reproductive tract with an increased risk of abortion. Work done by Allen (1974) and Voss and Pickett (1975) produced no evidence of such an association in pony mares. In fact, abortion of a twin by squeezing of (pinching out) the foetal sacks through the uterine wall per rectum at Day 40 requires considerably more relative force than ordinary rectal palpation, but is only reported to run a minimal risk to the remaining foetus. However, Osbourne (1975) demonstrated that myometrial activity within the uterine walls does increase at palpation and that increased levels of stress are associated with abortion. The consensus these days is that any association between rectal palpation and abortion is due to stress of handling rather than the technique per se.

Blood Tests

Blood samples are taken by a veterinary surgeon and used to assess the concentration of one or several hormones within the mare's circulatory system. These tests can be very accurate but have the major disadvantage of there being a delay before the results are available, as most have to be sent away to specialist laboratories for analysis. Another disadvantage is the high costs involved.

Pregnant Mare Serum Gonadotrophin

As discussed in Chapter 3, between days 40 and 100 of gestation the placental endometrial cups produce the hormone pregnant mare serum gonadotrophin (PMSG), which can be detected in the mare's blood for the period of time it is produced. Unlike some of the other hormones of pregnancy it is prevented from passing into the urine by its large molecular size which cannot pass through the kidney's filtration system. The presence of PMSG was traditionally detected by biological tests involving the injection of a blood sample into female mice. Observation of the mouse's ovaries and uterus after 48 hours would indicate the presence of PMSG, which caused enlargement and activation of the mouse's reproductive tract.

In the 1960s a new immunological or antibody test called the haemagglutination inhibition test was developed. This is the PMSG diagnostic test now used (Wide and Wide 1963; Allen 1969) and is more commonly known as the MIP test (mare immunological pregnancy test). During the period Day 40–100 post coitum the test's accuracy is 97%. This MIP test was further refined in the 1980s by De Coster et al. (1980) as an agglutination test with anti-PMSG immunoglobulins. The blood serum to be tested is added to a solution of PMSG antibodies and then to PMSG-coated sheep red blood cells. If PMSG is present in the blood sample, it binds to the PMSG antibody and no agglutination between the sheep blood cells and the anti-PMSG occurs. A precipitate is formed, evident as a doughnut shape in the bottom of the test tube indicating a positive pregnancy result. If there is no PMSG in the blood sample, agglutination between anti-PMSG and the sheep red blood cells occurs and no precipitate is formed.

As stated, this test is very accurate in detecting the presence of PMSG in a blood sample. However, the presence of PMSG is not a guarantee that the mare is carrying a viable foetus. After spontaneous or induced abortion the endometrial cups continue to secrete PMSG and, therefore, an erroneous positive result can be given (Mitchell and Betteridge, 1973).

Progesterone
Progesterone is the hormone responsible for the maintenance of pregnancy and therefore, as might be expected, elevated levels are indicative of the presence of a foetus. When testing very early on in pregnancy before a mare's possible return to service, elevated levels, ie greater than 1.5 ng/ml on Day 16–17 post coitum, are an accurate indication of pregnancy. At this time progesterone levels should be declining in non-pregnant mares. However, delayed oestrus can be mistaken for pregnancy. It is, therefore, essential to know if the mare being tested shows habitual delays in her return to oestrus. In such cases rectal palpation can be used as a back-up diagnosis after Day 21 (Hunt, Lein and Foote, 1978), but then one of the major advantages of progesterone testing is lost, that of early detection prior to a mare's expected first return to oestrus.

Progesterone can be tested not only in blood but also in the milk of mares with foals at foot. Milk samples taken between Day 18 and 19 give the most accurate results. Blood testing is most commonly used but milk testing is an alternative for mares that object violently to blood sampling.

Early Pregnancy Factor
An early pregnancy factor (EPF) has been isolated in some animals as early as 6 hours post coitum (Shaw and Morton, 1980). Such a

factor has not as yet been isolated in the mare though there is no reason to doubt that it will be eventually. Hopefully this will provide the ideal pregnancy test allowing the mare to be prepared for covering on her next natural oestrus or at an artificially accelerated return to oestrus. If diagnosis of fertilisation ever becomes routinely possible as early as 6 hours post coitum, this will even allow the mare to be re-covered at the same oestrus giving her another chance to conceive.

Oestrogen

Oestrone sulphate is one of the major oestrogens found in the mare's plasma during mid pregnancy (Cox, 1971; Raeside et al., 1979). Plasma oestrone sulphate concentrations can be measured by radio-immunoassay (RIA) procedures from Day 85 onwards (Terqui and Palmer, 1979), providing an accurate but late diagnosis.

There is a smaller rise in oestrogen evident Day 35–40 onwards and this can be used as a preliminary indication of foetal viability. However, there is evidence to suggest that oestrogen levels remain elevated for as long as 14 days after foetal death (Stabenfeldt, Daels, Munro, Kindahl and Lasley, 1991). This method is thus of no use for an early diagnosis.

Urine Tests

Oestrogens in the mare, being relatively small molecules, are able to pass unaltered through the kidney's filtration system and can therefore be detected in the urine. A biological test was used to detect their presence before Cuboni (1934) developed the Cuboni test by which detection is possible at Day 90 of gestation, although accuracy of diagnosis at this stage is low. By Day 150, accuracy improves significantly and is reported to be of the order of 95% or above (Boyd, 1979).

The urine Cuboni test is not very popular but does have its uses in non-lactating mares where rectal palpation or blood collection proves difficult (Evans, Kasman, Hughes, Coulo and Lasley, 1984).

Ultrasonic Pregnancy Detection

In recent years ultrasonic techniques have revolutionised the detection of pregnancy in many animals. Ultrasonics are based on the principle that ultrasonic sound waves are absorbed or reflected by objects or substances they hit. The relative amount of absorption or reflection depends on the density and movement of the object they hit and can now be transduced into a visual image. These methods have the advantages of being relatively cheap, giving an immediate result and being usually done on the stud as the equipment is fully portable.

Doppler System

Fraser, Keith and Hastie (1973) developed the use of the Doppler ultrasound. This machine allows the movement and hence beat of the foetal heart to be detected along with blood flow in the uterine arteries. The reflected signal is emitted from, and received by, a transducer placed on the mare's abdomen or in her rectum via a rectal probe and is heard as an audible sound. The foetal heartbeat can be heard as a distinct beat accelerated in comparison with the mare's from Day 42 of gestation onwards, though it is not consistently heard until Day 120. The enhanced blood flow through the pregnant mare's uterine artery can also be heard at the same time. It is present as a distinct whooshing noise at a slower beat than the foetal heart. This characteristic blood flow through the pregnant uterus is diagnostic in itself.

This method is very accurate at Day 120 but accuracy in earlier pregnancy is not guaranteed and thus it cannot be successfully used within the time scale required for the mare to be returned to the stallion at her next oestrus. The time span of use is similar to that of the test for PMSG, but it has the obvious advantage of detecting a viable foal.

Visual Echography

Hackelver (1977) developed the use of visual ultrasonic echography. The reflected ultrasonic waves in this system are transduced into a visual picture on a television monitor. Ultrasonic echography allows the ovaries and uterus to be viewed and, as discussed in Chapter 12, can be used to examine the mare's ovaries as an aid to the timing of service. Examination is per rectum by means of a probe carrying the ultrasonic emitter and transducer. The foetus can be detected from Day 14 as a discrete spherical sack. This typical spherical nature of the equine trophoblast and its characteristic position at the junction of the uterine horn and body is very fortunate and, as in the case of the human, allows detection at an accuracy reported to be greater than 92% at a very early stage. This method can also be used in the accurate detection of twins in early pregnancy, giving time for abortion to be induced and re-covering of the mare at her next oestrus (Plate 14.1).

The image produced by ultrasonic echography illustrates reflected sound waves, ie hard structures such as bone, as white, and illustrates absorbed sound waves, ie fluid, as black. There are obviously many variations between the two extremes depending on the relative reflection and absorption of the waves (Plate 14.2).

Ultrasonic Pregnancy Detection

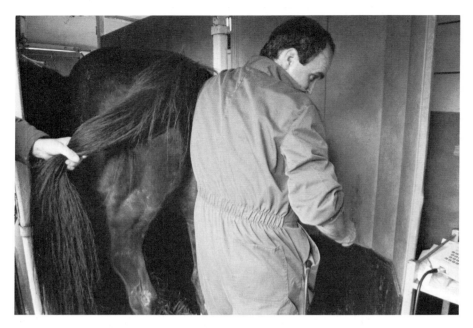

Plate 14.1

The transducer probe is placed in the rectum of the mare and angled down towards the uterus. The image produced is displayed on a television monitor.

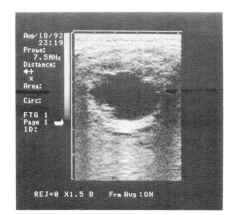

Plate 14.2

A typical scanning photograph showing a 20-day-old pregnancy as a white structure towards the bottom of the black fluid-filled sack.

Foetal Electrocardiography

A foetal heart electrocardiogram is obtained from electrodes strategically placed on the mare's body, which pick up the foetal electrical heart impulses. The readout can be used not only to detect the presence of a foetal heart beat but also to diagnose any abnormalities. Its popularity as a tool for diagnosing pregnancy is limited due to the complication of setting it up. However, it does have its uses in detecting foetal stress or cardiac abnormalities, especially at parturition during a difficult delivery. It can also confirm foetal viability at any stage in later pregnancy (Buss, Asbury, and Chevalier, 1980).

Pregnant Mare Management

The management of the pregnant mare is important in order to ensure that the offspring she is carrying is given every advantage for health and development at the start of its life. In addition, it is vital to try to maintain the reproductive capacity of the mare for her future life. Good fit condition throughout a pregnancy is an advantage for a mare and allows her to foal with maximum ease. In the intensive systems of today the mare is usually required to return to the stallion at her foal oestrus. For this to be successful she has to not only ovulate and show oestrus but be in optimum physical condition to embark on another pregnancy. An understanding of the developmental changes to both mare and foetus during gestation is essential to gear her management towards maintaining her in top condition for reproduction. These developmental changes are detailed in Chapter 5.

There is little information on the environmental factors that affect foetal well-being. However, good management of the pregnant mare is clearly one, and can be reviewed under five main headings: exercise, nutrition, parasite control, vaccinations and teeth and foot care.

Exercise

Exercise and nutrition are closely connected and together are the major determining factors of body condition. As with most things, extremes can prove harmful, and so in the case of exercise a happy medium is to be aimed for. The absolute level depends on the individual mare and her past history. A moderate exercise regime may be provided in the form of self-exercise by regular turnout in an open field or gentle ridden exercise in early pregnancy. Mares accustomed to being ridden can be ridden with increasing gentleness until 6 months of pregnancy, but again it depends on the individual mare.

Gentle exercise promotes and enhances the mare's blood circulatory system. As discussed in Chapter 5, the foetus depends entirely on its mother for its nutrient intake and waste output. This transport system is provided by the blood circulatory system, especially the utero-ovarian vein and artery. Blood transport through this system is enhanced by exercise, and hence oxygen and nutrient supply to the foetus are enhanced too. Exercise reduces water retention, or oedema, often associated with mares who are kept inside for prolonged periods of time. Exercise helps to maintain body condition and reduces the levels of fat. Hence the chances of complications at parturition are reduced and the maintenance of muscle tone will enhance delivery (Plate 14.3).

Exercise must be gentle and will depend on the mare's previous activity. Excessive exercise will result in exhaustion and has been shown to be associated with high abortion rates. Excessive exercise may also cause stress, and so may sudden movements, travelling, sale rings and even low flying aircraft. Stress is known to increase

Plate 14.3 *Exercise is essential in order to maintain a mare's fitness and body condition score of 3 and to prevent circulatory problems in the later stages of pregnancy.*

abortion rates via an elevation in the concentrations of prostaglandin F (PGF) metabolites. It is this association between abortion and stress that leads some people to believe that the stress of rectal palpation may be associated with an increased abortion risk. In most farm livestock minimising stress is especially important at the time of implantation. There is no direct evidence that this is the case in the mare, but it seems highly probable. Therefore, as a rule stress should be minimised throughout gestation but especially between Day 40 and 70 (the period of implantation) and during the last 6–8 weeks of pregnancy.

Nutrition

The nutritional requirements of a pregnant mare depend mainly upon whether she has a foal at foot and is lactating, or whether she was previously barren and so has only to satisfy the requirements for her own maintenance. The length of lactation in the mare depends today on man's interference and influence, but normally runs for 6–8 months. It is, therefore, during this time that the two groups of mares with or without foals at foot vary in their requirements. In an ideal system, and providing there were enough mares to justify it, these two classes of mares should be kept and fed separately. During pregnancy a mare's body condition and weight should be carefully monitored to ensure that neither excess fat is laid down, nor does she have to mobilise her own body reserves to supplement inadequate nutrition. Her feeding management should vary accordingly. The use of a weigh pad is ideal to monitor her weight on a regular basis. However, this is financially beyond the reach of many establishments and a weigh band can be just as good. Assessment of body condition by eye and feel can also be undertaken (Plate 14.4).

The following discussion concentrates on the nutritional management of the non-lactating mare, as the management of the lactating mare is addressed in some detail in Chapter 16. The general principles of nutrition apply to feeding the mare as to any formulation of a ration for a horse. The important nutrients are proteins, energy (sugars and fat), vitamins and minerals, and they should all be balanced according to need. As in the case of other animals, feed must be of a good quality, especially in late pregnancy. Fungal contamination can cause abortion. General poor nutrition has been associated with prolonged gestation, developmental abnormalities and decreased birth weights. These problems are exacerbated if low nutrition levels are evident in late pregnancy (Comline, Hall, Lavelle and Silver, 1975; Silver, Barnes, Comline, Fowden, Clover and Mitchell, 1979).

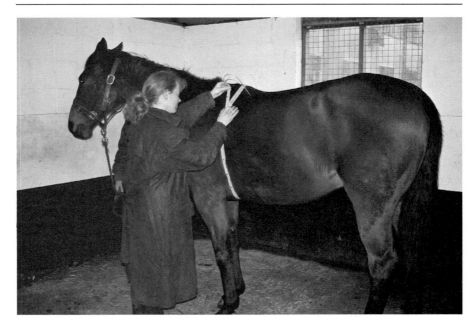

Plate 14.4 *Use of a weigh band is a quick and convenient method of monitoring a mare's weight, especially when used in conjunction with assessment of body condition by eye and feel.*

As discussed in Chapter 5, the vast majority of foetal growth occurs in the last 3 months (90 days) of pregnancy. Prior to this, nutritional requirements are similar to those for barren mares. During the last 3 months the pregnant mare requires more energy, protein, calcium and phosphorus in order to satisfy the foetus's increasing requirements and allow some laying down of fat within the mare and foetus for use as a temporary store to be mobilised, if needed, after parturition or during lactation.

Protein
Good-quality hay or grass will provide adequate protein for early pregnancy providing that calcium and vitamin A levels are adequate; if not, a supplement should be used. Levels of crude protein in the order of 10.0% (100% DM) are required in early pregnancy. However, as pregnancy progresses into the last 90 days, protein intake needs to increase in parallel with requirements. All changes in diet must be made gradually to help prevent digestive upsets. Levels of protein in the order of 10.6% (100% DM) are required in the last three months of pregnancy (Tables 14.1–14.4). This level may be achieved by some good fresh grasses, as dried grass or hay tends to lose protein in the

Table 14.1 The average composition of feeds commonly used in horse diets.

Feed	DM (%)	DE (Mcal/kg)	Crude protein (%)	Lysine (%)	Fibre (%)	Ca (%)	P (%)	Mg (%)	K (%)	Na (%)	S (%)	Vitamin A (IU/kg)
Alfalfa (hay)	91.0 / 100.0	2.07 / 2.28	17.0 / 18.7	– / –	25.5 / 28.0	1.24 / 1.37	0.22 / 0.24	0.32 / 0.35	1.42 / 1.56	0.11 / 0.12	0.26 / 0.28	41900 / 46000
Barley (grain)	88.6 / 100.0	3.26 / 3.68	11.7 / 13.2	0.40 / 0.45	4.9 / 5.6	0.05 / 0.05	0.34 / 0.38	0.13 / 0.15	0.44 / 0.50	0.03 / 0.03	0.15 / 0.17	817 / 922
Clover (hay)	88.4 / 100.0	1.96 / 2.22	13.2 / 15.0	– / –	27.1 / 30.7	1.22 / 1.38	0.22 / 0.24	0.34 / 0.38	1.60 / 1.81	0.16 / 0.18	0.15 / 0.16	9727 / 11005
Fish meal	91.7 / 100.0	2.93 / 3.20	62.2 / 67.9	4.74 / 5.17	0.7 / 0.8	5.01 / 5.46	2.87 / 3.14	0.15 / 0.16	0.71 / 0.77	0.41 / 0.44	0.53 / 0.58	– / –
Grassmeal (high protein)	88.0 / 100.0	2.31 / 2.62	16.0 / 18.2	0.8 / 0.9	22.0 / 25.0	0.6 / 0.7	0.23 / 0.26	– / –	2.1 / 2.3	– / –	– / –	– / –
Maize	88.0 / 100.0	3.41 / 3.88	8.50 / 9.65	0.26 / 0.29	2.5 / 2.8	0.02 / 0.02	0.30 / 3.34	– / –	0.31 / 0.35	– / –	– / –	– / –
Milk (Dried skim)	94.1 / 100.0	3.81 / 4.05	33.4 / 35.5	2.54 / 2.70	0.2 / 0.2	1.28 / 1.36	1.02 / 1.09	0.12 / 0.13	1.60 / 1.70	0.51 / 0.54	0.32 / 0.34	– / –
Molasses	77.9 / 100.0	2.65 / 3.40	6.6 / 8.5	– / –	0 / 0	0.12 / 0.15	0.02 / 0.03	0.23 / 0.29	4.72 / 6.06	1.16 / 1.48	0.46 / 0.60	– / –
Oats (grain)	89.2 / 100.0	2.85 / 3.20	11.8 / 13.3	0.39 / 0.44	10.7 / 12.0	0.08 / 0.09	0.43 / 0.38	0.14 / 0.16	0.40 / 0.45	0.05 / 0.06	0.21 / 0.23	44 / 49
Pea (seeds)	89.1 / 100.0	3.07 / 3.45	23.4 / 26.3	1.65 / 1.86	5.6 / 6.3	0.21 / 0.14	0.41 / 0.46	0.12 / 0.14	0.95 / 1.06	0.22 / 0.25	– / –	285 / 320
Sorghum (grain)	90.1 / 100.0	3.21 / 3.56	11.5 / 12.7	0.26 / 0.29	2.6 / 2.8	0.04 / 0.04	0.32 / 0.36	0.15 / 0.17	0.37 / 0.41	0.01 / 0.01	0.03 / 0.15	468 / 520
Soyabean (extracted meal)	89.1 / 100.0	3.14 / 3.53	44.5 / 49.9	2.87 / 3.22	6.2 / 7.0	0.35 / 0.40	0.63 / 0.71	0.27 / 0.31	1.98 / 2.22	0.03 / 0.04	0.41 / 0.46	– / –
Sugar beet (pulp)	91.0 / 100.0	2.33 / 2.56	8.9 / 9.8	0.54 / 0.60	18.2 / 20.0	0.62 / 0.68	0.09 / 0.10	0.26 / 0.28	0.20 / 0.22	0.18 / 0.20	0.20 / 0.22	88 / 97
Timothy (hay)	88.9 / 100.0	1.77 / 1.99	8.9 / 9.7	– / –	30.0 / 33.8	0.43 / 0.48	0.20 / 0.23	0.12 / 0.13	1.61 / 1.82	0.01 / 0.01	0.12 / 0.13	18964 / 21340
Wheat (bran)	89.0 / 100.0	2.94 / 3.30	15.4 / 17.4	0.56 / 0.63	10.0 / 11.3	0.13 / 0.14	1.13 / 1.27	0.56 / 0.63	1.22 / 1.37	0.05 / 0.06	0.21 / 0.24	1048 / 1177

Table 14.2   Daily nutrient requirements of pregnant mares of varying weights.

Pregnant mares	Weight (kg)	DE (Mcal)	Crude protein (g)	Lysine (g)	Ca (g)	P (g)	Mg (g)	K (g)	Vitamin A (10³ IU)
9 months	200	8.2	361	13	16	12	3.9	13.1	12
	500	18.2	801	28	35	26	8.7	29.1	30
	700	23.6	1039	36	45	33	11.3	37.8	42
	900	26.8	1179	41	51	38	12.9	42.9	54
10 months	200	8.4	368	13	16	12	4.0	13.4	12
	500	18.5	815	29	35	26	8.9	29.7	30
	700	24.0	1058	37	46	34	11.5	38.5	42
	900	27.3	1200	42	52	38	13.1	43.6	54
11 months	200	8.9	391	14	17	13	4.3	14.2	12
	500	19.7	866	30	37	28	9.4	31.5	30
	700	25.5	1124	39	49	35	12.3	40.9	42
	900	29.0	1275	45	55	41	13.9	46.3	54

Table 14.3   Nutrient concentrations in total diets for pregnant mares (100% DM). Values assume a concentrate feed containing 3.3 Mcal/kg and hay containing 2.0 Mcal/kg dry matter.

Pregnant mares	DE (Mcal/kg)	Diet proportions Conc (%)	Hay (%)	Crude protein (%)	Lysine (%)	Ca (%)	P (%)	Mg (%)	K (%)	Vit A (IU/kg)
9 months	2.25	20	80	10.0	0.35	0.43	0.32	0.10	0.35	3710
10 months	2.25	20	80	10.0	0.35	0.43	0.32	0.10	0.36	3650
11 months	2.40	30	70	10.6	0.37	0.45	0.34	0.11	0.38	3650

Table 14.4   The expected feed consumption by pregnant mares (% body weight).

Pregnant mares	Forage	Concentrates	Total
Early and mid pregnant mares	1.5–2.0	0–0.5	1.5–2.0
Mares in late gestation	1.0–1.5	0.5–1.0	1.5–2.0

drying process. Legume hay, for example alfalfa, tends to have a higher protein content, and even after drying may have adequate levels for the late pregnant mare. If in doubt, it is essential to have your feed analysed. Any good stud will as a matter of routine have all batches of hay, haylage, legume hay etc analysed to ascertain the dry matter content, protein, energy, vitamins and minerals. This allows accurate balancing of feeds available to meet the requirements and identifies the appropriate supplements needed.

Protein can be supplemented by the addition of animal products or plant products. Appropriate animal by-products include fish meal, bone meal etc. Such products tend to be expensive and have suffered a drop in popularity in the United Kingdom and parts of Europe in the light of the recent bovine spongiform encephalitis (BSE) scare in cattle. The other alternatives are plant products such as soya bean meal, linseed meal etc. These tend to be cheaper and hence more popular. Both kinds can be just as easily utilised by the horse's digestive system.

The total protein content of a diet is not the only important factor in satisfying the protein requirements of the late pregnant mare. The component parts of all proteins are amino acids, some of which are essential and others which can be manufactured by the horse's own metabolism from other amino acids. The latter are termed non-essential amino acids. Certain protein-rich supplements may be lacking in certain essential amino acids, so that even though the *total* protein content is high, the use of this to the foetus is limited. Barley, oats and linseed are all high in total protein content but are lacking in one or more essential amino acids. Soya bean meal, on the other hand, contains all the essential amino acids required for the development of the foetus and is, therefore, a very useful protein supplement for late pregnant mares.

Energy

Energy requirements also increase significantly in late pregnancy from levels required by the mature horse. Again, the energy demand in early pregnancy may be met by good-quality forage, levels of 2.25 Mcal/kg (100% DM) being required. However, if the mare is a poor doer, or being ridden, her diet may need to be supplemented by concentrate feeds to ensure that she can maintain her body condition. Again, analysis of all feed is a valuable tool to rationing horses appropriately. During early pregnancy the mare's weight gain should be minimal, though, as she progresses into the last third of pregnancy, she should be expected to gain in the order of 0.25 kg/day. This level ensures that weight gain is due to an increase in foetal weight and not excessive deposition of internal body fat.

Overfat condition in mares during late pregnancy causes excessive pressure on the internal organs at parturition as well as limiting uterine size and so reducing foetal birth weights and possibly post natal viability. The other extreme of allowing no body weight gain means that the mare does not lay down any fat at all. A limited amount of fat laydown in late pregnancy is essential to act as a temporary store for mobilisation in an emergency, especially during early lactation. Failure to gain weight may even mean that the mare's limited fat reserves of pregnancy are already being mobilised, reducing the energy available to the foetus and hence its growth in utero and final birth weight.

During this period of increased energy demand, energy supplementation will be required. This may be achieved in theory by an increase in hay intake. However, in late pregnancy the increase in uterine size begins to limit the capacity of the digestive tract. Energy levels of 2.40 Mcal/kg (100% DM) are required during this period. Good-quality feeds, low in bulk but high in nutrients, are best fed. It is, therefore, advised that in late pregnancy the mare's roughage intake (ie hay) is reduced and partly replaced by increasing levels of energy-rich concentrates, being careful at the same time to ensure that the required protein intake is still maintained (Tables 14.1–14.4).

Vitamins and Minerals
Vitamins and minerals are classified as micronutrients. The specific effects of deficiencies of many micronutrients on the pregnant mare, and indeed on general equine welfare, are as yet unknown. However, the importance of calcium, phosphorous and vitamin A is appreciated. The calcium:phosphorous ratio is of special importance and the involvement of both minerals in bone growth is well documented. In addition, these minerals are essential for the adequate function of nervous impulses, muscle contraction and relaxation. They are essential components of fat and protein and hence the development and laying down of soft tissue. In the pregnant mare, not only are these micronutrients important to the mare herself, but also to the foetus. Calcium and phosphorous are normally stored within the bones of the horse, much of which is a temporary store and can be mobilised to satisfy demands elsewhere. If the pregnant mare's dietary intake of calcium or phosphorous is inadequate, especially during late pregnancy, the mare will mobilise her own stores from within her bones to satisfy the foetus's demand. If the dietary deficiency is great, then her bones will suffer and become brittle and possibly unable to take the strain of the increased weight in late pregnancy or the stresses of parturition. The foals of such mares can also suffer from

deformities in tissue and bone growth and general weakness at birth.

The levels of calcium and phosphorous required in the first 8 months of pregnancy are in the order of 0.43% and 0.32% per day respectively. During the last 90 days the demand increases to 0.45% and 0.34% per day respectively (Tables 14.2 and 14.3).

Not only are the absolute levels of these two minerals important but also the ratio of relative amounts. Excess phosphorous interferes with calcium absorption and leads to an effect similar to calcium deficiency. A ratio of calcium:phosphorous of 2:1 is ideal, though ratios between 1:1 and 6:1 are considered acceptable. It is important to know that straights in the form of grains tend to be relatively high in phosphorous and, therefore, in late pregnant mares, when concentrate intake should be increased, calcium should also be supplemented if grains are fed. Calcium may be supplemented in the form of bone meal, but this, as mentioned previously, can be expensive and is not very popular these days. An often-used alternative is ground limestone flour or milk pellets, available on the market today. Because of its very high concentration of phosphorous, bran should be avoided in late pregnancy except when used in small quantities as a laxative.

Other minerals should also be supplemented, though their exact function is unclear: for example, potassium, magnesium, salt, iodine, manganese, copper etc. These may be easily supplemented by one of the commercially available vitamin and mineral blocks, providing the mares have free access to them (Tables 14.2 and 14.3).

Vitamin A, as mentioned, is also important as an essential component of the epithelium and is, therefore, indicated as important in reproductive function, cell regeneration and development. Adequate vitamin A levels can best be ensured by feeding fresh green forage. Beware of excess vitamin A supplementation, as this may cause abnormalities, especially associated with bone growth (Tables 14.2 and 14.3).

Water
As with all equine rationing, water is an essential but often forgotten component. The pregnant mare requires large amounts of water, approximately 50 litres/day, depending on the dry matter content of her ration. This water must be clean, fresh and available at all times to allow consumption in small but frequent amounts.

Parasite Control

Parasite control is of significant importance. A high parasite count is often the reason why some mares seem to be bad doers, and parasites in essence cause a large amount of the food fed to the animal to be wasted. If a mare's internal worm burden is allowed to get out of hand, the damage caused may become permanent, condemning that mare to being a bad doer for the rest of her life and making it very difficult for her to take in enough nutrients in times of high demand as in late pregnancy. As with all equids, worming should take place regularly at least every 12 weeks during spring and summer. Less frequent worming is required in winter once frosts appear; if temperatures do not fall below freezing, continual 12 week worming throughout the year should be practised.

The development of resistance to wormers (anthelmintic treatments) can be avoided by rotating the wormer types used, ie those based on the thiabendazole group followed by those based on the pyrantel embonate group. This does not allow the parasites time to develop resistance to a specific wormer type. Care must be taken in worming pregnant mares, as not all wormers are suitable. Wormers based on cambendazole, for example, are not suitable for very early pregnant mares, ie within the first 12 weeks, as its use during this period has been associated with an increased chance of abortion. At the other end of pregnancy organophosphate wormers used to control bot flies are not recommended as they may disrupt and trigger smooth muscle contraction, and induce late abortions due to uterine contractions. Wormers based on pyrantel panoate or febendazole are relatively safe for pregnant mares, though it is recommended that no wormers are used in the last month of pregnancy because of the risk of inducing premature delivery. The importance of clean grazing must not be overlooked and a combination of worming treatment and clean grazing gives the best results.

Delousing powder may be administered to pregnant mares, but again its use should be avoided in late pregnancy. Ideally, pregnant mares should not be allowed to get into such a condition as to need treatment.

Vaccinations

The vaccination programme required by a mare depends upon the endemic diseases present and hence the country in which she lives. In the United Kingdom vaccination against tetanus and influenza is automatic and is normally administered 4 weeks prior to parturition to allow the titre of antibodies to be raised within the mare and transferred on to the colostrum waiting in the mammary gland and, therefore, readily available to the foal at birth.

Equine rhinopneumonitis, which is caused by equine herpes virus type 1 (EHU1), is a problem in the United States and is becoming increasingly so in the United Kingdom. This infection causes abortion normally in the last trimester in up to 70% of infected mares. The most recent vaccination available utilises an inactivated virus vaccine and can be administered at any time during pregnancy, ie if an outbreak is suspected or routinely in months 3, 5 and 7 of pregnancy. The older live vaccines run the risk of inducing abortion.

Equine viral arteritis (EVA) also causes abortion. A modified live vaccine is now available against it. Although EVA is a problem in parts of Europe and the United States, it is not common at present in the United Kingdom, but we cannot become complacent, as the 1993 outbreak demonstrated.

Teeth and Foot Care

Finally the teeth of the pregnant mare should not be neglected, as regular rasping ensures that all the plates are level and can efficiently grind food during mastication. Rough, uneven teeth lead to inefficient chewing action and subsequent inefficient food digestion, reducing the nutrient value of the food to that mare. This will be of utmost importance when her system is under stress, for example during late pregnancy. All mares should have their teeth rasped regularly once a year and as soon as any suspicions of teeth problems are identified.

The mare's feet should also not be neglected. Regular filing if the mare is not shod should be carried out. Poor feet cause pain, and this pain can be exacerbated with the increased weight burden of late pregnancy. Such mares will be reluctant to exercise themselves, bringing on all the associated problems previously discussed. Shod mares must have any shoes removed prior to parturition to prevent any accidental damage to the foal.

Conclusion

Accurate pregnancy detection followed by correct management of the pregnant mare is essential if her reproductive potential is to be realised.

References and Further Reading

Allen, W.R. (1969) An immunological measurement of pregnant mare serum gonadotrophin. J.Endocr., 43, 581.
Allen, W.E. (1974) Palpalable development of the conceptus and fetus in welsh pony mares. Eq.Vet.J., 6, 69.

Boyd, H. (1979) Pregnancy diagnosis. In Fertility and infertility in domestic animals, 3rd Ed. Laing, J.A. Ed, pp. 36–58., Bailière Tindall, London.

Buss, D.B., Asbury, A.C. and Chevalier, L. (1980) Limitations in equine fetal electrocardiography. J.Am.Vet.Med, ss 177, 174.

Comline, R.S., Hall, L.W., Lavelle, R, and Silver, M. (1975) The use of intravascular catheters for long term studies of the mare and fetus. J.Reprod.Fert. Suppl., 23, 583.

Cox, J.E. (1971) Urine tests for pregnancy in mares. Vet.Rec., 89, 606.

Cuboni, E. (1934) Uber einer einfache und schelle chemische-hormonald schwanger schaftsekretian. Klin Wschr. 13, 302.

Day, F.T. (1940) Clinical and experimental observations on reproduction in the mare. J.Agric.Sci.Camb. 30, 244.

De Coster, R., Cambiaso, C.L. and Masson, P.L. (1980) Immunological diagnosis of pregnancy in the mare by agglutination of latex particles. Theriogenology, 13, 433.

Evans, K.L., Kasman, L.H., Hughes, J.P., Couto, M. and Lasley, B.L. (1984) Pregnancy diagnosis in the domestic horse through direct urinary estrone conjugate analysis. Theriogenology, 22, 615.

Fraser, A.F., Keith, N.W. and Hastie, H. (1973) Summarised observations on the ultrasonic detection of pregnancy and foetal life in the mare. Vet.Rec., 92, 20.

Hackeloer, B.J. (1977) The ultrasonic demonstration of follicular development during the normal menstrual cycle and after hormone stimulation. Proc.Int.Symp.Recent Advances in ultrasound diagnosis, Dubrovnik, 122–128.

Hunt, B., Lein, D.H. and Foote, R.H. (1978) Monitoring of plasma and milk progesterone for evaluation of post partum estrous cycle and early pregnancy in mares. J.Am.Vet.Med.Ass. 172, (11) 1298.

Mitchell, D, and Betteridge, K.J. (1973) Persistance of endometrial cups and serum gonadotrophin following abortion in the mare. Proc. 7th. Int.Congr.Anim.Reprod, and AI. Munich 1, 567.

National Research Council (1989) Nutrient Requirements for Horses. Bulletin No. 5 revised. National Acadmy Press.

Palmer, E. and Draincourt, M.A. (1980) Use of ultrasonic echography in equine gynecology. Theriogenology, 13, 203.

Osbourne, V.E. (1975). Factors influencing foaling percentages in Australian mares. J.Reprod.Fert. Suppl., 23, 477.

Raeside, J.I., Liptrap, R.M., McDonnell, W.N. and Milne F.J. (1979) A precurser role for DNA in feto-placental unit for oestrogen formation in the mare. J.Reprod.Fert. Suppl., 27, 493.

Shaw, F.D. and Morton, H. (1980) The immunological approach to pregnancy diagnosis. Vet.Rec., 106, 268.

Silver, M. Barnes, R.J., Comline, R.S., Fowden, A.l., Clover, L. and Mitchell M.D. (1979) Prostaglandins in maternal or foetal plasma and in allantoic fluid during the second half of gestation in the mare. J.Reprod.Fert. Suppl., 27, 531.

Stabenfeldt, G.H., Daels, P.F., Munro, C.J., Kindahl, H., Hughes, J.P. and Lasley, B. (1991) An oestrogen conjugate enzyme immunoassay for monitoring pregnancy in the mare: limitations of the assay between Days

40 and 70 of gestation. J.Reprod.Fert.Suppl., 44, 37.

Terqui, M., and Palmer, E. (1979) Oestrogen pattern during early pregnancy in the mare. J.Reprod.Fert.Suppl., 27, 441.

Van Niekerk, C.H. (1965) Early embryonic resorption in mares. A preliminary study. J.S.Afr.Vet.Med.Ass. 36, 61.

Voss, J.L. and Pickett, B.W. (1975) The effect of rectal palpation on the fertility of cyclical mares. J.Reprod.Fert.Suppl., 23, 285.

Wide, M. and Wide, L. (1963) Diagnosis of pregnancy in mares by an immunological method. Nature Lond. 198, 1017.

CHAPTER 15

Management of the Mare at Parturition

G ESTATION in the mare lasts on average 330–336 days, though considerable variation either side of these figures has been reported (Bos and Van der May, 1980; Rossdale and Ricketts, 1980). Average gestation lengths in Thoroughbreds and the larger riding type horses have been reported to be longer at 340.7 days on average (Badi, O'Bryne and Cunningham, 1981). The physical process of parturition and the endocrine control of parturition have been detailed in Chapters 7 and 8. This chapter will, therefore, concentrate on the management of the mare at parturition including the various means of artificial induction that may be used.

Pre Partum Management

About 6 weeks before the mare's estimated date of delivery, she should be introduced to her foaling unit or the paddock in which she is to foal. This will give her ample time to become familiar with the surroundings that will be her home at and immediately after foaling, reducing the stress of any sudden changes at such a time. Such familiarisation allows her to become accustomed to the particular management practices within the foaling unit, especially if she is to foal away from home. Changes in feed, exercise, housing and routine can be introduced in plenty of time to allow her to become accustomed to her new environment well before the foal is born.

In addition, if she is to foal away from home, a period of 6 weeks is required to allow her immune system to raise the necessary antibodies against any infections not present in her home environment. This will not only provide protection for the mare herself, but will also allow the antibodies to pass into the colostrum and provide the foal with immediate protection at birth. Bearing this in mind it is advised in Britain that all mares should be vaccinated against tetanus and influenza either as her annual booster or as a new programme at 4 weeks prior to parturition. This again allows her antibody titre to rise and pass into colostrum and hence transfer on to the foal.

During this period prior to parturition a regular routine for the mare is all-important. This helps to reduce stress, allowing her to relax and hence makes her easier to manage and eases the birth. Exercise is also very important. Regular free exercise in a paddock or field will be adequate for most mares and will help maintain fitness for foaling and reduce the chances of filled legs due to water retention in the later stages of pregnancy. Diet should, within reason, be on the laxative side as many mares suffer from constipation in late pregnancy, especially if exercise is limited. Components such as bran may be added in small quantities to the diet as a laxative, but must be used with care so as not to upset the overall nutritional balance of the diet, especially the calcium:phosphorus ratio, which is also very important at this stage. Lastly, but by no means least, clean and fresh water must always be available. Many mares demand large quantities of water in late pregnancy.

Throughout the preparation of the mare for foaling her udder and perineum should be watched for the characteristic signs of imminent parturition. These are detailed in Chapter 7 but in summary include the increase in size of the mammary glands, the possible secretion of milk and general relaxation of the whole abdominal, pelvic and perineal area.

By this time, except in the case of an emergency, it will have been decided whether the mare is to give birth naturally or is to be induced. If birth is to occur naturally, then as soon as any signs of imminent parturition are noticed the mare must be put into her foaling box if she is to foal inside, or put into a small quiet paddock if she is to foal outside. She should then be watched.

If she is to foal inside, the box provided must be at least 5 m × 5 m with good ventilation but free of draughts. Traditionally the floor covering was a deep bed of straw, which provides a soft, warm, dust-free surface on which the foal can be born (Plate 15.1).

In recent years, however, mainly as a result of research and development in North America, rubber matting is increasingly popular as flooring at larger studs (Plate 15.2).

This matting is expensive but provides a clean, insulated and dust-free floor that can be easily washed and disinfected. Some of the older rubber matting floors tend to raise off the concrete base and can prove very dangerous but the newer ones are much more successful and durable.

The foaling box should also be free of any protrusions, which may

Plate 15.1   *The traditional floor covering for the foaling box was straw that provides a good, deep, soft bed for the foal to be born on to.*

Plate 15.2   *Rubber matting as a flooring for foaling boxes is becoming increasingly popular. It is dust-free, warm and can easily be hosed down and disinfected.*

cause damage to the mare or foal. Ideally, it should have rounded corners to reduce the risk of the mare getting cast. Hay nets should not be used as the foal can get itself caught up in anything left dangling. Hay should be fed off the ground but some people use high hay racks to avoid the wastage of feeding off the floor. Unfortunately, this runs the risk of the mare and foal getting seeds in their eyes and ears. In an ideal purpose-built unit each box will have two doors, one to the outside for horse access and another facing into a central sitting area for human access and viewing (Plate 15.3). A closed-circuit television is also a good method of providing a means of viewing mares with minimal disturbance. The provision of radiant heat lamps in each box is an advantage for weak foals.

Plate 15.3   An ideal arrangement of the foaling unit allows the foaling boxes (through the doors in the background) to be viewed from a central sitting area (in the foreground). (The National Stud, Newmarket)

Once the mare has settled into her foaling quarters it is a case of careful watching and patient waiting. Many mares have the wonderful ability to give birth the minute your back is turned. However, careful observation can minimise the time from the first signs of trouble to action and can, therefore, be critical in saving lives.

Induction of Parturition

The mare shows a much wider variation in gestation length than other farm animals in which induction of parturition is practised. As discussed in Chapter 8, the natural length of gestation is affected by several factors, including foetal genotype, foetal sex, seasonal effects, nutrition and occasionally the number of foetuses. When considering inducing parturition artificially, it is essential that the date is as close as possible to the estimated delivery date. Premature induction will result in foals with all the classical symptoms of prematurity, including breathing difficulties, late to stand and a delay in the normal adaption mechanisms that occur immediately post partum. Such foals may, if they survive, suffer long-term effects, and there will be a dramatic increase in your own labour and veterinary bills. Induction of parturition in horses should, therefore, be carried out with great care, and its value as a routine technique is questionable.

Induction is normally carried out as an aid to management to ensure that all facilities and staff are available and ready when required. It is particularly useful if limited experienced labour is available and only a few mares are involved. Experienced labour and veterinary assistance can be organised for the time of the induced birth. Alternatively, induction may be used as a matter of urgency in the case of prolonged gestation, preparturient colic, pelvic injuries, ventral rupture or painful skeletal or arthritic conditions which get unbearable in late pregnancy. Finally it may be used as a research or teaching tool.

There are several methods of inducing parturition (Hillman and Lesser, 1980), and these will be discussed in turn.

Corticosteroids

Induction of parturition has been practised for many years in several species, and in ruminants the most successful method is with the use of corticosteroids (Adams and Wagner, 1970). Foaling can be successfully induced in large mares within a week, using 100 mg of dexamethasone, a corticosteroid administered daily for 4 days from Day 321 of gestation. The resultant parturition is reported to be normal, and live, healthy foals are born (Alm, Sullivan and First, 1974, 1975).

Other workers using pony mares have had less and more varying success. Some reports indicate that in pony mares such treatment results in stillbirths and placental retention, whereas others report success. However, it must be remembered that the number of mares used in most work is small (Rossdale and Jeffcott, 1975; Van Niekerk and Morgenthall, 1976; First and Alm, 1977). This apparent lack of success in mares when compared with ruminants is possibly due to the difference in the endocrine control of parturition. Chapter 8 details the possible control mechanisms governing parturition in the mare, and as discussed the answer is as yet unclear. However, even if corticosteroids are involved in the initiation of parturition in the mare, the naturally occurring levels within the foetus and the mare's circulatory system are not as high as in other ruminants and, therefore, their importance and hence effect are less. Corticosteroid administration will not induce parturition in ovarectomised mares and, therefore, the action of the corticosteroids is assumed to be via the ovaries.

Progesterone

Progesterone administration over a period of 4 days in late pregnancy will induce parturition over approximately a week (Alm, Sullivan and First, 1975). This does not occur in ruminants, but if the endocrine control of parturition is similar in the mare to that in the human, it is possible that the foetal adrenals are responsible for the conversion of progesterone to corticosteroids and hence its induction of parturition.

Both these methods have the disadvantage of being relatively inaccurate in the timing of the reaction to their administration which can be over a period of a week. A more accurate and more immediate induction agent is, therefore, required.

Prostaglandins

The use of prostaglandins gives a more immediate result than progesterone and corticosteroids; 250 µg of fluprostenol administered intramuscularly is sufficient to cause parturition within 2 hours in a mare at full term. The mare will show the initial signs of parturition, sweating, uneasiness etc within 30 minutes, the foal usually arrives within 2 hours and the afterbirth again some 2 hours later. However, if the timing of prostaglandin administration is not correct and the mare is not at full term it will not have any considerable action, though repeated injections over a longer period of time, several hours, may induce parturition/abortion. Incorrect timing of prostaglandin administration results in problems at birth, including weak foals and insufficient cervical relaxation (Rossdale, Pashan and Jeffcote, 1979).

In the natural course of events the rise in prostaglandin coincides with the delivery of the foal. Fluprostenol (Equimate) presumably imitates this. Its use too early though has little effect as elevated levels at this time are out of synchrony with other endocrine changes. Surprisingly, the use of natural prostaglandin has no effect on parturition (Alm et al., 1975; Cooper, 1979). The reason why the synthetic prostaglandin can be used as an induction agent is unclear, but it is known that the activity of prostaglandin is not via the ovaries, as it will work in ovarectomised mares, but, therefore, presumably via the uterus inducing the strong uterine contractions of parturition. In humans synthetic prostaglandins have a more potent stimulatory effect on uterine contractions than the natural prostaglandins, and it is possible that a similar situation is evident in the mare. It is interesting to note that natural prostaglandins used with glucocorticoids are also effective (Van Niekerk and Morgenthal, 1976).

Oxytocin

A further agent used to induce foaling is oxytocin. In the natural course of events oxytocin levels rise markedly during the second stage of delivery, causing the rapid contractions associated with foal delivery. When injected intramuscularly or intravenously, low levels of oxytocin, 5 iu, can induce foaling, in some instances even in mares not immediately at term. The exact amount of oxytocin required depends on the endocrine balance within the mare at administration (Pashan, 1980; Hillman and Ganjam, 1979). It is known that in many mammals there is a synergistic relationship between oxytocin and prostaglandins and that such an association does occur in mares. Oxytocin administration at a dose of 1-5 iu near term results in an immediate release of prostaglandins, equivalent to the natural release, followed by parturition (Pashan and Allen, 1979). High levels of oxytocin (60–120 iu) were initially used, but providing the timing is correct, levels as low as 1 iu can induce parturition successfully.

Oestrogen

Oestrogen can also be used as part of an induction regime. It is not successful when used alone, but aids the dilation of the cervix when used in conjunction with oxytocin, and hence eases the process of parturition.

It cannot be emphasised enough that if induction of parturition is being considered, accurate records of the mare's date of service and hence expected date of delivery are very important indeed, along with close observation for signs of parturition. Inappropriate use of induction agents can have disastrous consequences. Most of the danger is to the foetus rather than the dam, though the risk to both increases with the asynchrony between induction and the natural course of events.

Inappropriate timing of induction or drug administration may cause overexertion of the uterine contractions risking rupture or the development of ring contractions. Such complications may lead to injury to the foal, causing rib and bone fractures because of overstrong contractions and passage through a pelvic opening that has not relaxed completely prior to birth. Premature separation of the placenta may also cause foetal problems due to oxygen starvation. Finally, one of the biggest risks of premature delivery is that the foal will not be physically mature enough to survive in the extra-uterine environment.

Assessment of the state of the pregnancy is, therefore, very important to maximise the chance of a successful delivery and a resultant healthy foal. Various signs can be used to aid this assessment. These include mammary development. Presence of colostrum within the mammary glands is a very good indication of full-term development. Mammary secretions change in the last few weeks of pregnancy from a clear, dilute straw or amber colour liquid to cloudy and subsequently on to a yellow or yellowish white and viscous fluid at full term (Peaker, Rossdale, Forsyth and Falk, 1979). Cervical and pelvic relaxation is also a good sign of full term and therefore indicates the appropriate time for induction. The cervix softens and relaxes shortly before delivery. It has been suggested that a cervical dilation sufficient to insert two fingers indicates an appropriate stage for induction (Hillman, 1975). However, the amount of cervical relaxation and its timing vary considerably between mares and, therefore, cannot be accurately used as a standard sign. Pelvic ligament relaxation also occurs immediately prior to parturition, along with vulval relaxation. Pelvic relaxation is evident as hollowness of the mare's hind quarters (Colour plate 22).

Due to the variability in these signs within and between mares, induction of parturition in mares is a risky business and must be used bearing this in mind.

Management at Parturition

Whether the mare delivers naturally or is artificially induced, her management should be very similar. The main difference is that the time of delivery with induction will be known and, therefore, preparation can be better timed and organised. A mare may foal in a specifically built foaling unit or outside. Whichever system is chosen, and there are advantages in both, then the principles of the stages of labour and their management will be the same. Foaling mares outside is increasingly popular as the risk of disease is lower and the system is much nearer the natural situation (Colour plate 32). However, these days foaling out is normally restricted to pony or cob-type hardy mares or multiparous mares foaling later on in the season. Early foalers, maiden or difficult mares and those of great value are normally foaled inside, as this allows them to be observed more closely.

First Stage of Labour

First stage labour requires no special measures except close observation for the start of the second stage and for the identification of any potential problems. Some mares run to milk prior to, or during, the first stage of delivery; that is, they lose milk from the mammary glands (Colour plate 21). This milk is valuable colostrum, of which there is only a finite amount. If a mare shows considerable milk loss prior to parturition, she should be milked lightly and the colostrum collected into a clean sterile container and frozen. As soon as the foal is born this can be thawed and bottle fed to the foal, or tubed if necessary, to ensure that the foal receives the valuable antibodies it contains.

During the first stage of labour the mare will seem restless and may repeatedly get up and lie down several times. She will sweat quite considerably and be uneasy, looking at her flanks and grimacing. She may also dig up her bedding (Colour plate 33).

This continual moving around is thought to help position the foal within the birth canal. She may well show signs of discomfort followed by quiet. Thoroughbreds are notorious for this type of behaviour over quite a period of time. Whatever, her discomfort will increase as the frequency of contractions and pain increases, culminating in the breaking of the waters (release of allantoic fluid) at the cervical star. Excessively prolonged first stage labour may be a sign of problems, especially if the mare seems to be excessively distressed. It is very difficult to say how long first stage labour should last and at what stage you should call for assistance, as some mares will naturally show several false starts in the days preceding birth. However, as a general rule, the assistance of a veterinary surgeon should be obtained if the mare seems to be in prolonged discomfort, showing considerable agitation and profuse sweating, and before she is in any danger of becoming exhausted.

Second Stage of Labour

The management of the second stage of labour is more important and is marked by the breakage of the chorio-allantois at the cervical star and the resultant loss of allantoic fluid. In 90% of cases the mare will then take up a recumbent position, the most efficient for straining (Colour plate 34).

At this stage an episiotomy should be performed if required. Unfortunately, today many mares are routinely given a Caslick's operation, the suturing up of the vulva (see Chapter 1 for further details) because of inadequate perineal conformation. Such mares, along with those that are naturally too small, will need an episiotomy, cutting of the vulva, to allow passage of the foal; usually only a local anaesthetic is required for this (Plate 15.4).

As soon as this has been done, and within 5 minutes of the start of

Plate 15.4

An episiotomy should be performed at the beginning of stage two labour. This is routinely required in the case of mares having undergone a Caslick's operation.

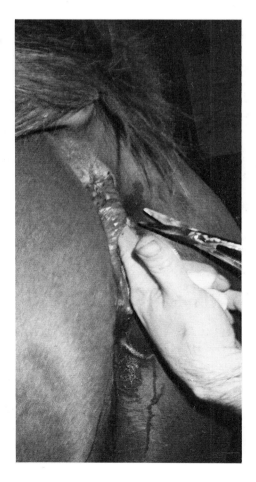

the second stage of labour in mares not requiring an episiotomy, a brief internal examination may be made to ensure that the foal is lying in the correct position for natural easy delivery (Colour plate 35). At this stage the allantoic sack should be evident as a white membrane bulging through the mare's vulva (Colour plate 36).

If the foal's forelegs and muzzle can be felt within the vagina, she should be left alone to deliver naturally. If there are problems, assistance should now be called. Care should be taken not to rupture the amnion during this process as it should be left to break naturally. When it does, the colour of the amniotic fluid should be noted for evidence of meconium staining, dark brown/green colouration, that may be indicative of foetal stress. If this is the case, then in order to minimise foetal stress, the delivery should be speeded up by traction as soon as the foal's head appears.

If all is well, all attendants should leave the box and allow the mare to foal unaided but observed quietly from a distance for signs of distress (Colour plate 37). Most mares lie down during second stage labour as this is the most efficient position for voluntary straining. Plenty of room is, therefore, needed to allow the mare to stretch her legs out during straining.

In most cases the mare will deliver on her own with no problems. However, the following complications which may be observed can be managed as follows. The chorio-allantois may appear abnormally thick and be evident as a red membrane forced to bulge out through the vulva. In such cases the membrane may be manually ruptured allowing the release of allantoic fluid and progression on to the second stage of labour. Abnormal presentation of the foetus has been discussed in Chapter 7. However, two of the most common complications can be easily corrected. If a foreleg is driven up into the roof of the vagina, there is the danger of its being forced into the rectum and causing a vaginal rectal fissure. This can be quite easily averted by drawing the foreleg down and out of the vagina. Secondly, if one foreleg is held back by the elbow lodging under the pelvic brim, gentle traction can alleviate this and bring both forelegs in line and ease delivery. General excessive straining by a mare with little result, or periods of prolonged inactivity between contractions, may be signs of problems. In this case gentle traction on the forelimbs may be used to aid the mare and to minimise her exhaustion (Colour plate 38). More complicated problems with abnormal positions require an experienced lay person or veterinary surgeon to correct. Further details on the potential problems encountered in foaling are given in Chapter 7.

After delivery the foal should be left with its hind legs still within its mother and the umbilical cord intact. The umbilical cord must not be broken prematurely as this will result in the potential loss of up to 1 litre of blood within the placental tissue. The umbilical cord will naturally constrict and degenerate at a point about 3 cm from the foal's belly and will break at this point when the foal first attempts to rise to its feet (Colour plate 39).

While the foal lies with its legs still within the vulva of its mother, there seems to be a tranquillising effect on the mare. As a result most mares will be reluctant to get up. The mare may turn to lick the foal or parts of the bedding that has become contaminated by allantoic fluid. This is the beginning of maternal foal bonding and recognition. Some mares, especially if stressed or a maiden, may get up and paw the ground at some danger to the foal.

Second stage labour should last on average 20 minutes, with a range of 5–60 minutes being acceptable. First-time foals tend to have a longer second stage labour and so do mares that have had a hard first stage.

If the foal has shown signs of distress and is limp and weak at delivery, the amnion should be broken immediately and the foal's head lifted to aid breathing. Occasionally the umbilical cord does not break after birth, despite drying up and constriction at the foal's belly. In such cases it may be broken by a sharp pull while placing the other hand on the foal's belly. Care must be taken to ensure that the umbilical cord has had enough time to break naturally, as premature breakage can result in the foal losing a considerable amount of blood left within the placenta.

Immediately Post Delivery

Immediately post delivery the foal can be dried off and the severed umbilical cord must be dressed with antiseptic powder or solution to prevent the entry of infection. The foal should be checked for its heart rate by a hand placed on the thorax (Colour plate 40). The rate immediately post delivery should be in the order of 40–80 beats/minute. Rates below or above this may be indicative of problems.

As soon as the mare stands up the placenta may be tied up to prevent the mare standing on it and ripping it out prematurely. The extra weight provided by the tying up also encourages its expulsion (Colour plate 41).

If an episiotomy was performed, then the vulva may be resutured at this stage. If this is carried out within 30 minutes of birth, no anaesthetic may be required because of the natural numbing of this area at delivery.

Third Stage of Labour

Post second stage labour, providing all is well, the mare should be left alone to recover and start the all-important bonding process with the foal. It is best for the mare to remain recumbent for as long as possible, ideally until after the end of the third stage of labour, the expulsion of the afterbirth, in order to reduce the inspiration of air into the vagina taking bacteria with it (Colour plate 42). Such contamination of the vagina increases the chance of post partum endometritis and delays uterine involution and any return to oestrus. Temporary Michel clips may be used to hold the lower lips of the vulva together and reduce air contamination. These clips can be easily removed 24–48 hours after the third stage of labour.

During this time the mare will appear restless again showing signs similar to first stage labour, those of sweating, pain and discomfort. The average duration of this third stage of labour is 60 minutes but there is wide variation. Occasionally the placenta may be expelled immediately after the foal and still attached to the umbilical cord. At the other extreme it may take several hours. Third stage labour in excess of 10 hours is indicative of problems, which are discussed in Chapter 7.

As soon as the third stage has been completed and the placenta expelled, it should be removed from the box and examined for completeness (Colour plates 29, 30 and 43). A good method to detect holes and therefore any missing fragments is to tie off both uterine horn ends of the placenta and fill it with water through the cervical star. Leakage of water indicates a break which should be examined to ensure that no membranes have been retained. If this is suspected, then a veterinary surgeon should be called to ensure its complete removal.

Placental retention can lead to septicaemia and eventual death if not treated as a matter of urgency. The placenta should be expelled in a matter of 10 hours, retention beyond this stage is indicative of problems and a veterinary surgeon should be called. The temptation to pull on the placenta in an attempt to release it must be resisted as this will invariably lead to rupturing of the placental membranes and the danger of fragments being retained. Only a very small fraction is required to set up a septicaemic infection.

During this period of third stage labour and afterwards, it is very important to leave the mare alone with her foal. The bond between mare and foal starts to develop now and can be irretrievably damaged by man's interference, however well intentioned.

There is no need to drag the foal to the mare's head and force her to stand up and move towards it as if left alone the vast majority of mares will reach around and move towards the foal without assistance (Colour plate 44).

Damage to a newborn foal by the mother is very rare, and if it does occur it is usually a result of her being disturbed or flustered by unwanted human interference. Occasionally mares will nibble or gently bite their foals to encourage them to move. Within reason this may be allowed, but if the mother is too aggressive a muzzle may be used. Very occasionally, usually in maiden mares, real aggression towards the foal may be evident involving kicking and vicious attacks. Such mares may be tranquillised for a short period of time. Tranquillisers may also be used on mares that will not stand to allow the foal to suckle. After a while most mares will get used to the foal and interference will not be required. A twitch should not be used for the first 24 hours post partum as this runs the risk of internal haemorrhage.

Foals that appear to be weak and unable to get adequate colostrum from the mare may be bottle fed or tubed colostrum. This ensures that they receive sufficient essential antibodies before they are 3 hours old, the most efficient time for antibody absorption.

Foals used to be given enemas using either medical paraffin or warm soapy water as a matter of course. This treatment is no longer deemed necessary unless the foal shows signs of meconium retention after 24 hours. Foals should be given the opportunity to suckle and pass meconium naturally before the decision is made to use an enema. The enema technique causes the foal unnecessary stress and rough handling at a very early age when delicate adaptation to the extra-uterine environment is still occurring and it is therefore best avoided if possible.

The mare should be given a feed about an hour after the birth. A light, easily digested and slightly laxative food is best along with fresh hay and water. Water should be given only under supervision initially unless automatic water feeders are installed in order to minimise the risk of the foal drowning, especially if it is on the weak side.

Conclusion

Management of the foaling mare is of utmost importance in ensuring the birth of a healthy foal. Inappropriate management or the failure to call in professional help when required can have disastrous consequences. Induction can be an aid to the management of the foaling mare, but it must be used with the utmost care, as inappropriate timing can have fatal consequences.

References and Further Reading

Adams, W.M. and Wagner, W.C. (1970) Role of corticosteroids in parturition. Biol.Reprod., 3, 223.

Alm, C.C., Sullivan J.J. and First N.L. (1974) Induction of premature parturition by parenteral administration of dexamethasone in the mare. J.Am.Vet.Med.Ass. 165, 721–722

Alm, C.C., Sullivan, J.J. and First, N.L. (1975) The effect of corticosteroid (Dexamethasone), progesterone, estrogen and prostaglandin F2α on gestation length in normal and ovariectomised mares. J.Reprod.Fert.Suppl. 23, 637.

Badi, A.M., O'Bryne, T.N. and Cunningham, E.P. (1981) An analysis of reproductive performance in thoroughbred mares. Ir.Vet.J., 35 (1), 1–12.

Bos, H. and Van der May, G.J.W. (1980) Length of gestation periods for horses and ponies belonging to different breeds. Livestock Production Science 7, 181-187.

Cooper, W.L. (1979) Clinical aspects of prostaglandins in equine reproduction. Proc. 2nd Equine Pharm. Symposium, 225–231.

First, N.L. and Alm, C. (1977) Dexamethosone induced parturition in pony mares. J.Anim.Sci., 44, 1072.

Hillman, R.B. (1975) Induction of parturition in mares J.Reprod.Fert. 23,641.

Hillman, R.B. and Ganjam, V.K. (1979) Hormonal changes in the mare and foal associated with oxytocin induction of parturition. J.Reprod.Fert. Suppl., 27, 541–546.

Hillman, R.B. and Lesser, S.A. (1980) Induction of parturition in mares. Vet.Clin.No.Am. (Large Anim.Pract.) 2, 333.

Pashan, R.L. (1980) Low doses of oxytocin can induce foaling at term. Equine Vet. J., 12, 85.

Pashan, R.L. and Allen, R. (1979) The role of the fetal gonads and placenta in steroid production, maintenance of pregnancy and parturition. J.Reprod.Fert.Suppl., 27, 499–509.

Peaker, M., Rossdale, P.D., Forsyth, I.A. and Falk, M. (1979) Changes in mammary development and the composition of secretion during late pregnancy in the mare. J.Reprod.Fert.Suppl., 27, 555.

Rossdale, P.D. and Jeffcote, L.B. (1975) Problems encountered during induced foaling in pony mares. Vet.Rec., 97, 371–372.

Rossdale, P.D., Pashan, R.L, and Jeffcote, L.B. (1979) The use of synthetic prostaglandin analogue (fluprostenol) to induce foaling. J.Reprod.Fert., Suppl., 27, 521.

Rossdale, P.D. and Ricketts, S.W. (1980) Equine Stud Farm Medicine. 2nd Ed. Ballière Tindall, London.

Van Niekerk, C.H. and Morgenthall, J.C. (1976) Plasma progesterone and oestrogen concentrations during induction of parturition with flumethasone and prostaglandin. Proc.8thInt.Cong.Anim.Reprod. and A.I. (Krackow) 3, 386–389.

CHAPTER 16

Management of the Lactating Mare and Young Foal

Foal Adaptive Period

IMMEDIATELY post partum the foal has to undergo substantial changes in its adaptation in order to survive in the extra-uterine environment. These adaptations involve changes to the foal anatomically, functionally and biochemically. In the normal foal adaptive changes can be identified right up to puberty and even until the achievement of mature size. However, in considering the true adaptive period for survival outside the uterus, the first four days of life are the most crucial. It is within this period that the majority of adaptive problems can be identified and, hopefully, rectified. If all is well then and the foal satisfies all the normal criteria at this age, it has a very good chance of survival.

Anatomical Adaptation

In a normal, straightforward birth the foal is born on its side lying with its hocks still within its mother and the umbilical cord intact (Plate 16.1). Normally the foal should be seen to breath within 30 seconds of final delivery. It may well have taken a few sharp intakes of breath as its muzzle first reached the air during passage through the birth canal, but a rhythm is normally established within 1 minute of final delivery. This initial breathing rhythm is normally 70 breaths/minute with a tidal volume (volume of air inspired/breath) of 520 ml, resulting in a minute volume (volume of air inspired/minute) of 35 litres. This initiation of breathing results in a significant increase in blood oxygen levels. During this first minute of life the foal's heart rate should be in the order of 50 beats/minute and its body temperature 37.5°C.

The significant increase in blood oxygen concentrations as a result of the first breaths activates the first reflexes and muscle

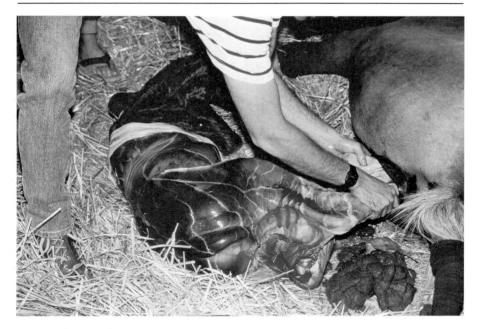

Plate 16.1 *At the end of second stage labour the foal lies outside the mother with its hocks still within the vagina and the umbilical cord still intact.*

movements. Within 5 minutes of birth the foal will respond to pain and it shows evidence of the reflexes associated with rising to its feet, in the form of raising its head, extending its forelimbs, blinking and possibly giving a whinny. It will also demonstrate the suckling reflex if given a finger or bottle to suck.

The next major event is the breaking of the umbilical cord, usually within 9 minutes of birth. As previously discussed, it is very important that the cord is left to break naturally. This occurs at a constriction that develops about 3 cm from the belly of the foal. The umbilical artery and vein degenerate and constrict naturally at this point and blood circulating through these vessels slows, allowing a clean break, with little blood loss to occur, reducing the chance of infection. If the cord breaks prematurely the foal may lose up to 1 litre of blood which is still circulating in the foetal side of the placenta, with its associated oxygen and nutrients diffused from the dam. This extra boost, or safety net, is invaluable, especially in foals that may have initial problems in breathing and, therefore, in obtaining enough oxygen on their own.

The mare may be recumbent post partum for up to 40 minutes and therefore, the cord is usually broken by the movements of the foal in its first attempts to raise itself. The increase in tension as a result of movement causes a break at the weakest point, the constriction, and, therefore, the umbilical cord breakage occurs naturally after the muscle reflexes of the foal have begun, on average 5–9 minutes after birth.

Around the time of umbilical cord breakage the foal makes concentrated efforts to raise itself to its feet (Plate 16.2). The series of movements is the same as in the mature horse: stretching forward of the head and neck, extending the forelegs, flexing the hind legs and raising the front end first off the ground, followed by the hind quarters. Many initial unsuccessful attempts are usually made and are all part of the process of developing the reflexes and muscle control. At this stage the foal is at risk of damage from projecting objects such as buckets, automatic water feeders etc. It is advisable to make sure that the foaling box is free from all such potential hazards. The foal's legs, muzzle and head are most at risk as it plunges around trying to stand, and fortunately damage is normally only superficial. Successful standing normally occurs in ponies at 35 minutes post partum but may take up to half an hour longer in Thoroughbred foals (Plate 16.3), successful standing in Thoroughbreds not often being achieved until 1 hour after birth. Failure to stand within 2 hours is indicative of problems and veterinary assistance should be sought (Jeffcote, 1972). This remarkable ability to stand and walk so quickly is the result of the evolutionary development seen in plain dwelling animals, such as the horse, to enable fleeing from potential predators soon after birth.

Though the suckling reflex is developed within 5 minutes or so post partum, actual successful suckling can obviously not occur until after standing, location of and movement to the udder. The actual reflexes involved in suckling are elicited by contact with soft warm surfaces; hence foals are seen to suckle and nuzzle their mother's flanks while searching for the udder. As soon as the foal stands it demonstrates directional movement towards the udder, located by the dark and warmth. This process of locating the mare's udder can be easily disrupted by human interference. It is very tempting to try to help a foal locate the udder and frustrating to watch it suckling at the hock, seemingly unable to locate the teat. However, it is much better to leave the mare and foal alone and resist the temptation to interfere. The mare will often assist the foal by gently nudging it and moving her hind leg away from her body to allow the foal easier access. Occasionally maiden mares may need to be held to allow the foal to reach the udder, as some mares seem initially to be ticklish and object to the foal's attentions. However, she will soon settle down and should be left alone. In Thoroughbreds successful suckling is achieved normally within 1.5 hours,

Plate 16.2   *Soon after the end of second stage labour the foal makes concentrated efforts to get to its feet, and at this time the umbilical cord breaks.*

Plate 16.3   *Successful standing in the pony foal takes on average 35 minutes compared to up to 1 hour in the Thoroughbred.*

and in ponies tends to occur within 65 minutes (Plate 16.4). At suckling a real affinity between mother and foal is set up and once it has begun develops into a very strong bond. Man's interference may well disrupt this initial bonding process. The foal, throughout the first few days of life, will suckle at 30–60 minute intervals; later on the interval period lengthens. If during the first few days of life a foal is not seen to suckle for 3 hours or so, problems should be suspected.

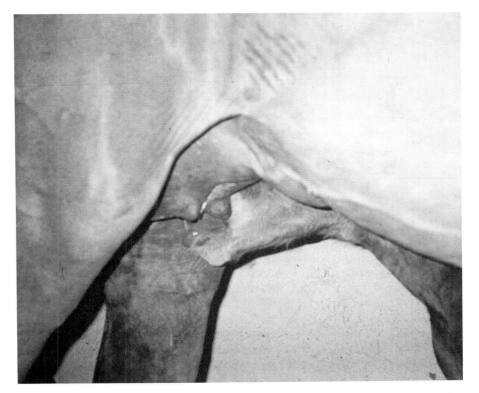

Plate 16.4 *In ponies suckling normally occurs within 65 minutes, whereas in Thoroughbreds it normally takes 1.5 hours.*

During this period of initial udder location, the foal's breathing rate falls to 35 breaths/minute with a slightly larger tidal volume at 550ml and a minute volume of 13 litres. Its body temperature rises slightly to 38°C and its heart rate to 120–140 beats/minute largely as a result of the exertion of initial standing and movement.

During the first 12 hours the foal should be seen to pass meconium, its first bowel movement. Meconium consists of the glandular secretions of the bowel during the foal's inter-uterine life, along with digested amniotic fluid and cell debris which are passed through the foal's digestive tract in utero. Meconium is stored in its colon, caecum and rectum ready for expulsion after birth. Premature expulsion may occur under stressful conditions immediately prior to, or during, delivery. This is why meconium staining of the amniotic fluid released during the first stages of labour should be noted as a sign of stress. Meconium should be brown or greenish-brown in colour and is usually all expelled within the first 4 days, followed by the characteristically yellow-coloured milk dung. The foal should also pass its first urine within 12 hours; colts tend on average to pass urine earlier than filly foals (Jeffcote, 1972; Roberts, 1975).

Functional Adaptation

The functional changes that occur in the foal in adaptation to the extra-uterine environment involve changes to the pulmonary, cardiac and circulatory systems along with temperature control mechanisms.

Pulmonary Ventilation

Prior to Day 190 of gestation the foetal lungs are very under-developed, incapable of expansion, with little superfactant, and are still developing and branching to form alveoli. At about Day 190–210 the lining of the lungs becomes progressively covered with a lipoprotein substance termed superfactant. This superfactant is not complete until 300 Days, and in some cases not until after delivery. This superfactant lining to the tubes, branches and alveoli of the pulmonary system reduces surface tension and, hence, increases the efficiency of oxygen and carbon dioxide transfer. Maximum efficiency is, therefore, not possible until superfactant development has been completed. This creates a problem encountered in premature deliveries. After birth, continued development of the alveoli occurs, induced by lung expansion and stretching of the bronchi with the first breaths. At birth the efficiency of gaseous exchange is low. Hence the rapid breathing rate, which reduces as the bronchi are stretched and alveoli development continues, results in an increase in surface area/air ratio, allowing more efficient gaseous transfer (Gillespie, 1975).

The foal's first breath possibly occurs in utero as a practice run for the real thing. The first proper extra-uterine breath is stimulated by low levels of oxygen and increased levels of carbon dioxide along with cold shock from the atmosphere and tactile stimulus associated with birth, and normally occurs within 30 seconds of the foal's hips appearing through the birth canal. The first breathing rate is 70 breaths/minute dropping to 50/minute after 1 hour and 34/minute after 12 hours (Dawes, Fox and Ledue, 1972; Gillespie, 1975; Stewart, Rose and Barko, 1984; Ousey, McArthur and Rossdale, 1991).

Cardiac and Circulatory Systems
In utero the blood can pass from the pulmonary artery via the ductus arteriosus to the aorta, bypassing the lungs and going directly to the placenta. In addition, the foramen ovale allows blood to pass from the right atrium direct to the left atrium and hence immediately around the body via the aorta, rather than to the right ventricle. The blood then enters the placenta via the two umbilical arteries and leaves via the single umbilical vein, to pass to the liver and then back to the right side of the heart. Immediately post partum the circulatory system of the foal changes dramatically to redirect blood through the pulmonary system to collect oxygen and rid itself of carbon dioxide, rather than through the umbilical veins and arteries to the placenta. The trigger for this is unknown but results in the closure of the ductus arteriosus and foramen ovale within the heart. Blood is now pumped from the right atrium via the pulmonary artery to the lungs for oxygenation and back via the pulmonary vein to the heart for pumping around the body. Closure of the foramen ovale prevents blood passing from the right to left atrium (Figures 16.1 and 16.2).

Immediately post partum the foal's heart rate is in the range of 40–80 beats/minute rising to 150 beats/minute with the exertion of rising to its feet. After this time the heart rate varies with activity but is in the region of 70–95 beats/minute when the foal is resting. Many newborn foals suffer initially from arrhythmia, irregular heart beats, but this soon settles down naturally. Many foals show signs of asphyxia during the second stage of labour with a reduction in blood flow and, therefore, oxygen to the head, resulting in a blue tongue and mucous membranes of the eyes. This is probably due to constriction of the head, neck and chest on passage through the pelvis. It is of no long-term significance providing the foal continues to be delivered normally and parturition is not delayed. After birth the mucous membranes may remain blue/grey in colour but should be the normal pink colour within 2 hours. The best place to take the heart rate of a newborn foal is by placing a hand over the left side of the chest in the area of the heart.

The red blood cell count at birth is considerably elevated compared to later on in life. The cause of this is thought to be foetal stress during birth as levels are further elevated in foals born with difficulty. Within 2 hours of birth red blood cell counts drop off and

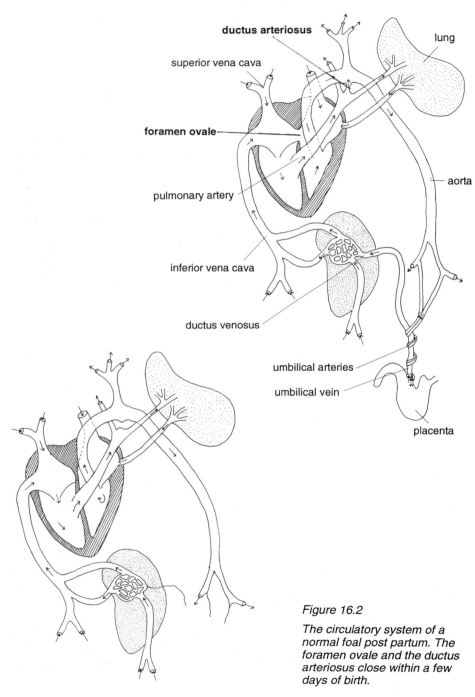

Figure 16.1   The foetal circulatory system in utero. The blood passes through two openings—the foramen ovale in the atrium wall and the ductus arteriosus—hence bypassing the lungs.

ductus arteriosus

lung

superior vena cava

foramen ovale

aorta

pulmonary artery

inferior vena cava

ductus venosus

umbilical arteries

umbilical vein

placenta

Figure 16.2

The circulatory system of a normal foal post partum. The foramen ovale and the ductus arteriosus close within a few days of birth.

allow white blood cell counts to rise to normal levels (Chavatte, Brown, Ousey, Silver, Cottrill, Fowden, McGladdery and Rossdale, 1991).

Temperature
The foal is born with a well-developed temperature control mechanism, unlike many other mammals, especially man, that cannot effectively control their body temperature for several weeks after birth. At birth the foal can maintain a steady body temperature of 37–37.5°C (100°F) (Rossdale, 1968). The exact mechanism by which it maintains this steady body temperature is unclear but is likely to be via shivering, which is evident in newborn foals during the first 3 hours post partum. In addition, muscular activity and the strain of the foal's first movement will also contribute to maintaining body temperature. The foal also has insulating layers of fat plus a hair coat. The foal does not have brown adipose tissue as seen in humans, as it has this ability to shiver, and so does not require brown fat heat-producing metabolism. The presence of brown fat is associated with neonates (newborn animals) not able to shiver and with less fine control over their body temperature (Ousey, McArthur and Rossdale, 1991).

Immune Status
As discussed in Chapter 5, the equine placenta is termed epitheliochorial and consists of 6 cell layers, 3 on the maternal side and 3 on the foetal side. As such, it presents a considerable barrier to the passage of blood components from mother to foetus, especially those of large molecular size, for example maternal antibodies. In the foal, therefore, the attainment of antibodies in utero is limited and so colostrum is vitally important for achieving adequate immune status for survival in the extra uterine environment. The small intestine is permeable to large protein molecules such as antibodies immediately post partum and for the following limited period of 36 hours (Jeffcote, 1971). Equine immunoglobulins can be divided into IgG, IgM and IgA. The most predominant in colostrum is IgG and it is these that are most evident in the circulation of the young foal (McGuire and Crawford, 1973). Adult levels of immunoglobulins are not immediately present in the foal and are reached only after the foal starts to actively produce immunoglobulins itself. Absorption of protein antibodies in the first 36 hours of life is via the intestinal epithelial cells by the process of pinocytosis and the absorption of proteins is non-selective. The ability to absorb whole proteins is seemingly enhanced by other components of colostrum and controlled in part by cortisol, though the exact means is unclear. Intestinal epithelial cells gradually lose

their ability to absorb these antibody protein molecules over the first 36 hours of life. It is, therefore, essential that newborn foals receive colostrum, 500 ml at least, within the first 36 hours of life, and preferably within the first 12 hours when absorption is at its most efficient. This ensures that the foal obtains maximum protection from infection via maternal antibodies.

In addition to receiving antibodies via colostrum, the foal becomes increasingly able to manufacture its own. The foal's immune system starts to function to a very limited extent during pregnancy, but does not reach its maximum capacity until the foal reaches 3–4 months of age. Colostrum, therefore, provides protection until the foal's own immune system is up and running. If the mare is immunised during late pregnancy (in the United Kingdom this is routinely carried out for tetanus and myxoinfluenza virus), then the antibodies raised within the mare pass to her colostrum and on to the foal, providing it with essential temporary protection. Such immunisation is advised at 6 weeks prior to expected delivery. Tetanus antitoxin may be given to the foal immediately post partum to ensure protection, especially in areas where tetanus infection is a particular problem.

Biochemical Adaptation

The foal's metabolic system at birth undergoes dramatic alterations from a dependent to an independent status. While in utero it depended upon the maternal system and hence placenta for all its needs. Post partum this dependency is removed and replaced by gastrointestinal absorption to meet its nutrient requirements.

At birth the foal goes through a transitional period after the severing of the maternal connection and before suckling. This period of time is one of considerable stress and exertion for which energy is required. To meet these energy requirements glycogen is stored within the liver during the last stage of gestation. This glycogen is mobilised in the form of glucose immediately post partum and supplies, in an easily convertible form, the energy required for those activities associated with the transition period. Glucose levels can be measured in the plasma of newborn foals and used to indicate the availability of these glycogen stores. Immediately post partum glucose concentrations are in the order of 50–70 mg/100 ml blood. After suckling these levels increase and in a normal 36 hour old foal reach values of 100–110 mg/100 ml blood. Bicarbonate levels rise steadily over the first 36 hours of life from approximately 23–28 mmol/litre (Jones and Rolph, 1985; Fowden, Mundy, Ousey, McGladdery and Silver, 1991). Lactate levels rise immediately after birth and fall off to normal adult levels by 36 hours. The initial increase in lactate coincides with a fall in venous pH and may be as a result of the change in circulatory flow on the energy demands during the transition period from placental to gastrointestinal dependency. This fall in pH rectifies itself within 12 hours. Plasma cortisol levels also rise significantly immediately post partum. The significance of this is uncertain but has been postulated in some mammals to be associated with the final maturation of internal organs, such as the respiratory system.

The biochemical changes apparent in the newborn and very young foal can only be assessed via blood sampling and vary considerably. They must, therefore, be viewed with a certain amount of caution when used as a diagnostic aid to identifying post partum problems.

Post Partum Foal Examination

In summary, within 1 hour of birth it should be reasonably obvious whether things are going properly or whether problems are suspected and a veterinary surgeon should be called to attend the foal. A physical examination of the following will give a very good indication of the situation:

Heart rate
Respiration
Ability to stand
Vigour
Ability to suckle
Straight legs
Body weight

Other possible routine procedures include:

Injection of 1500 iu tetanus antitoxin
6–8 ml antibiotic (penicillin/streptomycin)
Blood tests
Blood cell count
Isoerythrolysis test (mare/foal compatibility test)
Immune status test

The Orphan Foal

The inability of a mother to bring up her foal due to her death, illness or injury can be a serious setback for the individual foal. Such foals are classified as orphans. The newborn foal depends on its mother for a variety of things from vital colostrum and nourishment to maternal affection and psychological development, and the replacement mother has to meet all these needs. Survival rates in orphan foals were at one stage very low. However, today's research and development have allowed the needs of the foal when changing from newborn foal to weanling to be understood and the management of the orphan foal geared to suit these needs.

The treatment depends on when the foal was orphaned and to a certain extent under what conditions. For example, foals that have received no colostrum or a very limited amount need to be fed colostrum immediately. If death of the mare occurs during parturition her colostrum may be milked from her and fed to the foal. Foals that have been orphaned after 48 hours or so and have been able to suckle should have received enough colostrum for adequate protection. Colostrum may be collected and stored frozen for up to 5 years, though evidence suggests that it starts to degenerate after 12 months. Some studs routinely collect and freeze extra colostrum from mares that produce excess amounts or from those that lose their foals. Although there is nothing to beat the real thing, colostrum substitutes are available, and colostrum from other animals may be used but with caution as digestive upsets may be caused by different component concentrations. Serum transfusions may also be given to provide the foal with immediate antibody protection.

In addition to classifying a foal according to its colostrum intake, its state of health and strength should also be assessed. A rough birth due to presentation difficulties or gestational problems in addition to the death of the mother will result in not only an orphaned foal but also a very weak animal requiring immediate intensive care. Such foals will require stomach tubing to ensure adequate colostrum is received, plus antibiotic treatment if an infection is suspected. Strong orphans may readily suck from a bottle and will not require stomach tubing. In any assessment of an orphan foal all the post partum parameters of a normal foal should also be ascertained to determine the extent of the orphan foal's problems.

As mentioned, a foal may be orphaned because of the mare's inability to raise it. This inability may be due to a number of factors, and the details of the particular situation are very important in deciding the appropriate management. A mare may be unable to feed her foal herself through inadequate milk production, due to mastitis or a physical abnormality. In the case of mastitis the inability to produce milk may only be temporary and the foal can receive supplementary feed to keep it going until the infection has cleared.

If a mare seems physically unable to produce enough milk, her foal may be supplemented to a certain extent but left hungry enough to encourage him to suckle his mother, and so stimulate her mammary glands to increase their milk production. Occasionally foals may need supplementing for the first few days of life, especially if they are premature, to give the mammary gland time to catch up and secrete adequate milk. In such cases the foal need not be removed from its mother, and so there is no separation preventing the mother–foal bond to develop: psychological damage to the foal is thus limited.

Occasionally, if the birth has been particularly stressful, or in the case of some maiden mares, the foal may be rejected by its mother. This rejection may vary from disinterest to actual attack with the risk of physical damage. Such mares may be sedated temporarily or the procedures used for fostering foals, that will be discussed later, may be followed. After awhile the mare will often accept the foal and bring it up normally. During such incidences of weak mare–foal bonding man's interference must be minimised as any external stimulus will only serve to worsen the situation. Man's intervention must be used only if the foal's wellbeing is at risk.

Whatever the cause of orphaning, the foal must receive adequate colostrum either by bottle or tube within the first 36 hours of life. Without this its long-term survival will be in jeopardy, due to the added risk of infection and the psychological problems of having no mother.

The vast majority of orphan foals are at a disadvantage as far as health status is concerned, due to the additional strains on the system. Health care for such foals is, therefore, very important. All the routine parameters for the newborn foal should be checked at birth and it is also advisable to check temperature and heart rate twice daily for the first few weeks to allow rapid identification of abnormalities or problems, so that treatment or further investigation can start immediately. A tetanus injection may be given at birth, especially if it is doubtful how much colostrum the foal has received. The foal may also be given an injection of vitamins and antibiotics, especially if it shows signs of weakness.

Orphan foals are more susceptible to the ordinary infections that normal foals take in their stride and deal with easily. They should be watched carefully and kept in a scrupulously clean environment. Diarrhoea is a common complaint of bottle-fed foals, especially if fed artificial diets. The diarrhoea may be a direct result of the diet but may also be a sign of infections, and persistence after 24 hours should be considered as serious, as dehydration can bring a foal down very quickly. Respiratory diseases can also be a problem in orphan foals, so their housing should be draught- and damp-free while providing good ventilation. Meconium retention may also be a problem, especially in foals that have not received colostrum, as colostrum acts as a laxative. Such foals may require an enema to clear the bowels and should be watched for subsequent constipation.

As discussed, the loss of a mare deprives the foal of nutrition and psychological security. A foal can survive without the psychological stimulus of a mother, though this may affect his long-term behaviour. It is not, however, able to survive without nutrition. There are several ways of providing that nutrition.

Fostering

The fostering of a foal on to a mare that has lost her own foal is the best solution. This provides the foal with a source of nutrients and psychological security, and provides the mare with a substitute for the foal lost. The mare used should be as close in her stage of lactation as possible to the foal's natural mother. For example, if the foal has been orphaned at birth, it should ideally be fostered immediately on to a mare that has lost her foal at birth within the last 48 hours. This is ideal, as the nutritional components of the mare's milk, as discussed in Chapter 9, vary with time of lactation and are coordinated with the foal's developing requirements.

If no such foster mother is available, but there is a mare that has lost her foal in the last few weeks, then she can be used as a companion to the foal and will be an invaluable support to its psychological development. In such a case the foal should be supplied with artificial foal milk, designed for the very young foal, in addition to being allowed to suckle his foster mother if she will let it. If it shows signs of poor health, it can be prevented from suckling its foster mother and fed the specially formulated foal diet for its age, but it will still have her psychological support.

Within the United Kingdom and in other countries there are foaling banks which can be contacted in the case of losing a foal or mare. Orphaned foals and mares that have lost their foals can be teamed up appropriately. Multiple suckling, the introduction of an orphan foal to a mare that has plenty of milk with her own foal of a similar age still on her, is not very successful in horses. This technique is often practised with cattle, where milk yields are artificially high. In horses multiple suckling tends to lead to two poor foals and may lead to resentment between them. It is possibly worth a try with a suitable mare under close supervision. In North America nurse mares are available as foster mothers. These mares are exceptionally good milkers and are kept primarily for leasing out as foster mothers after their own foals have been weaned off them early on.

Once a foster mother has been found it can be quite a tricky business persuading her to accept an orphan. She is much more likely to accept a foreigner if it smells of her or her own foal. Tricks such as rubbing the placenta of the foster mother over the foal or skinning the dead foal and placing the skin over the orphan work quite well, but depend upon getting you, the mare and the foal all together at the right time. Later on, the orphan foal can be rubbed with the mare's urine, manure or milk, especially in the head, neck, back and navel region, again to mask its own foreign smell. Other tricks are used, such as rubbing strong smelling substances on the mare's nose and over the foal in an attempt to disguise their smells.

Once prepared the foal should be introduced to the mare with great care. One person must hold the mare and at least one hold the foal. They should be introduced to each other at the mare's head end and her reaction noted very carefully. If all goes well and she shows no objection, the foal can be allowed to slowly explore around the mare, making sure that plenty of assistance is available if the mare should object. The foal can then be introduced to the mare's udder and allowed to suckle, providing again the mare shows no signs of objection. Rarely will things go according to plan and the mare will often initially show some signs of annoyance or objection to the orphan, in which case it should be removed and reintroduced to her head end again. Eventually, most mares will let the foal suckle and once this has occurred it can be left with her unrestrained for a period of time in the stable.

The area should be relatively confined so that they remain in close contact with each other and the foal does not become isolated. Throughout this period observation at all times is very important and immediate action to remove the foal taken if the mare starts to object. Slow, patient progress will pay off and once the foal has suckled several times the mare will rarely object to it. After a couple of days they can be turned out into a small paddock alone to help develop the bond between them over a distance before introducing them to the melee of other mares and foals.

Problems with fostering do occur. Some mares, regardless of all persuasion, will not accept a foal and it should be given up as a bad job before humans and horses lose patience and the foal ex-periences yet another rejection. There are various techniques to help persuade reluctant foster mothers, including the use of a crate in which the mare is held unable to turn around. The sides of the crate should be solid with a hole at the end nearest the udder through which the foal, who is outside the crate, can suckle. This can work quite well, and after the foal has suckled several times the mare can be removed from the crate and she will often accept the foal. Other tricks are used to elicit the mare's protective response:

the introduction of a dog or another mare within sight can induce a protective response in the mare towards the foal that often leads to acceptance, but the applicability of such means depends on the individuals concerned.

If all else fails a nanny goat may be used as a foster mother. It is placed on an elevated platform for suckling, and its continual presence provides company for the foal. However, care must be taken as goat's milk is not of the same composition as horse's milk, though it is nearer than cow's. It has two-thirds the sugar content and three times the fat content of mare's milk and tends to cause gastrointestinal upsets, especially associated with gas retention.

Artificial Diets

If it is not possible to find a foster mother or if the foal still requires supplementary feeding, there are specifically formulated equine milk substitutes available on the market. These are formulated to mimic the mare's natural milk components. Foal-Lac, for example, is specifically formulated for young foals and can be fed from birth. All these are dry powders to be mixed under sterile conditions with water and fed at 37.5°C. Other formulas are used based on cow's milk or dried cow's milk with added components to make up the shortfalls in cow's milk compared to mare's milk. These are not very successful, though still popular, and often result in gastrointestinal upsets, causing diarrhoea, dehydration and if not rectified a rapid decline in foal development and growth. If diarrhoea does occur, then milk substitutes should be replaced by a 50% glucose solution made up in sterile water for 1–2 days, with the slow reintroduction of the milk substitute. The incidence of diarrhoea can be reduced by ensuring regular small feeds are fed rather than fewer large ones. Specially designed mare's milk substitutes are widely available nowadays and there is no excuse for feeding other formulas except in real emergencies.

During the first few days of life the foal should also receive antibiotics plus a broad spectrum vitamin injection to give it an extra boost over, what is to it, a very stressful time. The foal should be watched carefully and help called at any signs of trouble. Orphan foals tend to go downhill much quicker than those with their mothers. For the first few weeks the orphan foal will need to receive its milk via a bottle, and a large plastic squash bottle with a lamb teat works quite well. It is very difficult to prescribe how much and how often it should be fed, as there is no hard and fast rule. However, as a rough guide, 110–220 g of milk should be fed every hour in the first week, decreasing to once every 2 hours in the second week, once every 3 hours in the third week and once every

4 hours in the fourth week. After 4 weeks it may be fed just 4 times/day. The exact timing and amount depends on the introduction and acceptance of milk in a bucket and/or concentrate feed. After about day 3–5 made-up milk powder can be placed in a bucket and hung at a level giving easy access for the foal. This will allow it to wean itself slowly off the bottle and on to the bucket. Some foals move over to bucket feeding very quickly, while others need more encouragement. It must be ensured that the milk is always clean and fresh; otherwise the foal will be discouraged from taking it. Warm milk and encouragement by licking off human fingers can help a reluctant foal accept bucket milk.

Some foals, instead of moving over to bucket feeding, are kept on bottle feeds by use of an automatic feeder. The more sopisticated automatic feeders mix and regulate the amount and temperature of the milk provided, and the foal sucks through a conveniently placed teat in its pen wall. Less sopisticated machines require human filling up with pre mixed milk. Either, if accepted, reduces the labour, especially at night if there are several orphan foals to be fed. It is very important that such feeders are regularly stripped down and disinfected as bacteria build up rapidly in such an ideal medium and can cause the foal immense problems.

Introduction of Solids

Many individual breeders recommend the introduction of solids as early as 1 week of age though the foal's intake at this time will be very limited. Introduction at an early age gives the foal time to become accustomed to the solid feed and gradually wean itself off milk. This progression from milk to solids can be helped by introducing foal milk pellets as a halfway house. As soon as the foal is eating a regular amount of concentrates then the milk pellets fed can be reduced, persuading it gradually to go over to the concentrates completely. Ad lib access to fresh green grass, or alfalfa as a substitute, will also encourage a foal to eat. By 2 weeks of age at the latest, the foal can be turned out with a companion, weather permitting, into a small safe paddock. This will introduce it to fresh grass and allow it room to stretch its legs and experiment with movement and investigate its new environment.

By 2 months of age most foals are consuming enough concentrates to allow them to be completely weaned off milk. The exact time will obviously depend on the foal's wellbeing, and final weaning should not be attempted unless the foal is fit and healthy, as it will always result in a slight set-back.

Many propriety creep feeds are available on the market, specially formulated to ensure the foal gets the adequate nutrients for a healthy start in life. If you mix your own, there are a few considerations to bear in mind. Protein should be relatively high in foal diets when compared to adult diets. In particular these proteins should be digestible and contain the ten essential amino acids for horses: lysine, methionine, leucine, isoleucine, histidine, arginine, tryptophan, valine, phenylalanine and threonine. Many straights tend to lack lysine and soya bean meal and linseed meal can be added, within reason, as good suppliers of lysine. Other dietary components of special importance in growing animals are calcium and phosphorous. A ratio of these two minerals of 2:1 should be aimed for. These minerals are essential for healthy growth of bones, cartilage, tendons and joints. Excess calcium or phosphorus can cause problems. Excess phosphorus causes calcium to be mobilised from the foal's bones in order to maintain the ideal 2:1 ratio, causing weaknesses in bones and epiphysitis, delaying growth. Many people supplement calcium in the form of limestone flour or equivalent as many straights have a relatively high concentration of phosphorus.

In addition to concentrates, foals should have access to fresh green forage. Alfalfa is good as it is relatively high in digestible protein and calcium. Hay may be fed, but it must be of a good quality, with no evidence of dust, mould, dampness, etc. Best of all is free access to fresh grass which provides an ad lib supply of continually fresh material.

By three months the foal should be completely weaned and fed on a ration of concentrate to supplement fresh grass and forage. A close watch should be placed on the foal's body weight and its diet should be changed to accommodate any significant rise or fall in condition. Under-condition will obviously retard growth and development, and overweight or too good condition can be as detrimental, putting undue strain on developing joints and ligaments.

Discipline and the Orphan Foal

One of the problems encountered with orphan foals, and not to such an extent in foals brought up by their mothers, is bad behaviour. The presence of the mare acts to discipline the foal, teaching it respect for other horses and humans and passing on, it is hoped, good behavioural characteristics. Hand-reared foals can become overfriendly with, and lack respect for, human beings, treating them very much as they would other foals and horses with nipping, kicking, chasing etc. If you let this get out of hand, such animals can be extremely hard to handle in later life.

To avoid this, foals should be allowed sight of other horses as soon as possible. Direct contact may be dangerous at a very young age but sight and smell help. Once they are 1–2 months old they can have a gentle companion, one that is placid and will not bully the foal. Shetland ponies and/or donkeys are often used for this job. Once they are 4–5 months old they should be able to cope with other weanlings and can be all run out together. As such they will get rid of their high jinks and play with other compatriots. All the same, discipline while in the company of humans should be always strict and consistent.

Early Foal Management – The First 6 Weeks

Management of the lactating mare and foal depends to a certain extent on whether she is to be returned to the stallion or not. Her management as far as covering by the stallion is concerned has been discussed in detail in Chapter 12. If the mare is to foal at stud, her early management and that of the foal will be largely up to the general stud management and practice, and because of its importance in determining the foal's long-term prospects, this is an area that should be discussed during the selection of a stud.

Immediately after foaling the mare and foal should be left in peace with a supply of fresh hay or freshly cut grass and clean water in the foaling box. After an hour or so the mare may be fed a nutritious concentrate feed. Apart from this and regular unobtrusive observation the mother and foal should be left in the box. If the weather is particularly cold, a heat lamp may be used to provide a warm spot in the stable for the foal to move to.

Exercise

For the first 3 days of life the eyesight of the foal is not good enough for it to be allowed out of the stable. After 3 days it should have developed enough appreciation of distance and depth to be turned out with its mother for an hour or two during the day, providing the weather is good (Plate 16.5).

Avoid turning out any young foals if the weather is wet, damp or very cold and windy. Such weather will easily soak the foal through, and as both mare and foal will be reluctant to move around in such weather, it defeats one of the main objectives of having them out, that of exercise. The paddock provided for foals should be quite small, half an acre being fine to start with. It should have strong, well-constructed fences, ideally post and three rails with no wire. There should be no protruding objects, old machinery, wire, holes, low branches etc as these can prove death traps to young foals still not sure on their feet. An alternative

*Plate 16.5   After 3 days a foal's eyesight should have developed enough to
allow him to be turned out with his mother.*

system used in Europe, especially France, is electric fence paddocks. Providing the mares are accustomed to electric fences, large
fields can be divided up into small paddocks with electric tape,
allowing association with other mares but security in the first
weeks (Plate 16.6).

Water should be provided in a bucket. Streams or large water
troughs can also prove lethal for a very young foal. The paddock
should have plenty of good grass as this will encourage the mare to
eat, which is especially important in mares whose appetite has
fallen off after parturition.

Persuading the foal to leave its stable for the first time can be
difficult, but should be made as free from trauma as possible;
otherwise the nervousness will perpetuate itself. During the first 3
days of life the foal should have been handled gently, stroked all
over and got used to having arms put around it. When the big day
comes, therefore, it should be used to human contact.

The best time to turn a foal out is a nice sunny day, but not too hot
or flies will be a problem. However, the bright sunlight may
discourage the foal from leaving the much darker stable environment. There should be at least two handlers. The mare should be led
ahead slowly by one person, and another handler should cradle the
foal in their arms, one arm behind its hindquarters and the other
around its chest, and encourage it to follow its mother (Plate 16.7).

Plate 16.6   Large fields can be divided up into small single mare and foal paddocks using electric fence tape, providing contact with other mares and foals along with security from them during early life. This example was photographed in France.

Plate 16.7

The foal may be encouraged to lead for the first few times by cradling in your arms, one arm behind its hindquarters and the other around its chest.

Some foals will follow easily; others prove more difficult and more than one handler may be needed. A foal should never be pulled from the head by means of a halter, as this may seriously damage its neck and head. A soft twisted cloth/bandage or thick rope can be put around its neck after awhile when it is getting used to following its mother, and later a soft leather or webbing halter can be introduced and used to lead it. Leather is preferred as it will stretch and eventually break under strain. Some people like to leave head collars on foals while they are out; this can be very convenient for catching them and gives them time to get used to them. However, the collar must be very well fitting to ensure that the foal cannot catch it on anything or get its feet caught up in it (Plate 16.8).

Plate 16.8

Leather is preferred for a first time halter as it will stretch and eventually break under strain.

Handling

At this stage the foal must start to learn how to be tied. This is best done by using a round pole with no projections on it so that the foal cannot get itself all twisted up or get caught on fences. A rope can then be tied on to its head collar and on to the pole. As mentioned, it is not good to pull a foal by a rope attached to its head, as this can cause damage to its head and neck. An alternative can be to loop the rope around the foal's girth and up through its head collar and on to the pole. This method of restraint means that all the pull is taken on the girth and not the foal's head. Foals will soon learn that they cannot escape and that it is easier to stand still.

Once the foal has learnt to tie, the general stroking and handling can progress onto grooming and attention to its feet and eventually travelling. These are particularly important if you intend to show the foal. Grooming can develop slowly from use of a soft body brush on to the main comb and dandy brush. The foal will soon come to enjoy it, providing all progression moves slowly and patiently (Plates 16.9, 16.10 and 16.11).

The foal's feet should need little attention in early life unless they have a deformity. Nevertheless, picking up the feet, picking out the hooves and grooming the legs should be done regularly, and these, along with ensuring a general acquaintance with the blacksmith when the mare's feet are attended to, will ease work on the foal's feet in later life. Regular inspection of the feet will allow examination for injury and damage that may need treating. Providing the weather is good a foal can be bathed, again of great use if it is to be shown. The weather must be warm and it should be bathed at the beginning of the day to avoid it catching a chill.

Introduction to a trailer or lorry can also be done in the first 6 weeks of life. The mare can be used to encourage it in and in fact many foals take to it quite easily, providing the mother is a good loader. If she is not, there is danger of the foal picking up her bad habits or her fear. In such cases leaving the trailer in its paddock with the door open and feed in the top end can encourage it to investigate and get used to going in and out at will. If this is done the foal must be watched at all times to ensure that it does not hurt itself. Once the mare and foal have been successfully loaded and unloaded a couple of times they can be taken for a short ride. A foal will sometimes travel better if there is no central partition dividing the trailer. If there is a top door, it should be closed or a bar or cover used to prevent the foal from trying to escape over the tail board if it panics.

Behaviour

During the first 6 weeks of life the foal shows quite significant development in behaviour and social interaction. Initially the foal's whole world and social experience involves just its mother. Play also involves the mother alone, consisting of rubbing her mane and tail and kicking. Through this the foal begins to learn how far it can push her before being reprimanded. Once the foal has developed more steadiness on its feet, normally after about a week, it will start galloping around its mother, but never straying far. Over the next few weeks the circle gets bigger and it spends more time away from its mother, investigating and playing alone (Plate 16.12).

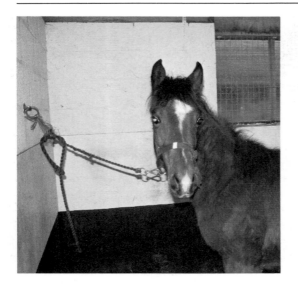

Plate 16.9

A foal should learn to tie as one of its first lessons.

Plate 16.10

Once the foal has learnt to tie it can be groomed.

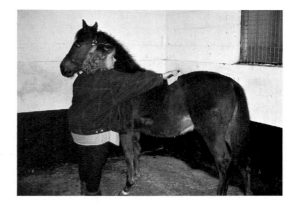

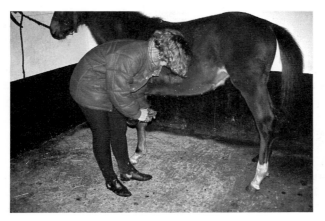

Plate 16.11

The foal should also learn to have its feet picked up, though they should need little attention at this stage.

Plate 16.12   The foal's circle of play becomes increasingly larger over the first few weeks of life as it gains independence.

If at this stage the foal has access to other foals and begins to interact with them, play will gradually include them rather than its mother. By 8 weeks it spends up to 50% of its time playing with other foals and only 10% playing around its mother. If, however, the foal has no contact with others it will play with its mother much longer and may try to play with older horses present or even dogs or other animals regularly in its company. If its mother is particularly possessive or shy, these characteristics can be passed on to the foal and it will not integrate with other foals so soon.

Apart from play, which takes up approximately 50% of its time, the foal spends a significant amount of time lying down resting (Plate 16.13). These are normally short periods of rest, particularly in warm sunlight, between periods of play.

The remainder of its time is spent suckling. These periods of suckling start off at regular intervals of 30–60 minutes. They become longer, but less frequent, as the foal matures (Plate 16.14).

Introduction of Solids

The foal may first be seen investigating concentrate feeds, and even ingesting some, at as early as 3 days of age. Investigation of the mare's feed by the foal at an early age is to be encouraged, as the mare's milk is naturally short of iron and copper, which invariably

Plate 16.13   Apart from play and suckling, the foal spends a large proportion of his time lying down resting.

Plate 16.14   As the foal grows up it spends less time suckling its mother, the intervals between suckling getting larger.

causes anaemia in very young foals. Copper and iron are vitally associated with red blood cell function and haemoglobin levels, and adequate levels of copper and iron can be achieved by the foal picking at the mare's feed. Anaemia will persist in foals too weak to nibble hay or concentrates until they are treated or are able to eat (Plate 16.15).

Plate 16.15

A foal will soon be seen investigating its mother's feed. This is to be encouraged as it provides the foal with essential copper and iron.

In theory, until peak lactation, week 6–8, the mare's milk provides the vast majority of the nutrients required by the developing foal. Gradual investigation and ingestion of its dam's feed and hay provides an increasing amount of nutrients, but this is not significant until after week 6. The amount of extra creep feed that a foal will require in the first few weeks depends largely on the mare's milk yield. In addition to concentrates, the foal must be introduced to roughage in the form of grass or hay, as a diet of concentrates and milk can cause diarrhoea.

Creep feed can be introduced as early as week 1 but the foal should never be forced to eat it. The progression from milk to solid food must be gradual and can start slowly at an early age. Once the foal starts to pick at its mother's food, care must be taken that when its intake becomes significant, she receives enough to meet her demands. Foal and mare should really be fed individually by now.

Dentition

Providing the teeth of the foal erupt as expected and at the correct angle, there is no need to do anything with its teeth in the first 6 weeks. Most foals are either born with the central incisors or have them by 8–9 days, and the middle incisors should erupt between 6–8 weeks.

Immunisation and Parasite Control

Foals can be immunised against tetanus as a routine in the first few days of life, though colostrum from a suitably immunised mare should provide the foal with adequate protection against numerous diseases including tetanus. In some countries it may be worth considering immunisation against strangles or rabies depending on the situation and the prevalence of the disease. Worming can be first done at 7 days of age, after which the foal should follow a regular worming regime.

Mare Management in Early Lactation – The First 6 Weeks Post Partum

During the first few weeks post partum, the mare must be watched carefully to ensure that she has not suffered any detrimental long-term effects as a result of the birth. Many studs perform a routine internal examination within 3 days of parturition to identify any problems and check that uterine involution is progressing properly. Any problems at this stage can be treated in time for covering at either foal heat or her first return to service. At between 72 and 76 hours after delivery the mammary gland starts to produce milk rather than colostrum. Details of composition and milk quality plus further details of the physiology of lactation are given in Chapters 9 and 10. During the start of lactation it is important that the mare looks healthy and fit in herself, as any infections or disease will pass easily on to the foal. The general wellbeing of the foal should also be used as an indication of milk production. If the mare is not producing enough milk the foal will appear tucked up, the mare's teats may be sore from continual unsuccessful suckling and she may begin to object to the foal suckling, risking the development of a perpetual situation making the condition worse. Low milk yields may be due to a physical inability, low nutritional intake, poor body condition or mastitis.

Mastitis, an infection of the mammary tissue, is relatively rare in the mare compared to the cow. It is characterised by a hot, swollen and painful udder with oedematous swellings developing along the abdomen and up between the hind legs. Milk secretion tends to be thick and clotted and should not be fed to foals. Mastitis can occur in early lactation but is very rare then. Normally if it does occur it appears post weaning, especially if the mare is still producing large amounts of milk. Mares that have had foals prematurely weaned are at particular risk and should be watched carefully. Treatment is similar to that in cows, by the injection of antibiotics directly into the mammary gland via an intramammary tube inserted up into the streak canal, repeated for several days. Alternatively, intramuscular injection of systemic antibiotics can be used, as many mares violently object to having a mastitic mammary gland touched.

There is a relatively rare condition, in which the mare has become isoimmunised against its foal, that is she has raised antibodies against the foal, and this results in a fatal reaction within the foal if it ingests the mare's colostrum. Alternative colostrum must be fed and the foal fed supplementary feed for 36 hours or more post partum until its intestinal system loses the ability to ingest large protein molecules.

Occasionally immediately post partum, the mare fails to produce any milk at all. This is due usually to a failure in the milk ejection reflex or in the production of milk. Failure of the milk ejection reflex is the most common, and this is thought to be caused by high adrenaline levels in the mare's blood system, due largely to stress and anxiety, which inhibit the action of oxytocin on the udder. Injection of oxytocin and/or warm compresses applied to the udder, along with a quiet calm atmosphere, will often rectify the situation. Failure of the udder to produce any milk, ie galactopoiesis or lactogenesis failure, is a serious condition about which little can be done. Many foals born to such mares have to be classified as orphan foals and brought up accordingly. Such mares should not be bred again, as the condition invariably repeats itself.

Teat abnormalities can also cause problems. Inverted nipples or conical nipples are very hard for the foal to suckle, though the condition may not affect milk production itself. Super-numerary teats may also be present, but are rare, and do not seem to affect milk ejection from normal teats. Fluid collection occasionally occurs within the udder in mares with high milk yields. Udder oedema, as it is termed, results in fluid accumulation around the udder and along the abdomen. As a result the udder becomes too painful to allow the foal to suckle and the condition predisposes the mammary gland to infections.

A lactating mare should be managed in much the same way as other stock with common sense and an eye for potential problems.

Nutrition

One of the specific areas to note in the management of the lactating mare is her nutrition. A lactating mare has higher nutrient demands than any other equid, even one in heavy work. As parturition approaches, her nutrient demand increases to meet the demands of the growing foetus. What is often forgotten is that after parturition, she still has to provide all the demands of that foal. As it gets larger, so do its requirements, until it starts to find nourishment elsewhere. In addition, the efficiency of the transfer of nutrients from mother to foal decreases from the transfer via blood in utero, to transfer via milk post partum. After parturition the mare's nutrient requirements increase by 75% to account in part for this decrease in efficiency. This is illustrated by the fact that the efficiency of energy transfer in milk is only 60%, as 40% of the energy intake does not appear as energy in the milk (Tables 14.1 (page 253), 16.1–16.3).

Protein and Energy

Protein and energy are important components of a lactating mare's ration. If her diet is nutritionally low in protein, she will not have enough to satisfy demands for her milk. Milk production will then fall off (Martin, McMeniman and Dowsett, 1991). Low dietary energy will result in mobilisation of the mare's own body reserves in order to try to maintain her milk production. If low dietary energy persists, milk yield will fall off and the mare will become emaciated (Pagen and Hintz, 1986). Some loss in weight, especially in mares that milk well, is inevitable but should be minimised by appropriate feeding. Weight loss in mares that are to be returned to the stallion can affect conception rates (Sutton, Bowland and Ratcliff, 1977; Gill, 1985). Hence, the protein and energy content of a lactating mare's ration are very important to her reproductive efficiency. Digestible energy concentrations of 2.35 Mcal/kg and crude protein at 12% is required in order to ensure adequate nutrition for maximum milk production (Tables 14.1, 16.1–16.3).

Calcium and Phosphorus

In addition to protein and energy, calcium intake is also very important. Calcium concentrations at 0.47% are required to ensure that the foal gets adequate calcium for bone and tendon growth and development. As previously discussed the ratio of calcium to phosphorus is important, excess phosphorus causing a drain of calcium from the mare's bones in an attempt to redress the balance. Phosphorous concentrations in the order of 0.3% are required in early lactating mares. Extra vitamins and minerals are

Table 16.1   Daily nutrient requirements of lactating mares of varying weights.

Lactating mares	Weight (kg)	DE (Mcal)	Crude protein (g)	Lysine (g)	Ca (g)	P (g)	Mg (g)	K (g)	Vitamin A (10³ IU)
Foaling–3 mths	200	13.7	688	24	27	18	4.8	21.2	12
Foaling–3 mths	500	28.3	1427	50	56	36	10.9	46.0	30
Foaling–3 mths	700	37.9	1997	70	78	51	15.2	64.4	42
Foaling–3 mths	900	45.5	2567	89	101	65	19.6	82.8	54
3 mths–weaning	200	12.2	528	18	18	11	3.7	14.8	12
3 mths–weaning	500	24.3	1048	37	36	22	8.6	33.0	30
3 mths–weaning	700	32.4	1468	51	50	31	12.1	46.2	42
3 mths–weaning	900	38.4	1887	66	65	40	15.5	59.4	54

Table 16.2   Nutrient concentrations in total diets for lactating mares (100% DM). Values assume a concentrate feed containing 3.3 Mcal/kg and hay containing 2.0 Mcal/kg dry matter.

Lactating mares	DE (Mcal)	Diet proportions Conc (%)	Hay (%)	Crude protein (%)	Lysine (%)	Ca (%)	P (%)	Mg (%)	K (%)	Vitamin A (IU/kg)
Foaling–3 mths	2.60	50	50	13.2	0.46	0.52	0.34	0.10	0.42	2750
3 mths–weaning	2.45	35	65	11.0	0.37	0.36	0.22	0.09	0.33	3020

Table 16.3   The expected feed consumption by lactating mares (% body weight).

Mares	Forage	Concentrates	Total
Early lactation	1.0–2.0	1.0–2.0	2.0–3.0
Late lactation	1.0–2.0	0.5–1.5	2.0–2.5

also required during lactation. Deficiency will cause lacklustre and ill thrift in both the mare and foal and inefficient use of other nutrients.

In order to satisfy the demands for milk production the mare will require concentrate feed in addition to good-quality forage. Phosphorus and protein tend to be deficient even in very good-quality forage diets for lactating mares. The concentrate ration should not, however, be more than 50% of the total diet. Not only should protein quantity be high to account for the inefficiency in conversion of nutrient protein to milk protein, but protein quality is also important. Lysine is often a limiting essential amino acid in conventional grain and grass forage diets. The introduction of alfalfa or soya bean meal can be used to increase lysine levels.

As stressed in earlier discussions it is imperative that all home-grown or bought-in straights and forages are analysed. An appropriate ration for a lactating mare can be ensured using this information and the analysis on commercial concentrates (Tables 14.1, 16.1–16.3).

Vitamins
Vitamins should also be considered, Vitamin A and D being of particular importance. Vitamin A is available in fresh green forage and vitamin D from exposure to sunlight. Lastly, but by no means least, free access to clean fresh water at all times is essential, as 90% of milk is in fact water.

It is not appropriate to be prescriptive on exactly what a lactating mare requires, as requirements will vary with individuals. Mares that produce more milk will obviously require more feed and will probably require higher levels of concentrate in order to ensure appropriate nutrient intake. However, some mares seem to be better doers and more efficient as milk producers. In such mares obesity may prove to be a problem and, therefore, limiting their concentrate intake should help. The yardstick to go by is the body condition of the mare: she should be fit not fat, with a body condition score of 3. From 6 weeks of lactation onwards the foal, as far as nutrition is concerned, becomes increasingly more independent. The mare's milk yield then decreases, having reached a maximum at approximately 6 weeks.

Foal Management – 6 Weeks of Age Onwards

From 6 weeks onwards the foal becomes increasingly independent of its mother.

Handling

The foal's handling at this stage should develop on from that started in the first 6 weeks of life, remembering patience and reward. Halter breaking should have been started by now (Plate 16.16). Late halter breaking can lead to confrontation and make the job harder than it need be. Leading lessons should develop and the foal should learn to lead without resistance. This process can be aided by a rope around the foal's hind quarters that can be pulled to encourage it to walk forward. By the time of weaning the foal should be happy to be led without resistance, both behind its mother and away from her. This can only be achieved by continual and patient training, using short and frequent lessons (Plate 16.17). It should also become further accustomed to travelling, leading on to travelling alone.

Behaviour

From 6 weeks of age onwards the foal continues to develop its independent traits, spending more and more time away from its mother and playing with other foals. This is invaluable in

Plate 16.16   By 6 weeks of age halter breaking should have been started.

Plate 16.17   *The foal should learn to lead from an early age without resistance.*

developing social awareness and learning about hierarchies in a group system in preparation for survival alone with its peers, without its mother for protection.

Nutrition

From 6 weeks of age a creep feed becomes increasingly important to the foal as a source of nutrients. From this stage onwards milk quality declines as an encouragement to the foal to seek nutrients elsewhere. The quality of creep feed must be carefully assessed to ensure that it provides all the nutrients required for optimum growth and development. However, it should be borne in mind that optimum growth is required, not maximum growth. As a rough guide a foal destined to make 15–16 hh may gain up to 2 kg/day but by 1 year of age it should not weigh over 80% of its expected mature weight. Excess weight gain causes strain on muscles, tendons, joints, the circulatory system etc. It is especially important when these structures are still developing that undue stress does not cause permanent deformity.

It is important that the mare does not have access to the foal's creep feed as this may discourage the foal from feeding. Specially designed creep feeders permit the foal to feed alone without danger that the mare can have access to the feed. The amount of food per foal should also be controlled and it is best if foals are fed individually if several are run together. This may be difficult to organise but will ensure that each foal is fed according to need and monitored against weight gain. Free access allows greedy foals to gorge themselves at the expense of smaller, less dominant individuals.

Several creep feeds specifically designed to meet foal requirements are available on the market. The protein requirement of the foal for growth is high and it becomes the first limiting factor as far as nutrient supplied by milk is concerned. Creep feeds should contain 20% protein in a highly digestible form, 0.8% should be calcium and 0.6% phosphorus. A calcium:phosphorus ratio within the range of 1:1 and 3:1 is acceptable.

As far as quantity of feed is concerned, 0.5 kg/day at 2 months is adequate, gradually increasing as weaning approaches. As a guide foals of 3 months of age should be taking 0.3 kg/100 kg body weight/day. If foals are fed outside in a creep feeder, food should be checked regularly to avoid mould developing. Free access to fresh grass or alfalfa is essential and access to a mineral supplement is good practice.

Feet

Foot problems can be identified and correction attempted within the foal's first year of life. In addition to regular trimming and handling this can ensure that minor faults and problems can be identified before training begins. An overzealous attack on a foal's feet, in attempts to correct leg problems, can cause further difficulties, and indeed if left alone many deformities often prove to be self-correcting.

Corrective trimming should be done only by a trained farrier or veterinary surgeon with experience in this area. A deformity that can be simply corrected by specific trimming is an incorrect hoof:pastern angle, which can be corrected within reason by changing the length of the horse's heel. Toes out or in result in uneven wear of the hoof wall and corrective and compensatory trimming can alleviate the problem. Excessively long toes or the opposite club foot can be corrected by ensuring that any trimming done is adequate, not overzealous.

Parasite Control

Worming from 2 months of age should be regularly performed. Wormers against Strongyloides, westeri and ascarids should be used primarily, as these affect young horses in particular (Plate 16.18).

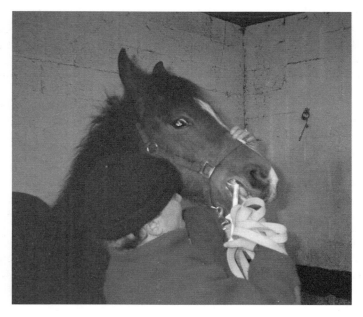

Plate 16.18   Regular worming should start from 2 months of age and continue as a routine for the rest of the foal's life.

Management of the Lactating Mare – 6 Weeks Post Partum Onwards

From 6 weeks post partum onwards lactation yield decreases, along with the quality of milk produced. The mare's udder shrinks as the amount of milk suckled reduces. As her milk yield decreases so do her nutrient requirements. Concentrates should then be slowly reduced to ensure that she does not become overfat. Protein concentrations can be gradually reduced to 10% and digestible energy to 2.2 Mcal/kg by 3 months post partum. Calcium and phosphorous are still important, concentrations of 0.33 and 0.2% respectively being required (Tables 16.1–16.3).

Exercise at this stage is essential to build up the mare's fitness again. Indeed, exercise, along with a decrease in nutrient intake, helps to dry up the mare's milk production slowly. The mare and the foal should be turned out for as long as is possible. If they are of a hardy pony type, they can be turned out all day and night providing the weather is good. If the weather is too hot it may be suitable to bring them in during the day to avoid the flies and to be out at night.

Worming regime, immunisation programmes, attention to teeth and feet should be maintained and not neglected to ensure that the mare remains in tip-top condition, especially if she is in foal again. Particular attention should be paid to the mare's body condition, as mares often put on extra condition in the summer months with decreasing demands for milk by the foal. It is not appropriate to put a mare in foal on a diet later on in the autumn to try to lose excess weight gained during the summer months.

References and Further Reading

Adams, R. (1993) Identification of Mare and Foal at High Risk for Perinatal problems. In: Equine Reproduction. Ed. A.O. McKinnon and J.L. Voss. Lea and Feibiger, Philadelphia & London, p. 985.

Adams, R. (1993) Neonatal Disease: An Overview. In: Equine Reproduction. Ed. A.O. McKinnon and J.L. Voss. Lea and Feibiger, Philadelphia & London, p. 997.

Adams, R. (1993) The Musculoskeletal System. In: Equine Reproduction. Ed. A.O. McKinnon and J.L. Voss. Lea and Feibiger, Philadelphia & London, p. 1060.

Asbury, A.C. (1993) Care of the mare after foaling. In: Equine Reproduction. Ed. A.O. McKinnon and J.L. Voss. Lea and Feibiger, Philadelphia & London, p. 976.

Blanchard, T.L, and Varner D.D. (1993) Uterine Involution and Post partum Bleeding. In: Equine Reproduction. Ed. A.O. McKinnon and J.L. Voss. Lea and Feibiger, Philadelphia & London, p. 622.

Chavatte, P., Brown, G., Ousey, J.C., Silver, M., Cotrill, C., Fowden, A.L., McGladdery, A.J. and Rossdale, P.D. (1991) Studies of bone marrow and leucocycte counts in perepheral blood in fetal and newborn foals. J,Reprod.Fert.Suppl., 44, 603.

Dawes, G.S., Fox, H.E. and Leduc, B.M. (1972) Respiratory movements and rapid eye movement sleep in the foetal lamb. J.Physiol.Lond., 220, 119.

Equine Clinical Neonatology. (1990) Ed. A.M. Koterba, W.H. Drummond and P.C. Kosch. Lea and Feibiger, Philadelphia & London.

Fowden, A.L., Mundy, L., Ousey, J.C., McGladdery, A. and Silver, M. (1991) Tissue glycogen and glucose-6-phosphate levels in fetal and newborn foals. J.Reprod.Fert. Suppl., 44, 537.

Gill, R.G. (1985) Post partum reproductive performance of mares fed various levels of protein. Diss.Abstr.Int.B.Sci.Eng., 45, 3670.

Gillespie, J.R. (1975) Postnatal lung growth and function in the foal. J.Reprod.Fert. Suppl., 23, 667.

Jeffcote, L.B. (1971) Perinatal studies in Equidae with special reference to passive transfer of immunity. Ph.D. Thesis, University of London.

Jeffcote, L.B. (1972) Observations on parturition in crossbred pony mares. Equine Vet.J., 4, 209.

Jones, C.T, and Rolph, T.P. (1985) Metabolism during fetal life: a functional assessment of metabolic development. Physiol.Rev. 65, 357.

McClure, C.C. (1993) The Immune System. In: Equine Reproduction. Ed. A.O. McKinnon and J.L. Voss. Lea and Feibiger, Philadelphia & London, p. 1003.

McCue, P.M. (1993) Lactation. In: Equine Reproduction. Ed. A.O. McKinnon and J.L. Voss. Lea and Feibiger, Philadelphia & London, p. 588.

McGuire, T.C. and Crawford, T.B. (1973) Passive immunity in the foal: measurement of immunoglobulin classes and specific antibodies. Am.J.Vet.Res. 34, 1299.

Martin, R.G., McMeniman, N.P. and Dowsett, K.F. (1991) Effects of a protein deficient diet and urea supplementation on lactating mares. J.Reprod.Fert.Suppl., 44, 543.

National Research Council (1989) Nutrient Requirements of Horses. 5th Ed. National Academy of Sciences, National Research Council, Washington D.C.

Ousey, J.C., McArthur, A.J. and Rossdale, P.D. (1991) Metablic changes in Thoroughbred and pony foals during the first 24 hours post partum. J.Reprod.Fert.Suppl., 44, 561.

Pagen, J.D. and Hintz, H.F. (1986) Composition of milk from pony mares fed various levels of digestible energy. Cornell Vet. 76, 139.

Roberts, M.C. (1975) The development and distribution of mucosal enzymes in the small intestine of the foetus and young foal. J.Reprod.Fert. Suppl.,23, 717.

Rossdale, P.D. (1968) Clinical studies on the newborn Thoroughbred foal. III thermal stability. Br.Vet.J., 124, 18.

Rossdale, P.D. and Ricketts, S.W. (1980) Equine Stud Farm Medicine. 2nd Ed. Ballière Tindall, London.

Schott II, H.C. (1993) Assessment of fetal wellbeing. In: Equine Reproduction. Ed. A.O. McKinnon and J.L. Voss. Lea and Feibiger, Philadelphia & London, p. 964.

Stewart, J.H., Rose, R.J. and Barko, A.M. (1984) Respiratory studies in foals from birth to seven days old. Equine Vet.J., 16, 323.

Sutton, E.I., Bowland, J.P. and Rattcliff, W.D. (1977) Influence of level of energy and nutrient intake by mares on reproductive performance and on blood serum composition of the mares and foals. Can.J.Anim.Sci., 57, 551.

Traub-Dargatz, J.L. Post natal care of the foal. In: Equine Reproduction. Ed. A.O. McKinnon and J.L. Voss. Lea and Feibiger, Philadelphia & London, p. 981.

Tyznik, W.J. (1972) Nutrition and disease. In: Equine Medicine and Surgery. 2nd ed. Ed. E.J. Catcott and J.R. Smithcors, p.239. Illinois, American Veterinary Publications.

Vaala W.E. (1993) The cardiac and respiratory systems. In: Equine Reproduction. Ed. A.O. McKinnon and J.L. Voss. Lea and Feibiger, Philadelphia & London, p. 1041.

Welsch, B.B. (1993) The Neurologic System. In: Equine Reproduction. Ed. A.O. McKinnon and J.L. Voss. Lea and Feibiger, Philadelphia & London, p. 1017.

CHAPTER 17

Weaning and Management of Youngstock

Weaning

WEANING is essential to allow the pregnant mare's mammary gland to recover in order to ensure an adequate milk supply for the new foal to be born. Naturally the foal would be weaned at 9–10 months of age, giving the mare's system a good month to recover before the birth of the new foal. By 9 months of age the foal will normally be consuming a large quantity of solid food and hardly taking any milk at all. The dry period after weaning and before lactation for the next foal allows the mare's whole system to recover and her body reserves to be built up. It also minimises nutritional strain during the most nutritionally stressful time of pregnancy and season. The mammary gland recovers and lactogenesis can begin in readiness for the foal in utero.

During the last 3 months of the natural lactation, the foal, now over 6 months of age, takes very little milk from his mother. Especially during the later stages, his intake of grass and herbage provides the vast majority of the nutrients he requires. It is, therefore, quite possible, with careful management, to wean foals routinely at 6 months, providing they are taking in adequate amounts of concentrated feed.

Most studs wean all foals at 6 months as a matter of course. Weaning, however, should not be considered unless it is certain that the foal is taking adequate concentrates and is in good physical condition. Weaning is a very stressful process indeed. Not only will the foal be forced to leave his mother and there will be no milk in his diet, but he will be introduced to strange horses and there will be more handling and contact with humans. Careful management can ease these stresses somewhat and, therefore, reduce the stress of weaning itself. The trauma of weaning along with disease can be devastating and, therefore, ways of minimising

316

these two aspects must be addressed prior to weaning.

Solid food intake by the foal must be adequate in order to allow it to take the sudden removal of milk from the diet. Sudden changes in diet at all ages can cause digestive upsets and in foals they can lead to a considerable setback in their growth and development (Plate 17.1). Some individuals advocate milking the mare for a few days after weaning in order to reduce the risk of mastitis. The milk collected can be fed to the foal, along with his concentrate diet. This helps to smooth the change in diet but it is quite time-consuming and some mares object violently to being milked by hand. The stress of new companions can be reduced by introducing the foal to his post weaning companions prior to removal from his mother. This, along with regular handling prior to weaning, as discussed in Chapter 16, will reduce the stress of increased human contact.

A foal being considered for weaning must be in good health. Any animals showing signs of ill-health such as runny nose, coughing, listlessness, starry coat, etc must not be weaned until their condition improves. Young animals can suffer quite dramatically from seemingly small problems, resulting in considerable setbacks to their development, with possible permanent damage. If in doubt, it is advisable to call a veterinary surgeon.

In exceptional cases, and only under veterinary supervision, foals suffering from ill-health may be weaned, as some medicines may be easier to administer and more effective in a foal not on a milk diet. Early weaning as soon as 4 months of age may also be practised if

Plate 17.1   *Intake of solid food must be adequate prior to weaning so as to minimise the upset within the digestive system as a result of the change from a liquid to a solid diet.*

the mare is suffering. Providing the foal has been well prepared and introduced to solids soon enough, and is in good physical condition, it should not suffer much as a result. Early weaning can cause complications as such foals cannot be run out with other foals, which will either not yet be weaned or will dominate the foal and not allow it adequate accessibility to concentrate feeds. An alternative companion has to be sorted out to provide psychological security and development, and a small quiet donkey or pony can provide an ideal companion for such foals.

Methods

Plans for weaning foals should be put into practice well in advance of the actual event in order to ensure that the foal's physical condition and nutrient intake are appropriate at the planned weaning date.

Traditionally foals were weaned suddenly and individually by removing the mare abruptly and leaving the foal in a stable or loose box out of earshot of his mother (Plate 17.2). A more popular alternative to this system is gradual weaning in groups. Mares and foals to be weaned are run together for a while prior to the projected

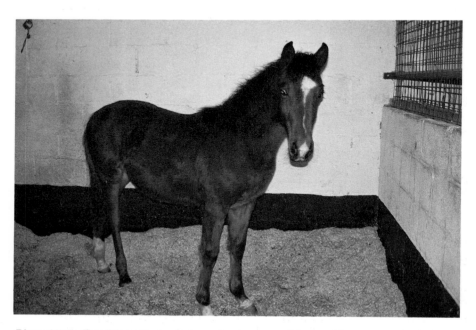

Plate 17.2   Traditionally foals have been weaned by sudden removal of the mare, leaving the foal in a secure stable.

Plate 17.3 *The stable used to leave the foal in after removal of the mare should be very secure and have an upper grill or mesh as well as a solid upper door.*

date of weaning and the mares removed either one at a time or all together for increasing lengths of time prior to complete removal.

Sudden Weaning

If sudden weaning is to be adopted, then the stable in which the foal is to be left must be free of any projections which may cause damage as the foal careers around when the mare is removed. Water buckets should either be fixed to the wall or not left unattended. Hay should ideally be in a hay rack off the ground. A hay net is not appropriate as a foal can strangle itself on it; hay on the floor is better but does tend to get wasted. The bed should be deep and ideally made of straw providing good protection as the foal launches around the box. The stable door should be very secure and have an upper grill or metal mesh door as well as a solid upper door (Plate 17.3). For the first few days the upper mesh door should be closed at all times to prevent the foal attempting to jump out and yet still provide ventilation. Foals are notoriously unaware of danger and will launch themselves at obstacles that an adult horse would not dream of attempting. They are, therefore, very prone to damage and extra care should be taken to avoid potential hazards: prevention is infinitely better than cure.

The mare and foal are separated by removing the mare from the loose box leaving the foal behind. The foal should be accustomed to handling by now and can be held in the stable while the mare is removed. She must be kept moving, even though she will be very reluctant, and the quicker she is removed and with the least fuss the better. As soon as she is out of the stable both solid doors top and bottom should be shut and a light left on in the stable. The foal should be relatively safe under these conditions for a short while as you sort out the mare.

The mare should be taken to a field out of earshot of the foal, with limited grass cover. The field should be very secure with safe boundaries. The mare may be very disturbed for the first few hours and can easily damage herself by careering around. She should be watched until she has settled down a bit and started to graze. The foal should then be checked. It should be given water as it will have invariably worked itself into a good lather. Hay should also be made available ad lib and a small feed given as soon as the foal has calmed down.

The foal should remain in its box for the first few days to allow it to calm down and get used to life alone. A large stable is, therefore, advantageous. Whenever the foal is being handled or mucked out, the top door should be closed on the stable as well as the lower to avoid the foal escaping. These first few days are very stressful for the foal and it is extremely susceptible to damage and disease; this is, therefore, one of the main disadvantages of this system. After a few days, providing the foal is calm, it may be turned out for periods of time in the day with a companion in a small secure paddock. The length of time of turnout can be increased gradually to all day and night if appropriate. In many systems foals are still brought in at night right through until the following spring as weaning does not occur until late summer or autumn with the increased chance of foals catching a chill if left out all night.

During this time the mare should not be neglected. Her udder may start to show signs of tenderness and discomfort due to the increase in pressure within from the build-up of milk. As mentioned earlier, a small amount can be milked out daily for the first few days in order to reduce the pressure and hence the chance of mastitis. The amount of milk removed daily should only be small and gradually reduced over 5 days. The removal of too much milk will only serve to prolong the problem. If milk build-up within the udder leads to excessive pressure, infection can result and mastitis caused. Post weaning mares must, therefore, be watched for such problems, especially in the first few days. Mastitis normally also results in elevated body temperature due to the development of a systemic infection, and can prove fatal, if not treated in time. Antibiotic treatment is usually successful, though there is always the danger of the infected half of the udder being lost.

The major disadvantages of this system are stress caused to the foal and the risk of mastitis to the mare.

Gradual Weaning

Gradual weaning is a newer and increasingly popular method which attempts to reduce the stresses of sudden weaning, but does itself have many disadvantages. There are two methods generally used, both are based on the same principles: that of gradual separation of mother and foal and the gradual introduction of other horses to replace the foal's mother as a companion.

In both systems careful planning well in advance is needed. It is not possible to practice gradual weaning with single foals or with foals of vastly differing ages. Ideally the foals should be born in batches within 2 weeks of one another and brought up together after the first couple of weeks. This allows them all to become accustomed to each other. In such systems some mares will be seen allowing a foal other than her own to suckle her, providing her foal is not also demanding milk. The foals get used to each other and a hierarchy is developed while their mothers are still around to dilute any aggression. Near the projected time of weaning all foals should be checked for physical condition and adequate solid food intake. In an ideal system on large studs there will also be other batches of younger foals born later following behind and any foals not ready for weaning in one batch can be transferred to the next one, giving them 2 more weeks or so to become prepared.

Once the foals are ready for weaning the mares can all be removed, initially for a short period of time, then increasing to a number of days until they can be removed altogether. This is, however, a labour intensive system. Alternatively, one mare can be removed on one day, followed by another the next, etc. until all have been removed. Again the mares must be taken well away from the foals and out of earshot. In the first system the time of separation from the mother is increased gradually over time and hence the stress of complete separation is much reduced. In the second alternative the foal that has lost his mother soon forgets due to the influence and security of his fellow foals and the other mares.

Again, the mares should be checked for mastitis, especially in the second system where individual mares are suddenly removed. When they are removed for increasing lengths of time this is not so much of a problem.

Plate 17.4 *In the gradual weaning system groups of mares and foals of similar ages are run together in preparation for the removal of the mares at weaning.*

As discussed, this system goes a long way to reduce the problems of weaning stress. However, it does have disadvantages. Having been weaned from its mother, the foal then has to be 'weaned' from its other companions at a later stage, prolonging the stress. Such systems are also not really possible with a single or a few foals, as batches are needed for it to work well.

Regardless of the system used, weaning is stressful and potentially dangerous. This stress can be reduced considerably by good management. Maintenance of familiar surroundings and routines as much as possible will considerably reduce the stress, in addition to a gradual preparation in diet and handling for the event. Damage to the foal can be minimised by ensuring that it is in safe and secure surroundings. Ideally a paddock is best to allow exercise, sunlight, fresh air, etc. However, it is easier to ensure that a stable is safe and 100% secure for expensive foals, and stables are, therefore, popular in the initial post weaning days, though not ideal.

Post Weaning Care of the Mare

As discussed the mare's udder should be watched for evidence of tenderness due to extreme back pressure from the build-up of milk within the udder. In order to minimise milk production post weaning, the mare should have all hard feed cut out of her diet and be put in a paddock with limited grass cover (Plate 17.5). This ensures that her nutritional intake is limited and that she is forced to exercise in order to obtain what grass she eats. Exercise helps to relieve pressure within the udder and uses up nutrients that would otherwise go towards milk production (Plate 17.6). Her paddock should be secure and free of hazards as the mare may be quite agitated initially due to the absence of the foal. The mare usually recovers quicker from the separation than the foal and in some cases may even seem relieved to be free of the extra burden.

Management of Youngstock

The management of youngstock in itself could well be the title of a book, hence this section will only attempt to give a summary of the principles and points to consider. The more specialised texts available should be consulted for in-depth details of the systems that can be used.

Nutrition

A large amount of the foal's development occurs in its first year of life and it continues its development until maturity at 4–5 years of age. Nutrition during this period is thus of the utmost importance. Nutrition must provide all the building blocks required by the body for growth and development, but not to excess, as this causes obesity, which in turn puts extra strain on young limbs, tendons and the circulatory system.

Prior to weaning the digestive system of the foal gets progressively more used to the digestion of solids and roughage, and at weaning the transition from fluids to solids is completed. In general, nutrient deficiencies in youngstock are more critical than in older animals, and so rations for weanlings should be designed with particular care.

Energy
Energy is the most important component of a youngster's diet. Low energy levels depress growth rates and at the extreme can cause stunting even if other diet components are appropriate. Energy levels in the order of 2.90 MCal digestible energy/kg (DE/kg) for youngsters at 4–12 months of age, decreasing to 2.80 MCal DE/kg by 12 months of age are required to achieve adequate growth. These relatively high energy levels can be obtained by feeding high

Plate 17.5   After weaning, mares can be turned out into a paddock with limited grass cover to help dry up their milk production.

Plate 17.6   At weaning, the mare's udder must be watched carefully for mastitis, though by this stage it should have begun to reduce in size as the amount of milk demanded by the foal has decreased.

concentrate diets. A good intake of roughage must also be maintained. A ratio of 3:7 of roughage to concentrates is ideal if the roughage has a DE content of 2.0 MCal/kg DM. This can be achieved by feeding legumes such as alfalfa. Most other roughages have energy concentrations less than 1.0 MCal/kg and so either the amount of concentrates fed or the energy concentration of the hard feed should be increased to compensate. If the amount of concentrate is increased it should not represent more than 70% of a 12-month-old youngster's diet (Tables 14.1 (page 253), 17.1–17.3).

Within these guidelines the body condition of the youngster should also be monitored and the level of feed should be altered to account for any over- or under-weight. Inappropriate energy levels are a prime cause of obesity in youngstock and are also thought to be contributory to conditions such as epiphysitis, structural deficiencies, contracted tendons, etc in the growing horse.

Protein
Protein is a necessity for bone growth and development, in addition to its involvement in energy utilisation. Protein concentrations in the order of 14.5% for youngsters up to 12 months of age are recommended. The total protein content is not the only important factor, as the quality of the protein provided is also important, ie its amino acid make-up. The adequate supply of essential amino acids is extremely important to youngstock. Lysine, usually the most limiting amino acid in most diets, is especially important in youngsters. Hence, even if the total protein content of the diet is appropriate, the horse may still suffer from protein deficiency due to the lack of an essential amino acid. This can be alleviated by feeding quality protein straights. Soya bean meal is a good source of protein, being highly palatable and high in lysine, and it is a popular choice as a component of a youngster's diet.

As mentioned, protein is involved in the utilisation of energy components within the diet. Inadequate protein concentrations are associated with signs of energy deficiency even if there are adequate DE levels within the diet fed. It is, therefore, usually recommended that protein levels slightly above requirements are fed in order to ensure that energy utilisation is optimised (Tables 14.1 (page 253), 17.1–17.3).

Vitamins and Minerals
As might be expected, vitamin and mineral requirements in growing youngsters are of immense importance. Vitamin and mineral deficiencies within the component feeds of a diet and deficiencies in feedstuffs are due to the environmental conditions under which they are grown, particularly soil deficiencies. For example grain

Table 17.1   Daily nutrient requirements of youngstock of varying weights.

(Tables 17.1–17.3 are adapted with permission from *Nutrient Requirements of Horses*, Fifth Revised Edition. Copyright 1989 by National Academy of Sciences. Courtesy of National Academy of Sciences, Washington, D.C.)

Animal	Weight (kg)	Mature weight (kg)	Daily gain (kg)	DE (Mcal)	Crude protein (g)	Lysine (g)	Ca (g)	P (g)	Mg (g)	K (g)	Vitamin A (10³ IU)
Maintenance	200	200		7.4	296	10	8	6	3.0	10.0	6
	500	500		16.4	656	23	20	14	7.5	25.0	15
	700	700		21.3	851	30	28	20	10.5	35.0	21
	900	900		24.1	966	34	36	25	13.5	45.0	27
Growing horses											
Weanling 4 mths	75	200	0.40	7.3	365	15	16	9	1.6	5.0	3
	175	500	0.85	14.4	720	30	34	19	3.7	11.3	8
	225	700	1.10	19.7	986	41	44	25	4.8	14.6	10
	275	900	1.30	23.1	1154	48	53	29	5.8	17.7	12
Weanling 6 mths											
Moderate growth	95	200	0.30	7.6	378	16	13	7	1.8	5.7	4
	215	500	0.65	15.0	750	32	29	16	4.0	12.7	10
	275	700	0.80	20.0	1001	42	37	20	5.1	16.2	12
	335	900	0.95	23.4	1171	49	44	24	6.2	19.6	15
Rapid growth	95	200	0.40	8.7	433	18	17	9	1.9	6.0	4
	215	500	0.85	17.2	860	36	36	20	4.3	13.3	10
	275	700	1.00	22.2	1111	47	43	24	5.4	16.8	12
	335	900	1.15	25.6	1281	54	50	28	6.5	20.2	15

Yearling 12 mths										
Moderate growth										
140	200	0.20	8.7	392	17	12	7	2.4	7.6	6
325	500	0.50	18.9	851	36	29	16	5.5	17.8	15
420	700	0.70	26.1	1176	50	39	22	7.2	23.1	19
500	900	0.90	31.2	1404	59	49	27	8.6	27.7	22
Rapid growth										
140	200	0.30	10.3	462	19	15	8	2.5	7.9	6
325	500	0.65	21.3	956	40	34	19	5.7	18.2	15
420	700	0.85	28.5	1281	54	44	24	7.4	23.6	19
500	900	1.05	33.5	1509	64	54	30	8.8	28.2	22
Yearling 18 mths										
Not in training										
170	200	0.10	8.3	375	16	10	6	2.7	8.8	8
400	500	0.35	19.8	893	38	27	15	6.4	21.1	18
525	700	0.50	27.0	1215	51	37	20	8.5	27.8	24
665	900	0.70	33.6	1510	64	49	27	10.9	35.4	30
In training										
170	200	0.10	11.6	522	22	14	8	3.7	12.2	8
400	500	0.35	26.5	1195	50	36	20	8.6	28.2	18
525	700	0.50	36.0	1615	68	49	27	11.3	36.9	24
665	900	0.70	43.9	1975	83	64	35	14.2	46.2	30
Two-yr-old 24 mths										
Not in training										
185	200	0.05	7.9	337	13	9	5	2.8	9.4	8
450	500	0.20	18.8	800	32	24	13	7.0	23.1	20
600	700	0.35	26.3	1117	45	35	19	9.4	31.1	27
760	900	0.45	31.1	1322	53	45	25	12.0	39.4	34
In training										
185	200	0.05	11.4	485	19	13	7	4.1	13.5	8
450	500	0.20	26.3	1117	45	34	19	9.8	32.2	20
600	700	0.35	36.0	1529	61	48	27	12.9	42.5	27
760	900	0.45	42.2	1795	72	61	34	16.2	53.4	34

Table 17.2 Nutrient concentrations in total diets for youngstock (100% DM). Values assume a concentrate feed containing 3.3 Mcal/kg and hay containing 2.00 Mcal/kg of dry matter.

Animal	DE	Diet proportions		Crude protein	Lysine	Ca	P	Mg	K	Vitamin A
		Conc	Hay							
	(Mcal/kg)	(%)	(%)	(%)	(%)	(%)	(%)	(%)	(%)	(IU/kg)
Maintenance	2.00	0	100	8.0	0.28	0.24	0.17	0.09	0.30	1830
Growing horses										
Weanling 4 mths	2.90	70	30	14.5	0.60	0.68	0.38	0.08	0.30	1580
Weanling 6 mths										
Moderate growth	2.90	70	30	14.5	0.61	0.56	0.31	0.08	0.30	1870
Rapid growth	2.90	70	30	14.5	0.61	0.61	0.34	0.08	0.30	1630
Yearling 12 mths										
Moderate growth	2.80	60	40	12.6	0.53	0.43	0.24	0.08	0.30	2160
Rapid growth	2.80	60	40	12.6	0.53	0.45	0.25	0.08	0.30	1920
Yearling 18 mths										
Not in training	2.50	45	55	11.3	0.48	0.34	0.19	0.08	0.30	2270
In training	2.65	50	50	12.0	0.50	0.36	0.20	0.09	0.30	1800
Two-yr-old 24 mths										
Not in training	2.45	35	65	10.4	0.42	0.31	0.17	0.09	0.30	2640
In training	2.65	50	50	11.3	0.45	0.34	0.20	0.10	0.32	2040

Table 17.3 The expected feed consumption by youngstock (% body weight).

Animal	Forage	Concentrates	Total
Maintenance	1.5–2.0	0–0.5	1.5–2.0
Young horses			
Nursing foal 3 mths	0	1.0–2.0	1.0–2.0
Weanling foal 6 mths	0.5–1.0	1.5–3.0	2.0–3.5
Yearling foal 12 mths	1.0–1.5	1.0–2.0	2.0–3.0
Yearling 18 mths	1.0–1.5	1.0–1.5	2.0–2.5
Two-year-old 24 mths	1.0–1.5	1.0–1.5	1.75–2.5

grown on land deficient in selenium will itself be selenium deficient. It is, therefore, ideal to have all rations fed to youngsters analysed for the major minerals, at least calcium, phosphorus, sodium and chlorine, and if necessary, an appropriate supplementary feed or a general vitamin and mineral supplement can be fed as routine so as to ensure no deficiencies arise. However the use of a general supplement may complicate imbalances if adequate minerals were originally present within the ration.

Calcium and Phosphorus
Calcium and phosphorus are essential for bone, tendon and cartilage growth and are, therefore, extremely important in the growing youngster. At 4 months the youngster requires ideally 0.68% calcium and 0.38% phosphorus in the ration; levels slightly above this are often fed to give a margin for error. These concentrations can decrease to 0.56–0.61% and 0.31–0.34% respectively for youngsters of 6 months of age (Tables 17.1–17.3). However, the ratio of calcium:phosphorus is as important as their specific concentrations, as excessive calcium in the diet binds to phosphorus and causes signs of phosphorus deficiencies within the body. A ratio of 1.5:1 is ideal for 6–month-old youngsters, though some variation either side of this can be tolerated. The older animal is able to tolerate more variation but even so the ratio of calcium:phosphorus should not exceed 3:1 as above this signs of phosphorus deficiency will become evident.

Sodium Chloride
Youngsters' rations should contain 0.9% salt, though the actual demand and intake of salt will be affected by work and environmental conditions. Ideally a salt lick or general vitamin and mineral supplement containing salt should be available to youngsters on an ad lib basis to allow them to supplement their salt intake as and when they require it.

Trace Minerals
Other minerals required by the growing horse include selenium, iodine, iron, copper, cobalt, manganese, potassium, magnessium, molybdenum, sulphur and fluorine. The amounts required are low, hence the term trace elements. Most feedstuffs will provide the adequate amounts of these; however, certain feedstuffs may be deficient due to low concentrations of the trace elements within the soil in which they were grown. A good example of this is selenium, that is deficient in soils in parts of mid Wales, United Kingdom, areas around the Great Lakes in the United States of America, and also parts of New Zealand. If horses are fed exclusively on feeds

grown within these areas they will show signs of deficiencies. However, most commercial diets are a combination of feedstuffs from numerous areas and as such normally contain adequate trace elements. Problems may be encountered in horses fed purely home-grown and mixed rations in deficient areas.

A free access vitamin and mineral lick or routine supplement to the diet can provide all the vitamins and minerals required by the growing youngster.

Water
Water must not be forgotten as an essential component of any diet, as a continual supply of clean fresh water is essential.

Environment

The nutritional requirements of any youngster are partly determined by its environment. The higher a horse's maintenance requirement, the higher his overall nutritional demand. If maintenance requirements are increased by adverse environmental conditions, ie wind, rain, cold, etc, then the overall ration must be increased to compensate. If this is not done, the growth and development of the horse will suffer.

In order to minimise maintenance requirements and, therefore, feed costs, many youngsters are kept housed for part, if not all, of the day, especially in adverse weather conditions. This also helps prevent damage and injury to the horse, which is of prime importance in show stock. However, such management has disadvantages as far as the psychological development of the horse is concerned. Isolation can lead to behavioural problems and such horses are also prone to obesity. An alternative that goes some way to solving the problem of isolation is to keep youngsters together in groups in a barn. Youngsters kept this way need time to adjust to each other and to develop a hierarchy, preferably in a large open field, before being confined within a barn. This system does also run the higher risk of damage to weanlings from close contact with fellows, and animals kept in this way must be watched closely for signs of bullying, and appropriate action taken.

Exercise

Exercise and social interaction with other horses are essential in the youngster for both physical and psychological development. Ideally, youngsters should be reared at pasture and this is becoming an increasingly popular method. Such animals can exercise at will, reducing the strain on growing bones and allowing muscles and tendons to develop and grow in strength and size. Social interaction with other horses allows a respect for fellows and it reduces boredom and the associated bad habits of confinement (Plate 17.7). The only major disadvantage of pasture-kept youngsters is the difficulty in maintaining concentrate intake unless each one can be fed individually. Most studs feed a standard ration to all pasture-kept youngsters all together, and only individuals that seem to be in under- or over-condition due to inappropriate concentrate intake are dealt with separately.

Handling

Before weaning all foals should be halter broken and be accustomed to being handled (Plate 17.8). It is much easier to teach foals basic manners at a young age than face a stroppy and potentially dangerous yearling.

A well-handled foal that is used to grooming, handling, feet trimming, clipping, boxing, etc, will be much easier to train in later life. Such an animal will have developed confidence that can be used and built upon in later training schedules. He is also much more likely to concentrate on the lesson in hand and, therefore, be easier to train, as he has confidence in his relationship with the human training him.

When handling any youngsters it must be remembered that they can be very unpredictable. They may be full of the joys of spring and so there is no guarantee that they will always react to things in the same way. As they develop, their interaction with, and therefore their reaction to, their environment changes and at times can be very unpredictable. Every precaution should be taken to ensure that if the unexpected occurs, no one is hurt and minimum panic ensues. A youngster is very impressionable and, therefore, panic or nervousness within the human handling him can easily be sensed and picked up by the youngster and detrimentally affect him. If this is allowed to get out of hand it can become self-perpetuating and result in an owner too frightened to handle the youngster and a horse that is a bag of nerves and increasingly unpredictable in behaviour. A complete outsider is often required to break such a downward spiral.

The youngster should always be treated with constant discipline. It gains security from predictable, consistent handling, even though it will try to push the limits of behaviour. Bad habits and bad behaviour should be disciplined immediately with the voice. Physical punishment is not necessary, except in extremes, and if not carried out with care, it can degenerate into a confrontation between the horse and handler.

Plate 17.7   *Youngsters should ideally be reared at pasture, allowing them to develop social interaction and respect for others, as well as providing exercise to the advantage of bone, muscle and tendon development.*

Plate 17.8   *Before foals are weaned they should be halter broken and used to being handled, which considerably eases their management post weaning.*

Youngsters should be taught not to bite, nip, push, barge, etc and to learn respect for their handlers. A foal should not be allowed to play with handlers. It is great to teach a foal to shake hands, but when it starts to kick out in front as a yearling or older horse, this can be dangerous. This is not the horse's fault, and he will find it very confusing to be told off for doing something it was once praised for.

Charging and pushing handlers, especially when leaving a stable or going through a gate, is potentially very dangerous. The foal should be taught to let the handler go first, and if it barges it should be halted with a short sharp tug on the lead rope. If that does not work, then a harder series of tugs should be given. A continual tug is not advised as this will encourage the horse to fix his neck and head and use his strength against the handler in the future.

Praise is as important as discipline and whenever a youngster does well or as he is told, he should be praised by a pat and also by voice. Throughout his handling he should get used to the human voice and certain clear commands. Initally 'Whoa' is very useful to get the horse to stand still, and it can be used later in lunging lessons. There are other commands for other actions, eg come to call etc, as appropriate. A horse will get used to any commands providing they are clear and consistently used. One of the most important discipline lessons to learn is to stand still. As long as a horse will stand still when told, you will always have control over him.

Initially a youngster will be very exuberant and full of energy, and at this stage it may be appropriate to overlook minor bad behaviour until the basics have been achieved. It can be very demoralising for a youngster, and handler, if the foal is always being disciplined and can seemingly do nothing right. The less desirable behaviour can be disciplined at a later stage providing it does not get out of hand. The handler must use his common sense and assess the situation.

Further handling of the youngster will depend on his destination in life and can be geared towards a particular aim, for example, racing, breeding, showing, hacking, etc. It is beyond the scope of this book to discuss the specific training of youngsters for specific aims.

Whatever the ultimate destiny for the youngster, exercise is of particular importance. Exercise allows muscles and tendons to grow and develop and good muscle tone and body condition to be maintained. During the first year free exercise is usually allowed by turnout into a paddock. If turnout has to be restricted, the period out will be determined by the environmental conditions. It is best to avoid turnout in hot weather, when flies will worry horses and also when the sun may bleach the coat of show horses. At about a year of age forced exercise can be introduced in the form of leading out, or lunging, and provides the grounding for further training especially in riding horses.

Regular exercise also provides conditioning and fittening of horses, ensuring that when training begins in earnest an initial period of fittening is not required. Training of unfit horses runs the risk of strains and sprains of unconditioned muscles and tendons, leading to a waste of time in resting injuries. A youngster that does not have enough free exercise can be very difficult to train as he will spend much of the time careering around to get his high spirits out of him before work can be commenced in earnest. Such animals are also very difficult to lunge or put in a horse walker and in fact they run a high risk of injury or damage. A combination of free and forced exercise allows agility and physical ability to develop along with stimulation of the youngster mentally. Development of both mental and physical faculties is essential as a highly sophisticated state is required in many disciplines.

Conclusion

Considerable thought should be put into the method of weaning and the handling of youngstock. Inappropriate treatment at this young age can have a long-term detrimental affect on the foal's physical and psychological development and affect its ability to fulfill its potential in later life.

References and Further Reading

Coldrey, C. and Coldrey, V. (1990) Breaking and Training young Horses. Crowood Press, Gipsy Lane, Swindon, Wiltshire, U.K.

Equine Research, Breeding Management and Foal Development. Equine Research Inc. P.O. Box 9001, Tyler, Texas 75711.

National Research Council (1989) Nutrient Requirements of Horses 5th Ed. National Academy of Sciences, National Research Council, Washington D.C.

Rossdale, P.D. and Ricketts, S.W. (1980) Equine Stud Farm Medicine. 2nd Ed., Ballière Tindall, London.

Wynmalen, H. (1989) Horse Breeding and Stud Management. Revised and enlarged by A. Leighton Hardman.

CHAPTER 18

Stallion Management

THE management of the stallion should not be neglected in the enthusiasm to get the management of the mare and foal just right. The management of the stallion has already been discussed at length in Chapters 11–13. However, it is worth considering the introduction of the stallion to his work as a breeding animal and the general training and management principles that should be borne in mind when keeping stallions.

Early Training

Early training, well in advance of the stallion's first introduction to a mare, is very important to reduce the chances of injury to both horse and handlers, and to reduce the risk of him developing bad habits which can prove dangerous.

It is very important that a stallion is taught discipline and respect for his handlers from a young age. Once a good grounding has been established this can be built upon. If the grounding is not there, it is near impossible to start disciplining a 3- or 4-year-old stallion without the considerable risk of injury, and it will inevitably lead to conflict and not respect. In training a stallion the handler's attitude and competence are also of extreme importance, as rough handling and incorrect and inconsistent training can cause many problems, from poor breeding behaviour and performance to dangerous habits.

Once basic discipline, including the acceptance of the handler, bridle, leading, obeying voice commands to halt, walk on, back up, along with boxing, shoeing, veterinary inspection and general handling, has been achieved, the stallion may be trained further in a specific area of discipline or maintained at this level for breeding. Further training for riding or driving is very good for the stallion as it provides him with a constructive outlet for his energies other than covering. This advanced level of discipline leads normally to greater respect and subsequently a stallion that is easier to handle.

335

Restraint

There are several means of constraint that can be used to control a stallion. The method used depends on the stallion's age, temperament, the facilities available and the handler's personal preference. The effect of the handler on the behaviour of the stallion cannot be overemphasised. A nervous and insecure handler will transfer these feelings to the stallion, who in picking them up is more likely to act uncharacteristically and unexpectedly. It is especially important that the handler dealing with young stallions is calm and confident and has had plenty of experience. The overuse of constraint or punishment to compensate for a nervous handler is a trap that can be all too easily fallen into. If a stallion needs reprimanding it should be immediate, quick and effective. Continually ineffective, half-hearted attempts at reprimanding, often due to a lack of experience or confidence in the handler, lead to resentment of the handler by the stallion.

Stallions are by nature proud and courageous, attributes much to be admired. However, they must be treated with care and respect in order to maintain these attributes and channel them into safe expression and not into conflict with handlers.

Ideally the stallion should be restrained in day to day management by a good strong leather or webbing head collar or halter and lead rope. This must be checked regularly as, due to his strength, a stallion can easily break away from inadequate restraint and has the potential to wreak havoc in a yard. He should have been taught acceptance of the halter and leading at a young age so that this is no problem. However, it is inevitable that at an older age he will become more boisterous, and if he has been inappropriately trained in early life he will need a more substantial means of restraint even for basic day to day management. This may consist of a halter plus chain passed over the nose from the ring at the top of the cheek piece.

A pull on the chain in such a arrangement applies pressure to the nose and provides extra restraint on the stallion. A more severe method along these lines is to pass the chain through the stallion's mouth or under his chin. This should be done with care as it is potentially very dangerous to the stallion's mouth and nose if extreme pressure is applied. Many stallions are restrained, especially for covering, by means of a snaffle or stallion bit to which a chain is attached in several positions. It may be passed through the left-hand ring and attached to the right-hand ring (Plate 18.1); attached to a separate chain that is passed through both rings; or, more severely, passed through the left-hand ring, then through the

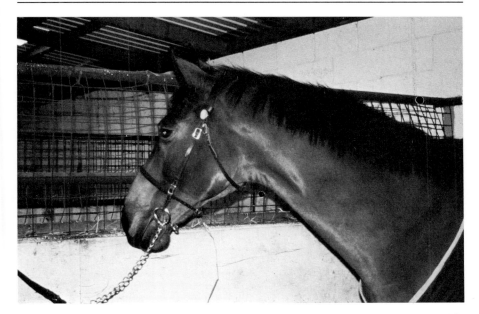

Plate 18.1 *This form of stallion restraint with the chain passed under the chin is thought to encourage rearing.*

right-hand ring and passed back to and attached to the left-hand ring. This arrangement of the stallion chain passed under the chin is thought by some to encourage rearing and so an alternative is to pass it over the nose, looped through the noseband. This method is reported to discourage rearing as it encourages the head to come down when pressure is applied on it (Plate 18.2). All these methods result in pressure being exerted on the lower jaw with varying severity and hence exert further restraint on the stallion.

It is best to keep a stallion's tack for varying occasions separate and different, so that when he is expected to behave and mind his manners he knows by the restraint used. The tack used will also let him know when he is expected to exhibit libido and cover mares.

The effective use of the handler's voice should not be underestimated. A clear, confident voice command is just as effective as a physical reprimand to a well-trained stallion. It is not a good idea to use the previously described more severe forms of restraint on a stallion as a matter of course; they should only be a last resort. A halter for everyday use and a snaffle bridle for covering, along with the effective use of the voice, are all a well-trained stallion requires (Plates 18.3 and 18.4).

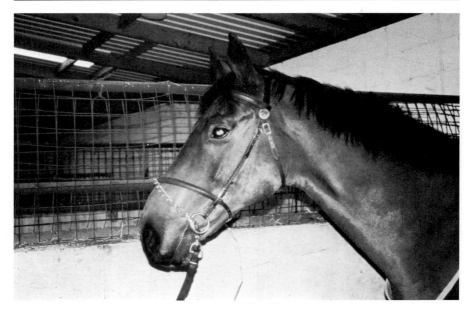

Plate 18.2   This alternative form of stallion restraint with the chain passing over the nose is thought to discourage rearing, as the pressure applied encourages the stallion's head to come down.

Plate 18.3

All a well-behaved stallion should require for restraint on an everyday basis is a halter.

Plate 18.4 A snaffle bridle, along with effective use of the voice, is all a well-mannered stallion should require when covering a mare.

Introducing the Stallion to Covering

A young stallion should not be expected to cover mares until he is at least 3 years old. During his first season he will only be able to cover a relatively few mares compared to when he is fully mature. He should be limited to 10–15 mares spread out over the season and he should not be expected to cover more than 1 per day. This is in contrast to the work rates that mature stallions can cope with, which are in the order of 60 mares per season and 1–3 mares on a single day, providing rest days are given. The owner of a young stallion should also be prepared to cancel further nominations in the first season if the stallion shows signs of losing interest and lacks libido or gets injured. All mares for young stallions should be individually picked, as only mature mares of quiet disposition well in season should be chosen. Such mares are normally offered a no foal, no fee service as even if a semen evaluation has been carried out, the stallion has as yet no proven fertility record.

A stallion's first covering should be in hand with an experienced handler who knows the stallion well. Even if the stallion is eventually to be used in pasture breeding or other non-in-hand breeding situations, it is advisable that his first cover is in hand to ensure that evasive action can be taken in the event of emergencies. This first cover is extremely important and its success can seriously affect a stallion's long-term ability and behaviour. It is essential that the mare to be covered is experienced and quiet and is in full oestrus. Maiden mares are not advisable as they themselves may be unpredictable and it is enough of a job watching an inexperienced and, therefore, unpredictable stallion without having the added complication of an unpredictable mare. Ideally, the mare should be slightly smaller than the stallion, making it easier for him to mount.

The stallion and mare should be prepared for covering as detailed in Chapter 12. The stallion should be familiar with the covering area, having been shown it beforehand, and extra care should be taken to ensure that the floor is non-slip and that there are no protrusions that may injure him or cause him to fall.

The handler should be experienced in the normal sequence of events when mating mature stallions as ultimately the novice will need to behave in a similar fashion. However, an inexperienced stallion cannot be expected to behave in such a way immediately. At the first covering it is best to more or less allow him his head to get on with the job. As discussed in Chapter 12, excessive interference by man is very off-putting for the stallion, and at this stage excessive guidance or discipline should be avoided and the stallion allowed to gain confidence in his ability, before he is taught manners as well. However, potentially dangerous habits such as kicking or biting should be corrected immediately, as if allowed to persist they could render the stallion unusable.

At his first few coverings the stallion may need prolonged teasing and may mount the mare several times before ejaculation is achieved. He should not be hurried or forced in any way, as this will only serve to upset him, put him off his stride and result in long-term problems. If he seems unable to ejaculate properly he should be taken away and returned to his box and either tried again with the same mare later on in the day or, better still, with another mare. There is no reason why such hiccups in the first few covers should have any effect on his long-term performance. This first season is all about building up the stallion's confidence in his ability to do the job and gradually instilling manners into him for the sake of safety. It is a gentle balancing act between the two aims, and the rate of progress very much depends on the individual stallion. You must always be prepared to suspend all attempts to cover mares if he has a bad experience, usually injury from a mare, and to start all over again at the beginning to restore his confidence. A bad experience during these initial coverings can affect the stallion for life. If serious it can permanently reduce his libido, if not it may mean he develops an aversion to a particular type of mare in regard to colour, size, age etc. In order to minimise problems of rejection by mares a young stallion must be introduced to his mares very carefully, allowing plenty of time for them to become acquainted, and for him to develop confidence in them.

The whole aim of the stallion's first season is to ensure that he associates his new job with pleasure, in a calm and secure atmosphere, in order to build up his confidence so that he will be able to deal with the occasional not so cooperative mare that he may well come across in later life. It is essential, therefore, that everything is carried out calmly and that any incidents are dealt with calmly and with confidence. Panic and insecurity in the handler will affect the stallion's attitude and performance. Confidence in his handler and surroundings can only serve to enhance his own self-confidence and, therefore, his ability at his job.

General Stallion Management

The general management of the stallion is extremely important to ensure that he is fit, able and willing to do his job. It also helps to prevent the acquisition of bad habits and increases the safety when using him. Stallion management can be divided into housing, exercise, nutrition, feet care, dental care, vaccination and worming programmes.

Housing

A stallion naturally would roam wide areas of land, migrating steadily over new pasture with his mares. He could, therefore, exercise himself at will and was always provided with fresh clean grazing. Domestication has largely put paid to this, except in some pasture breeding systems. We, therefore, have to compensate the stallion for the things that domestication has taken from him, in order to maximise his efficiency and avoid habits of boredom.

Ideally a stallion should be turned out in a large paddock. This is not always possible and the climate in many areas precludes the use of all-year turnout.

Most stallions are confined to a stable for some period of time at least. This stable must be large, ideally at least 5 m × 5 m for a 15 hh horse, and be light and airy. It should have a good strong secure door with both top and bottom sections, plus a top grid that can be shut to provide extra safety but still allow ventilation. Details of the stallion's pedigree can be displayed on the inside of the upper door as at the National Stud, Newmarket (Plate 18.5). This adds interest, especially to yards where visitors are catered for. The stable should also have a tie ring at the back to which the stallion should be tied (racked up) as part of his general routine each day, possibly when his bed is cleaned out and water and hay are replenished. It is a good idea to rack stallions up routinely as it makes them easier to handle if visiting mares are around in the yard and they can be easily caught if it is their turn to cover. Consistent routine enhances their discipline.

A stallion's stable should always be clean and free of flies. Regular cleaning of water troughs and feed mangers is also essential and helps to keep flies down.

In order to reduce boredom and the development of bad habits such as crib biting, weaving, stable walking, masturbation etc, the stable door should overlook a busy part of the yard but still be away from any mares. Chains, plastic bottles and swedes hanging from the ceiling, or a football or even a cat have been successfully used to provide the stallion with entertainment and, therefore, reduce his boredom. The box should have easy access to a paddock, which is normally exclusively for his use. Two acres per stallion is ideal and allows plenty of room for exercise. The cost of fencing can be minimised as just this paddock need have strong, high stallion-proof fencing. Fencing should be at least 2 metres high and be made of post and rail. An electric fence along the top or projecting into the field about 15 cm from the top of the fence would provide extra security (Plate 18.6).

The paddock has to be cleaned of manure at regular intervals to ensure it remains clean and fresh and does not become horse sick. It may also be provided with a field shelter to give the stallion protection if the weather turns nasty while he is out.

Exercise

Exercise, as for all horses, is essential for the stallion. It helps to reduce boredom and associated problems as well as helping to maintain basic fitness and muscle tone. Fitness is especially important in stallions as they must undergo short, sharp periods of extreme exercise when covering. Their system must, therefore, be able to cope with this. Exercise improves the cardiovascular system and reduces the chances of conditions such as azoturia or tying up and other infectious diseases that can affect overall performance (Dinger and Noiles, 1986). Exercise also helps aid digestion and promotes a healthy appetite. It can be in the form of either free or

Plate 18.5 *A stallion box at the National Stud, Newmarket, showing the stallion's pedigree displayed on the inside of the upper door.*

Plate 18.6 *Two metre high post and rail fencing is ideal for a stallion's paddock. An electric fence running along the top or inside the fencing provides extra security. (The National Stud, Newmarket)*

forced exercise, ie turnout or riding/lunging. Many stallions are unbroken and the only option in such cases is free exercise, so the stallion should be turned out for a large proportion of the day, weather permitting, or alternatively turned into an indoor school (Plates 18.7 and 18.8).

Free exercise is fine for stallions willing to exercise themselves. However, some refuse to move around the paddock, or at the other extreme charge around like mad and pace the fence, spending no time grazing, and, so, lose condition. The exercise of such stallions must be controlled by means of forced exercise, which also improves discipline and respect, especially in nervous and highly strung stallions.

Riding (Plate 18.9) and lunging are popular forms of forced exercise. Other forms include swimming (Plate 18.10), which is particularly beneficial to the cardiovascular system and for lame horses. Treadmills or horsewalkers (Plate 18.11) provide a good means of forced exercise but should be restricted to stallions accustomed to them.

Some stallions, especially of the native type breeds, can be turned out with mares and foals in a large field. This system has the added advantage that mares returning to heat after covering can be picked up and served by the stallion turned out with them (Plate 18.12). This system must be confined to use with well-behaved, older stallions which are well into their working season.

Whatever exercise is used it must be geared to the particular stallion. It must be closely monitored along with his nutrition to ensure that he remains in body condition score 3. He must not be overtired or he will not have enough energy for the real job in hand. Monitoring individual stallions allows their exercise to be tailored to suit their individual needs.

Nutrition

A properly balanced diet is essential for a stallion's well-being. Each stallion should be fed individually according to his size, condition, work load, temperament etc. He should be closely observed to ensure that his condition does not get too good or too poor. His feed should be carefully monitored not only during the breeding season but also out of season. If his nutritional intake is neglected during the non-breeding season, there will not be enough time to correct any over- or under-condition before the season begins. During the breeding season the work load of a stallion, in nutritional terms, is greater than that of a performance horse. As a general rule a stallion will eat 2–3% of its body weight daily and at least 50% of this should be good-quality roughage in the form of

Plate 18.7   Turnout into a field provides ideal exercise for a stallion, weather permitting. (The Welton Stud)

Plate 18.8   In bad weather turnout into an indoor school provides a good alternative free exercise. (The Derwen Welsh Cob Stud)

Plate 18.9

Riding provides ideal exercise for a stallion and also helps to improve discipline and respect for his handlers.

(Below) *Plate 18.10*

Swimming provides good exercise especially for the cardiovascular system and for stallions with lameness problems.

Plate 18.11   **Treadmills or horsewalkers are a good means of forced exercise but should be restricted to stallions that are used to them.**

Plate 18.12 *Stallions of a quiet disposition can be turned out with mares, especially towards the end of the season. This relieves boredom in the stallion and allows mares returning to service to be picked up.*

hay or grass. This will also meet a stallion's vitamin, mineral and protein requirements (Hintz, 1993).

Young growing stallions should be fed on a slightly higher proportion of concentrates, ie a ratio of 60:40 of concentrates : roughage to increase protein intake.

Protein and Energy
A protein concentration of 9.6% in the diet is recommended for a mature stallion and higher levels of 11–13% for young stallions. Energy levels of 2.4–2.6 MCal DE/kg similar to those for horses in heavy work, are recommended (Tables 14.1 (page 253), 18.1–18.3).

Vitamins and Minerals
Many breeders also feed a vitamin and mineral supplement on a free access basis regardless of feed analysis. This is not always necessary but can be used as a precaution. The only vitamin that is likely to be short in a well-balanced diet is vitamin A, important in the regeneration of germinal epithelium within the testis.

Inappropriate nutrition is one of the major causes of low libido and poor reproductive performance. Correct monitoring of a stallion's

Table 18.1   Daily nutrient requirements of stallions of varying weights.

Stallions	Weight (kg)	DE (Mcal)	Crude protein (g)	Lysine (g)	Ca (g)	P (g)	Mg (g)	K (g)	Vitamin A (10³ IU)
	200	9.3	370	13	11	8	4.3	14.1	9
	500	20.5	820	29	25	18	9.4	31.2	22
	700	26.6	1064	37	32	23	12.2	40.4	32
	900	30.2	1207	42	37	26	13.9	45.9	40

Table 18.2   Nutrient concentrations in total diets for stallions (100% DM). Values assume a concentrate feed containing 3.3 Mcal/kg and hay containing 2.00 Mcal/kg of dry matter.

Stallions	DE (Mcal/kg)	Diet proportions Conc (%)	Hay (%)	Crude protein (%)	Lysine (%)	Ca (%)	P (%)	Mg (%)	K (%)	Vitamin A (IU/kg)
	2.40	30	70	9.6	0.34	0.29	0.21	0.11	0.36	2640

Table 18.3   The expected feed consumption by stallions (% body weight).

Animal	Forage	Concentrates	Total
Mature stallion	1.00–2.5	1.0–1.5	2.0–3.0
Young stallion	0.75–2.25	1.25–1.75	2.0–3.0

Tables 18.1–18.3 are adapted with permission from *Nutrient Requirements of Horses*, Fifth Revised Edition. Copyright 1989 by National Academy of Sciences. Courtesy of National Academy Press, Washington, D.C.

condition and adjustment of nutrition and exercise accordingly cannot be overemphasised. Reducing or increasing nutritional intake immediately prior to the breeding season can have as detrimental an effect on performance as over- or under-nutrition per se.

Feet Care

The feet of a stallion should never be neglected. Lameness can severely reduce and restrict the stallion's ability to cover a mare. This is especially evident if the lameness is in the hind feet and is often first seen as uncharacteristically low libido. A stallion should not be shod during the breeding season as shoes can hurt and damage the mare's flanks and shoulders at mating, as well as inflicting more damage than unshod feet if the stallion should kick out. Regular trimming should be carried out to ensure that the feet remain clean, uncracked and unsplit. This should be done at 6 week intervals and any problems dealt with immediately to avoid long-term complications.

Dental Care

Uneven, rough teeth, along with lesions or abscesses, can reduce a stallion's appetite, sometimes seriously, due to pain when chewing. They can also cause food to pass into the stomach without full mastication, reducing the efficiency of digestion. If a stallion starts to lose condition for no obvious reason, one of the first things to check is his teeth and mouth.

A stallion's mouth should be checked annually to see whether rasping is required. Regular rasping will ensure that uneven grinding of the teeth does not result in sharp projections, which cause lacerations. Sharp points can be filed quite successfully by a veterinary surgeon, or a specialised horse dental surgeon.

Vaccination

Vaccination is an essential part of a preventative medicine routine that should be developed and implemented regularly for a stallion throughout his life. The vaccinations required depend on the country in which the stallion stands and on the prevalence of various infections. In the United Kingdom stallions should have up-to-date influenza and tetanus inoculations. Influenza injections should be up-dated every year, whereas for tetanus a booster every other year is adequate in most areas of the country.

Vaccination against EHV1 is also becoming more popular in Britain preventing passage of the virus to mares, in which it causes abortion.

Parasite Control

Worming is another essential part of a preventative medicine routine. As with all wormers in all horses the key to success is regular use and rotation of product to ensure that the parasite burden remains low and that resistance to a certain anthelmintic is not developed. High worm counts or, for that matter, any parasitic infection causes listlessness and, hence, low libido and reduced reproductive performance. Worming should be carried out regularly at 6 week intervals throughout the spring, summer and autumn. During periods of temperature consistently below freezing, ie during the winter in the United Kingdom, worming is not required. Some wormers are themselves reported to cause listlessness and reduced libido in stallions but only for a few days post administration. As a result some studs try to organise their worming regime so that stallions do not need to be wormed at the height of their covering season.

Stallion Vices

Largely because of the conditions that man keeps stallions in, for their safety and the safety of other stock, they are in danger of developing bad habits due to boredom. Prevention is infinitely better than cure and it is, therefore, essential that the stallion's management is geared towards reducing boredom. A regular routine of work, exercise, feeding, etc and a stable in an area of the yard where he can see activity goes a long way to relieving boredom. A frustrated and bored stallion releases his energies and tensions in the only way possible to him—by developing bad habits. These vices can be harmful to the stallion himself and dangerous to the handler, and they may also affect his reproductive performance. If prevention has failed and vices have developed there are certain practices that can help control them and/or their effects.

Windsucking and Crib Biting

Primarily caused by boredom, these habits are related and often develop one from the other. In crib biting the horse bites part of the stable structure or other convenient object and chews (Plate 18.13). The habit can be discouraged by removing all objects that can be grasped by the teeth, or by painting all structures with Cribox or an equivalent foul-tasting substance. Wind sucking can develop on from crib biting, and in this more serious condition the horse, while grasping a projecting structure, arches his neck and gulps air in. If

Plate 18.13 *Telltale signs of crib biting in a horse: protruding surfaces, commonly stable doors, showing signs of teeth marks*

the habit is allowed to continue unchecked, it can lead to colic and low appetite, as well as excessive wear and tear on the upper incisor teeth. A muzzle can be used to prevent a horse from crib biting. A cribbing strap, placed over the throat lash and so preventing the stallion from tensing the neck muscles used in windsucking, can also help.

A more extreme method of cure is the severing of the neck muscles attaching the hyoid bone to the base of the tongue. Alternatively the nerves serving these muscles can be severed. In the majority of cases such procedure will effect a cure, and in the remainder considerable improvement is obtained, but these are rather drastic solutions to a problem that is largely avoidable with the appropriate management.

Weaving

Weaving involves the swaying of the horse's head and neck rhythmically from side to side, often over the stable door. This can cause damage to the forelegs as the horse's weight is shifted from side to side. A chronic weaver may weave himself to the point of exhaustion. The condition can be alleviated by anti weave bars over the lower stable door (Plate 18.14).

Unfortunately chronic weavers will continue to weave within their boxes. Also often associated with weaving is stable walking, in which the stallion continually walks around his box, seemingly chasing his tail. There is little that can be done to cure this behaviour except to turn the horse out into a paddock to relieve the boredom, and such horses usually revert to walking as soon as they are stabled again.

Self-mutilation

This vice, unlike the others, is not normally an expression of boredom, but nevertheless can be extremely distressing to the stallion and his owner. It involves the stallion biting his own legs, shoulders and chest, causing himself some considerable damage. It is particularly evident after mating. If evident only then, thorough washing of the stallion after dismount reduces expression of the vice, which in this case is thought to be due to the smell of the

Plate 18.14   During the times that the top door of the stable is open, anti weave bars may be used to discourage the stallion from weaving over the bottom door and encouraging other horses to copy.

mare. If, however, the stallion is a habitual self-mutilater, there is very little that can be done to cure him, though the use of a muzzle or cradle can prevent him inflicting damage to himself.

Masturbation

Masturbation by a stallion is a further vice evident in a stallion suffering from boredom, especially those that are sexually frustrated. The penis extrudes from its sheath and the stallion rubs it along the under side of his abdomen. In extreme cases such masturbation will result in ejaculation and loss of valuable sperm, so reducing the work load the stallion is capable of. The problem of masturbation can be reduced by reducing boredom and also by ensuring that the stallion does not become sexually frustrated and is used to cover mares on a regular basis. Teasers can develop the habit if they are not allowed occasionally to cover a mare. Cure is difficult but good results have been reported from placing a ring around the end of the penis, thus causing pain when masturbating. The ring often cures the problem while in place but the stallion will go back to his old habits once it has been removed.

Aggressive Behaviour

Some stallions develop extremely aggressive behaviour which can be very dangerous for both handlers and any mares covered. Occasionally, this behaviour is associated with certain conditions or restraint or is directed towards certain people and can, therefore, be averted by avoiding such situations. However, more often than not, it is expressed generally and is due to mismanagement during his early formative years. If the behaviour is beyond control the stallion can be gelded. In 95% of cases gelding significantly reduces aggressive tendencies. If the stallion must remain entire, due, for example, to financial considerations, then certain measures can and should be used to help protect his handlers and mares. He may be muzzled to prevent him savaging his mares during covering, and he can be controlled by a pole attached to his bit giving his handler more control and preventing the stallion attacking him. Such stallions may be so bad that they are condemned to breeding by artificial insemination (AI) only.

In deciding whether to continue to use an aggressive stallion, it must be certain that his behaviour is management induced and not hereditary, as it is very important that such behaviour is not perpetuated.

Rearing and striking out with the front feet is a relatively common vice, which is potentially very dangerous to handlers. This should be corrected, especially in young stallions where the vice can be cured. To avoid being kicked, the handler should always stand to one side of the stallion, never in front. A long lead rein should be used so that contact can still be maintained from a distance if the stallion rears. As soon as the stallion starts to rear, his lead rein should be jerked and he should be told to get down. Backing the stallion at the first signs of rearing can also help to avert the situation. Use of a chain under the chin is one of the more popular forms of stallion restraint but has been reported to be associated with a higher incidence of rearing and hence should be avoided.

Finally biting is another relatively common vice evident in mares and geldings as well as stallions. Due to the stallion's naturally more aggressive behaviour he tends to bite more. Such a vice should be corrected at a young age by a short sharp jerk on the lead rein or a sharp hit on the muzzle as punishment. If allowed to continue, such a stallion can become almost impossible to handle. Some stallions only bite in certain situations and hence these should be avoided whenever possible. Many will bite when they are eating, after mating, when they are being groomed or when they are fed by hand, and these situations should be reduced to a minimum and any handlers should be made aware of the problem.

Conclusion

Management from a very early age has important implications on the stallion's reproductive ability and behaviour. Many problems encountered in stallions which either do not perform to their full potential or exhibit antisocial behaviour stem from man's management at an early age. One of the major problems encountered with stallions is boredom, which can directly affect his libido and hence performance and also behavioral characteristics. Unfortunately, boredom often becomes a self-perpetuating problem leading to a downward spiral which is very difficult to correct. One of the major rules in stallion management is consistent discipline and ensuring that the stallion has plenty of opportunity to be involved in activities other than covering.

References and Further Reading

Dinger, J.E. and Noiles, E.E. (1986) Effect of controlled exercise on libido in 2 yr old stallions. J.Anim.Sci., 62, 1220.

Dinger, J.E. and Noiles, E.E. (1986) Prediction of daily sperm output in stallions. Theriogenology, 26, 61.

Dougall, N. (1973) Stallions: Their Management and Handling. Devonshire Press, Torquay, Devon.

Equine Research. Breeding Management and Foal Development. Equine Research, P.O. Box 9001, Tyler, Texas 75711.

Hintz, H.F. (1993) Feeding the Stallion. In: Equine Reproduction. Ed. A.O. McKinnon and J.L. Voss. Lea and Feibiger, Philadelphia & London, p. 798.

Johnson, L. and Thompson, D.L. (1983) Age related and seasonal variation in the Sertoli cell population, daily sperm production and serum concentration of follicle stimulating hormone, luteinising hormone and testosterone in stallions. Biol.Reprod, 29, 777.

National Research Council Nutrient Requirements of Horses. 5th Ed. National Academy of Sciences, National Research Council, Washington D.C.

Pickett, B.W. (1993) Sexual Behaviour. In: Equine Reproduction. Ed. A.O. McKinnon and J.L. Voss. Lea and Feibiger, Philadelphia & London, p. 798.

Probett, B.W. and Voss, J.L. (1972) Reproductive management of stallions. Proc.18th.Ann.Conv.Am.Ass.Equine Practnrs. 501.

Rossdale, P.D. and Ricketts, S.W. (1980) Equine Stud Farm Medicine, 2nd Ed. Ballière Tindall, London.

Umphenour, N.W., Sprinkle, T.A, and Murphy, H.Q. (1993) Natural Service. In: Equine Reproduction. Ed. A.O. McKinnon and J.L. Voss. Lea and Feibiger, Philadelphia & London, p. 798.

Varner, D.D., Schumacher, J., Blanchard, T.L. and Johnson, L. (1991) Diseases and Management of Breeding Stallions. American Veterinary Publications, 5782, Thornwood Dr., Goletra, CA 93117.

CHAPTER 19

Infertility

INFERTILITY is a vast area, and this chapter can only aim to be an introduction to the subject and provide a basis from which further information can be sought. Infertility may have its cause in either the stallion or the mare and both of these aspects will be discussed in turn.

Table 19.1 Average fertility rates for various types of horses and ponies. (Day, 1939)

Breed	Average fertility rates
Heavy horses	59%
Light horses	52%
Thoroughbreds	68%
Wild ponies	95%
Feral ponies	23–80%

Expected fertility rates vary enormously and are affected by breed as seen in Table 19.1 As an average within the United Kingdom, 50–80% of mares covered deliver a live foal. However, in considering fertility rates management is of prime importance, as rates can be significantly affected by management prior to covering, at covering and during gestation. Failure to produce a live offspring in any single year can be due to extrinsic or intrinsic factors, and this is the case with both the mare and the stallion. Extrinsic is the term given to external factors affecting reproductive performance whereas intrinsic is the term given to internal, usually physiological, factors. When considering the stallion and the mare these two areas will be dealt with in turn.

Stallion Infertility

On average a stallion might be expected to cover 1–2 mares per day during the breeding season with a rest day every 7–10 days resulting in reasonable success in achieving fertilisation rates in the order of 60%. However, this figure is not only affected by the type and condition of mares presented to a stallion but also the characteristics of the individual stallion. This figure and work load is very much an average, with some stallions being capable of much heavier loads and some struggling with less. It is, therefore, one of the major responsibilities of the stallion manager to be aware of the limitations and expectations of his individual stallion and to work within these (Kenney, 1990).

Research into the causes of infertility in the stallion is as yet rather limited due to previous concentration on mare infertility. However, in any breeding programme 50% of the outcome is determined by the stallion and as such deserves fair discussion. The difficulty in obtaining standard figures for fertility and the reluctance of the majority of stallion owners, especially in the Thoroughbred industry, to select for reproductive performance and to assess for breeding soundness have led to relatively low average fertility rates that have shown little, if any, improvement in recent years. Before discussing the subject further the following glossary should be noted to prevent a confusion in terms:

Sterility   permanent inability to reproduce
Infertility   temporary inability to reproduce
Subfertility   inability, either temporary or permanent, to repro-
                 duce at full potential
Impotency   temporary or permanent inability to ejaculate
                 semen; sperm capable of fertilising an ovum may,
                 however, be produced

Extrinsic Factors Affecting Reproductive Efficiency in the Stallion

Extrinsic factors affecting the reproductive efficiency of a stallion include lack of use, the presentation of subfertile or infertile mares, poor mare management, poor stallion management and the imposition of an artificial breeding season. Each of these will be discussed in turn in the context of reproductive efficiency/infertility. It will be noted that many aspects have already been discussed in previous chapters, especially those concerning management.

Lack of Use
Reproductive efficiency in any animal reflects its use. A stallion may not be used in a certain year by design due to financial or management considerations of the owner. Disease may also preclude

a stallion from use for part or all of a season. The disease may preclude his use due to the risk of direct transfer to mares in the case of venereal or contagious diseases, or it may limit his ability to perform in the case of non-contagious diseases. Alternatively the stallion may have suffered from disease or infection during the previous year and is not to be used in the following season in order to allow full recovery, or because the long-term effects of disease on his reproductive performance deem it inappropriate to use him until he is fully recovered. Diseases of the stallion's reproductive tract will be discussed later under intrinsic factors.

Finally semen evaluation, which is part of good practice and should be carried out regularly at the beginning of each season, may have indicated very poor quality and the stallion may be taken out of use until the cause has been isolated and the problem solved.

Subfertile or Infertile Mare

Both mare and stallion are equally responsible for the production of an offspring. A stallion is only as good as the mare he is expected to cover and vice versa. It is essential that any mare presented to a stallion is capable of reproducing and does not suffer from any of the intrinsic factors affecting reproduction that will be detailed in the later sections on mare infertility. If the mare herself is subfertile or infertile due to infection or physiological abnormality, lack of success cannot be blamed on the stallion.

Poor Mare Management

Mare management is discussed in detail in Chapters 12–16. Inappropriate management in any of the areas considered will adversely affect the mare's ability to conceive and, therefore, the apparent fertility rates of the stallion she was put to. The most important management area as far as stallion reproductive efficiency is concerned is during the time of covering. Service at an inappropriate time because of failure to detect oestrus accurately will obviously be reflected in poor fertility rates. Failure in the detection of oestrus is due usually to either prolonged dioestrus preventing oestrus from being displayed or infrequent or inaccurate teasing, along with lack of records and observation of mares. In such cases veterinary examination by means of rectal palpation has been shown to significantly improve heat detection and, therefore, fertility rates.

Poor Stallion Management

Stallion management has been detailed in Chapters 12, 13 and 18. All aspects of a stallion's management will affect his ability to cover mares successfully. Stallion management as far as its effects on reproductive efficiency are concerned can be subdivided into the following sections.

Excess Work Load   As discussed previously the amount of work or number of mares a stallion may be expected to cover successfully during a season is highly variable, depending upon individuals. It is one of the responsibilities of the stallion manager to know the capabilities of his stallions and to work within these limits. The work load of a stallion depends largely on the ability of his testes to produce sperm. This is a function of testis size that can be assessed by calipers or ultrasonically (Graph 19.1).

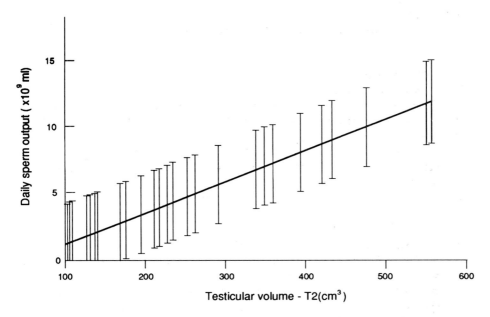

Graph 19.1 *The relationship between predicted daily sperm output and testicular volume for 26 stallions. (Love, Garcia, Riera and Kenney, 1991)*

Stallions with large testes have larger daily sperm outputs and can cope with a heavier work load than those with smaller testes. Testis size is also a function of age, which has an important bearing on fertility, especially at the extremes of youth and old age. It is a good idea, especially in new stallions, to carry out a semen analysis to give a guide to daily sperm production. This may be done in a number of ways, the most accurate being collection daily or every other day over a period of a week with evaluation for semen concentration and volume carried out on the last sample. It has been shown that 90% of the stallion's sperm reserves will have been used up after a week of regular collection, and subsequent daily collections will thus reflect daily production. So many collections are, however, very time consuming and, therefore, samples on 2 or 3 consecutive days may be taken. The first is of little use in determining sperm production but may be used for assessing other sperm characteristics. The following collections will give an indication of the sperm production in the previous 24 hours and will, therefore, indicate daily sperm production, from which the stallion's potential work load can be gauged.

Sperm concentrations are usually in the range of $100–300 \times 10^6$ sperm/ml and for successful fertilisation, assessed by the use of artificial insemination, $300–500 \times 10^6$ sperm are required (Pickett and Voss, 1972). On average 50–60% of the sperm collected can be classified as normal progressively motile sperm capable of fertilising an ovum (Pickett and Voss, 1975; Kenney, 1975). The average daily sperm production for a stallion is $600–2000 \times 10^6$ depending upon season, environment, age, etc (Pickett and Voss, 1972). From these figures it can be calculated that in theory the average number of successful services a stallion could be expected to perform per day is between one and three. There are, however, other considerations to take into account when looking at work loads.

It is interesting to note that total sperm production per week is the same regardless of whether a stallion is used daily or on alternate days. However, use on a daily basis results in a lower concentration of sperm per ml. This may be of no consequence in stallions with high daily sperm production, as concentrations will still be acceptable, but daily use of stallions with lower daily sperm production figures may have a detrimental effect on fertility rates.

Overuse will not only deplete the stallion's sperm reserves but will also result in the ejaculation of immature sperm. As discussed in Chapter 2, sperm have to spend a period of 24–48 hours within the epididymis of the testis in order to mature in readiness for fertilisation. If this period of time within the epididymis is reduced because of overuse, then the sperm ejaculated will have a limited fertilising potential.

Excessive work loads in a stallion may also result in a lack of libido. As a result of this the stallion will be slow to breed or may even fail to ejaculate. (This lack of libido is also encountered in teasers that are not occasionally allowed to cover a mare.) In such cases it is best to take the stallion out of work for a short period of time, and reintroduce him a week or so later. If libido is still low it may well be indicative of further problems.

In an ideal world a stallion should be used on alternate days and given regular periods of 1–2 days' rest every 10 days or so. However, there are many pressures, not least financial, that entice stallion managers to increase their work loads. Because of individual variation some stallions can cope with more than one mare per day or periods of excessive use in a busy season providing they are given adequate periods of rest.

Training Management As discussed in Chapter 18, early training is of utmost importance in the long-term ability of a stallion to perform to his full potential. A stallion brought up in a relaxed, unstressed environment with consistent and fair discipline and respect is much more likely to perform to his full potential in later life.

One of the major problems encountered as a carry-over effect from early life, which is often unappreciated, is that caused by stallion isolation. Many managers isolate stallions to ensure the safety of personnel, other stock on the yard and the stallions themselves. This is a self-perpetuating problem, as such treatment often results in excessive excitability and unpredictability in the stallion, which in turn results in further isolation in the interests of safety. A happy medium between safety and stallion participation in the general yard activities has to be achieved.

Breeding Discomfort Full physical examination of the stallion is essential before purchase to ensure that no abnormalities are present. Details of this are given in Chapter 11. It is also advisable that stallions undergo regular examinations at the beginning of each season to ensure that no problems have arisen that may cause pain at breeding. Pain associated with the act of covering a mare can cause a permanent reduction in stallion libido. Poor foot care can lead to laminitis, causing pain on mounting, especially if the problem is in the hind feet. Muscular or skeletal problems, including arthritis, may also limit the stallion's ability to mount due to pain. Irritation and soreness in the prepubital or sheath area may also cause pain at covering, especially if smegma has been left in this area or soap or antiseptic wash has not been rinsed thoroughly away. Breeding accidents involving inadequate erection at intromission, kicking by a mare and rough handling by handlers will discourage a stallion from covering because of an association with pain. Finally, when using an artificial vagina (AV) care should be taken that the internal temperature is not too hot. Temperatures above 40°–44°C will cause the stallion pain and will reduce his future willingness, not only to use an AV, but also in natural service. Further details on the use of an AV are given in Chapter 20.

Nutrition As discussed in Chapter 18, appropriate nutrition ıroughout the year is essential in order to ensure that the stallion ıs in tiptop physical condition in readiness for the season. Obese or excessively thin stallions suffer from low libido, and nutrition, along with exercise, is a major determinant of body condition. A body condition score of 3 is to be aimed for.

As far as specific deficiencies are concerned only limited research has been carried out. It is known that general severe nutritional deficiencies are associated with a delay in puberty, testicular atrophy and a reduction in sperm production. Severe deficiencies in vitamin A are specifically associated with a reduction in testis weight and spermatogenesis. Deficiencies in selenium have also been correlated with reduced spermatogenesis. In addition, low dietary intake of copper, iron and/or cobalt result in a reduction in appetite with accompanying loss in weight and anaemia.

Obesity will result in a loss of libido, and may also cause a reduction in spermatogenesis. Obesity is associated with excess fat deposition within the scrotum, which increases scrotal insulation and hence may cause an increase in testis temperature with an associated fall in the efficiency of spermatogenesis.

Chemicals   As discussed in Chapter 12, it is essential that prior to use, a stallion that has been on a drug regime must be given time to allow that drug to be cleared from his system.

Anabolic steroids are sometimes used in an attempt to improve male characteristics, for example, weight gain, muscle growth and performance in young horses. They have also been used to improve stallion libido. In man and other animals such use of anabolic steroids is associated with infertility and a similar association has been indicated in stallions, anabolic steroids being reported to result in a decrease of up to 40% in testicular size and weight. Spermatogenesis is also reduced, with fewer sperm/gram of testicular tissue being found, and the sperm produced also have lower motility rates.

Anabolic steroids thus are not recommended for stallions in breeding work. These drugs not only have an immediate effect, but may also have a long-term effect, at least until they are cleared from the stallion's system.

Testosterone therapy has been used to improve libido in stallions and as such is relatively effective. However, it does have serious potential side effects as far as fertility is concerned. Chapter 4 outlines the fine control and delicate hormonal balance controlling male reproductive functions. In summary the control is via the hypothalamus, pituitary and testicular axis. This axis in turn produces GnRH (from the hypothalamus), driving LH and FSH production by the pituitary, controlling testosterone and sperm production by the testis. Testosterone is largely responsible for sperm production and also feeds back negatively on the hypothalamus, hence indirectly controlling its own production and allowing the whole system to equilibrate at a steady rate. If, however, one of these components is added exogenously, for example testosterone, it affects the delicate balance of the system. Testosterone treatment reduces the release of GnRH, which in turn reduces the release of LH and FSH and hence reduces spermatogenesis. Testosterone therapy is, therefore, associated with low fertility due to low sperm counts and hence is not advised for use in stallions in work unless under veterinary supervision.

In addition, the use of any other drug or treatment that causes a drop in appetite, diarrhoea or lack of condition is ill-advised during the breeding season and should only be used under veterinary supervision. Some wormers have also been reported to be associated with a temporary decline in fertility, and thus many breeders arrange their parasite control regimes to ensure that stallions are not treated during the breeding season.

Imposed Breeding Season
Reproductive activity in the stallion, as in the mare, is naturally limited by a breeding season, though with enough encouragement some may cover mares out of season. However, as discussed in Chapter 4, season affects the number of sperm/ejaculate, total sperm number, number of mounts per successful ejaculation and reaction times. As a result, fertilisation potential out of season is significantly reduced. The natural breeding season, with its optimum fertilisation rates and libido, is nature's way of ensuring that foals are born during the spring and early summer so as to maximise their chances of survival.

Unfortunately, the natural breeding season does not coincide with the arbitrary breeding season imposed by man in an attempt to achieve foaling as near as possible to 1 January, which is the official registered birth day of all foals in several breed societies, the Thoroughbred being the most well known. The arbitrary breeding season in the northern hemisphere starts on 15 February as opposed to the natural breeding season that starts in April/May. In the southern hemisphere the imposed season starts 15 August as opposed to the natural season in October/November. Stallions are, therefore, expected to cover mares at a time of the year when their libido and fertilisation rates are naturally low and when they are unable to perform to their full potential. Indeed if stallions are overused during this period the effect is even more detrimental than overuse during the natural season. Occasional use towards the end of the non-breeding season can be quite successful but a full work load can in no way be expected. Exact performance depends on the individual animal but improved fertilisation rates and libido can be obtained by the use of lights and heaters in the stallion's stable to advance the reproductive system by mimicking the early onset of spring.

Intrinsic Factors affecting Reproductive Performance in the Stallion

Intrinsic factors affecting reproductive performance in the stallion include age, along with chromosomal, hormonal, testis, vas deferens, accessory gland, penis and semen abnormalities. These will be discussed in turn in the context of reproductive performance.

Age

Age is an important aspect in considering the potential fertility of a stallion. Young and old stallions may have problems with taking on a full work load with consistent success rates. A young stallion is not capable of producing enough sperm to perform consistently throughout the season. In addition, he is still learning the job and as such can easily be adversely affected by his handlers and/or management. He may, therefore, be slow to breed, mounting several times per successful ejaculation or even failing to ejaculate, ejaculating prematurely or exhibiting enlargement of the glans penis before intromission. Careful treatment and handling during this period is essential to ensure that any such behavioural problems are not perpetuated.

As far as physical capabilities are concerned, 3-year-old stallions, which have reached puberty, have been shown to be perfectly capable of fertilising a mare, but they have a limited sperm production capacity. By 4 years of age they are capable of producing adequate numbers of sperm to mate as many mares as an adult stallion but full fertilising capacity is not attained until 5 years of age on average (Johnson, Varner and Thompson, 1991).

At the other end of the spectrum old age may be a problem. Age related decline in fertility is associated normally with general age related problems rather than reproductive capacity per se. Arthritis causing pain on mounting is a major cause of lack of libido and, therefore, low fertilisation rates in older stallions. If such problems are encountered they may be alleviated, to a certain extent, by the use of breeding platforms or an artificial vagina, as well as allowing the stallion extra time to do his job. At over 20 years of age evidence suggests that degeneration of testicular tissue results in lower sperm counts and, therefore, a reduction in the stallion's ability to carry out a full season of work. However, the effect of age is very different in individual stallions, and older stallions should not be precluded from use, as such animals have had many years in which to prove their worth as far as their own performance and that of their progeny are concerned. Older stallions often tend to be more gentlemanly to handle, know their job well and good to use on maiden, shy or nervous mares, giving them confidence. If using an older stallion his semen should be evaluated regularly and monitored closely to allow a reduction in his work load if a drop in any of the semen evaluation parameters is detected.

Chromosomal Abnormalities
Chromosomal abnormalities may be the cause of infertility in stallions that otherwise seem fit. These may be associated with semen abnormalities or more obvious abnormalities of the genitalia. Hermaphrodites and genetic mosaics have been reported in stallions but are relatively rare. Other genetic abnormalities are associated with cryptorchidism (rig) and testicular hypoplasia (Bishop, David and Messervy, 1966). Cryptorchid stallions may still be fertile but only capable of reduced work loads. They are also reported to show increased sexual activity due to elevated testosterone production within the retained testis. Before purchase of a stallion, correct descent of both testes should be checked (Kenney, Kent, Garcia and Hurtgen, 1991). The majority of breed societies refuse to register rigs as stallions.

Other specific genetic deformities are umbilical and inguinal hernias. These may correct themselves naturally but the trait may well be perpetuated in succeeding generations. Other genetic factors causing abnormalities of the reproductive system will be discussed under the specific areas of the tract detailed below.

Hormonal Abnormalities
As mentioned previously in this chapter and in Chapter 4, endocrine control of the reproductive system is finely balanced. Circulating concentrations of testosterone have a direct effect on reproductive performance, on both libido and sperm production, and low testosterone levels are often blamed for poor fertility rates. Attempts to improve fertility by exogenous testosterone have, however, proved largely ineffective and, as discussed in the previous section, may lead to detrimental side effects. Treatment of a stallion showing low circulating testosterone levels with LH has been reasonably successful but research in this area is limited. Stimulation with GnRH has been tried with only limited success in attempts to increase testosterone concentrations, but such work as yet is still in its infancy (Blue, Pickett, Squires, McKinnon, Nett, Amann and Shiner, 1991; Roger and Hughes, 1991; Boyle,

Skidmore, Zhang and Cox, 1991).

Abnormal hormone concentrations are associated not only with low libido, but more seriously with feminisation of the external genitalia and mild hypothyroidism. Feminisation of the external genitalia is normally the result of a failure of the male embryo to react to testosterone in utero, resulting in the development of female rather than male genitalia. The offspring appears from the external genitalia to be a female, but internally has testes and is sterile.

Hypothyroidism, a rare inherited condition, results in the lack of thyroid hormone. A deficiency of thyroxine causes obesity due to excess fat deposition, muscle disorders, a thin coat as well as delays in puberty, decreased testis size, reduced sperm production and low libido. In severe cases complete reproductive breakdown can be seen. There is a suggestion that hypothyroidism is associated with high environmental temperatures and hence low thyroid production may be associated with summer infertility.

Testis Abnormalities

Testis abnormalities, as with most anatomical abnormalities, are caused either by disease or are inherited traits. It is reported that one in five stallions has an anatomical abnormality, the significance of which varies from life-threatening to a minor flaw that may be of little consequence as far as reproductive performance is concerned but may still reduce his market value.

Cryptorchidism

During gestation the male testes lie within the abdomen below the kidneys. At some time from just prior to birth until the foal is 18 months of age, the testes descend through the inguinal ring into the scrotum. The inguinal ring then closes to prevent the testes returning into the body cavity. The testes thus lie within the scrotum outside the body cavity to ensure optimum temperature for sperm production (Figure 19.1).

Occasionally only one testis descends and the other is retained within the body cavity; this one is, therefore, kept at body temperature rather than at the lower temperature usual within the scrotum. A stallion that shows a failure in the descent of one testis is termed a unilateral cryptorchid or a rig; a stallion in which both fail to descend is a bilateral cryptorchid. A testis that remains in the body cavity, not having descended at all, is termed abdominal cryptorchidism (Figure 19.2). Alternatively a testis may partly descend and be situated just above the inguinal ring, in which case it is termed inguinal cryptorchidism (Figure 19.3). Hormonal treatments have been used to correct the failure of testes to

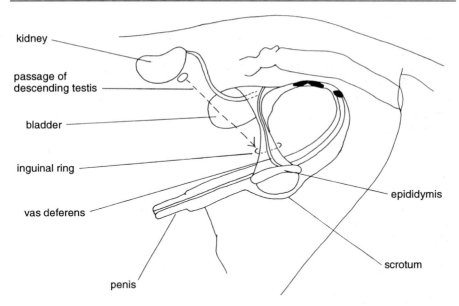

Figure 19.1   The normal passage of descent of the testis in the stallion.

descend. Success has, however, been limited, especially in cases that have been left until after puberty and in abdominal cryptorchidism. If left untreated, there is always the chance that any testes that have failed to descend may do so of their own accord up until the colt is 3 years of age.

When considering unilateral cryptorchidism it must be remembered that even though one of the testes remains within the body cavity, the stallion may well be still fertile. Leidig cells of the testis are not affected by the elevated temperature within the body cavity to the extent that Sertoli cells are. Therefore, testosterone production still occurs and sex drive and libido in rigs is often as high as that seen in stallions, but the ability to produce sperm is impaired. The longer the testes remain within the body cavity, the greater the degeneration of the seminiferous tubules which has a long-term effect. Hence, stallions successfully treated in later life for undescended testes may well be permanently subfertile.

Although the use of a unilateral cryptorchid as a stallion is possible, it is not advisable, as the condition is heritable. It is normally advised that stallions without the full descent of the testes by 12 months of age should be castrated in an attempt to eliminate the trait from the population. Most breed societies will refuse to register unilateral cryptorchids. The condition is also associated with higher incidences of cancer and testicular torsion.

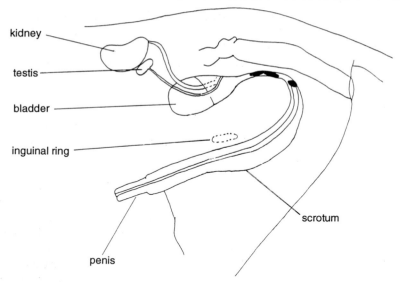

Figure 19.2   An abdominal cryptorchid stallion is characterised by the testis lying up within the body cavity. In a unilateral abdominal cryptorchid one testis has failed to descend; in a bilateral both remain in the body cavity.

Figure 19.3   An inguinal cryptorchid stallion is characterised by the testis having only partly descended and remaining associated with the inguinal ring. Again, failure of testis descent may be seen in both (bilateral) or only one (unilateral).

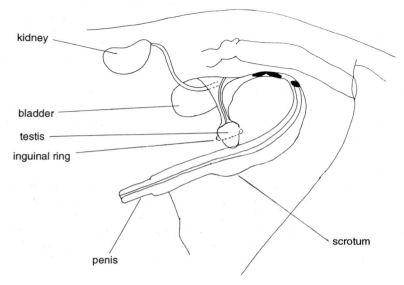

Incomplete descent of the testes in which they lie nearer the body than normal at the inguinal ring is also seen. In such cases the efficiency of the thermoregulatory system within the testis is reduced and as a result such stallions often have lower fertilisation rates.

Hernias

Scrotal hernias are caused by parts of the intestine and associated structures passing through the inner and outer inguinal rings along with the testes. This condition is seen normally in young foals, but may also be evident in older stallions. These stallions show larger inguinal rings that have not closed tightly enough after testes descent. This allows the intestine to pass through, especially as a result of trauma, though there is evidence that it may also be an inherited trait. Scrotal hernias in newborn foals may well correct themselves as the inner inguinal ring closes later than normal. Such hernias are truly called inguinal hernias as the intestine is not evident externally because it has only passed through the inner inguinal ring (Figure 19.4).

Scrotal hernias in older stallions may be evident as external swellings within the scrotum or just as a change in the gait. They may be associated with extreme pain in which case they must be treated immediately (Figure 19.5). One of the worst complications is torsion of the intestine severely restricting its blood flow and causing necrosis and death of that area of the intestine with potentially fatal consequences. Treatment may be by rectal palpation, especially in stallions under three years of age, by which means the intestine can be pulled out of the hernia allowing the inguinal ring to close. However, in more serious cases surgery may well be required, often accompanied by the removal of the testes, as natural full closure of the inguinal ring is not likely, and so the chances of repercussions are high unless the ring is closed artificially.

The intestine within the hernia causes an increase in temperature within the scrotum, and a further complication is an elevation in testicular temperature and an associated reduction in spermatogenesis.

Hypoplasia

Testicular hypoplasia is characterised by inadequate development of primitive sperm cells or spermatogonia within the testis of the foetus. The extent of hypoplasia does vary between individual sufferers and may be due to spermatogonia degeneration in later life. In the stallion, hypoplasia is evident as seemingly shrunk testes within the scrotal sack. This is due to small epididymides as

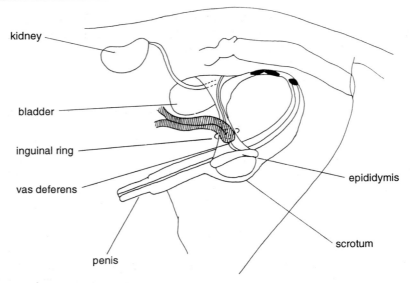

Figure 19.4   An inguinal hernia in the stallion in which a loop of the intestine folds through the inguinal ring.

Figure 19.5   A scrotal hernia in the stallion is a more extreme case of inguinal hernia, in which the loop of intestine has entered the scrotum and there is significant danger of complete ligation of the intestine, necrosis and death.

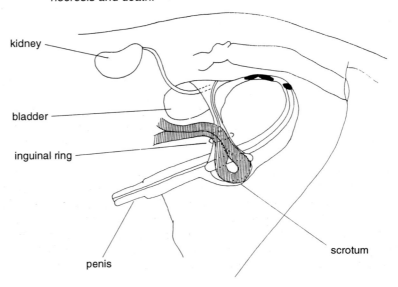

there are few if any sperm maturing within them. The sperm count, therefore, is also very low. The condition is rare, and sufferers seem to be predisposed to testicular tumours.

Incorrect Testicular Number
The complete absence of testes is rare. The condition is easily confused with cryptorchidism when the testes are exceedingly small and may be further degenerated by their position high up within the abdominal cavity. Additional testes are also uncommon, but if evident may not necessarily have a detrimental effect on reproductive performance. Possession of both male and female genitalia is again uncommon, but is evident in individuals carrying excess X or XY chromosomes. These animals often have under-developed testes and ovaries which do not function correctly. However, they may show significant stallion behavioural tendencies and as such may be used as a teaser to detect mares in oestrus. The external genitalia are also often an amalgamation of male and female with the presence of an overdeveloped clitoris in the form of a rudimentary penis along with underdeveloped testes.

Defects in testicular seminiferous tubules and epididymides are characterised by subfertility. Such stallions may be accepted and used but the condition should be identified as it can be an inherited trait and its perpetuation should be prevented.

Testicular Torsion
Testicular torsion may also be evident with results varying from little pain and few detrimental effects on sperm production to acute colic-like pain associated with obstruction of the testicular blood supply. Occasionally the torsion may only be transient and correct itself.

Testicular Disease
Before we consider testicular disease in particular it must be noted that any systemic infection or disease that results in an elevation of body temperature will cause periods of infertility due to the detrimental effects of increased temperature on sperm production. The extent of the infertility depends largely on the severity of the infection or disease and may vary from a temporary subfertility to infertility for a considerable period of time. The return of fertility rates to normal will be delayed after recovery from disease as it will take some time for the seminiferous tubules and associated Sertoli cells to recover and produce sperm for maturation within the epididymides. Such stallions should not be used immediately after recovery, as time should be allowed for the resumption of sper-matogenesis and maturation of adequate sperm numbers within

the epididymides. Systemic diseases or infection that may cause such testicular change include equine infectious anaemia (EIA), a viral infection causing weight loss, oedema (fluid collection) and anaemia, as well as pneumonia, *Actinomyces bovis* infection and peritonitis, a swelling of the peritoneal membrane lining the abdominal cavity. Infection by equine viral arteritis (EVA) virus and strongyle larvae can cause inflammation of the testicular artery and hence adversely affect the efficiency of the testicular heat exchange mechanism within the pampiniform plexus.

Prolonged systemic infection can result in the stallion being recumbent for prolonged periods of time and as such the lower testis, which is in closer contact with the body, will suffer more from the effects of elevated testicular temperature. In the majority of cases, disease of the testes is associated with orchitis, inflammation of the testicular tissue. Such inflammation may be irreversible, causing permanent physical damage to Sertoli cells and, therefore, permanent subfertility or infertility due to the elevated testicular temperatures associated with the inflammatory reactions.

Signs of orchitis are swollen and firm testes that feel warm to the touch and are extremely sensitive. Degeneration as a result of orchitis is characterised by soft and flabby testes with a poor sperm count. If inflammation, and hence degeneration, persist, then the testes may be felt as small hard structures within the scrotum. Infections that may cause orchitis include haemolytic streptococci, *Klebsiella pneumoniae*, *Pseudomonas aeruginosa*, *Streptococcus zooepidemicus*, *Streptococcus equi*, *Brucella abortus*, *Pfeiferella mallei*, *Salmonella abortus equi* and herpesviruses I and II. In these conditions the testes become swollen and painful and may be associated with epididymitis. In such cases the stallion usually refuses to cover any mares, and if he does, semen evaluation will reveal high leucocyte counts, in addition to poor sperm motility and morphology scores. If chronic orchitis develops, the prognosis for long-term recovery is poor, though acute conditions may be cured using systemic antibiotics. Cooling of the testes by hosing or cold compress may reduce the incidence of permanent testicular damage, especially that associated with elevated temperature. Most infections affect both testes and result in bilateral degeneration except as discussed earlier in cases of prolonged recumbency.

Testicular Tumours
Though rare, tumours of the testes and associated structures are not unknown. Those reported include testicular and epididymal tumours, dermoid cysts, especially evident in cryptorchids, teratomas, Sertoli cell and interstitial cell tumours and seminomas (malignant tumours of spermatogonia).

Tumours of the anterior pituitary may also occur causing deficient stimulation of the testes, via LH and FSH, resulting in low sperm and testosterone production. Finally melanomas of the scrotal surface may also be seen.

Vas Deferens Abnormalities
Inflammation of the vas deferens is often associated with inflammation of the epididymis and is often the cause of epididymitis. Infection of the vas deferens is characterised by swelling of the duct, possibly associated with fibrous growth and hence the chance of obstruction. Such inflammation is often associated with enlargement of the ampulla accessory gland detected by rectal palpation. The result of such infection is usually pain along with sperm of poor quality and low longevity.

Accessory Gland Abnormalities
As discussed in Chapter 2, the stallion has four accessory glands: the ampulla, the seminal vesicles, the prostate and the bulbourethral glands, all situated along the vas deferens or urethra and responsible for the secretion of seminal plasma. As mentioned above, swelling of the ampulla often accompanies inflammation of the vas deferens. The seminal vesicles are potentially susceptible to the infection termed seminal vesiculitis. Infection by *Corynebacterium pyrogenes* and *Brucella abortus* is the most common cause of infection to these glands. Infection is characterised by increased leucocyte concentrations within semen, especially in semen collected after rectal palpation. Rectal palpation will also reveal that the seminal vesicles are swollen and painful.

Penis and Prepuce Abnormalities
Abnormalities of the penis or prepuce are normally associated with trauma or injury to this area. The penis, especially when erect, is very vulnerable to traumatic injury from kicking by an unreceptive mare. This causes vascular rupture and/or haemorrhage making the return of the penis to within the prepuce very difficult and painful. Immediate application of a cold compress or cold hosing may alleviate some of the swelling and long-term effects. Haemorrhage of penile blood vessels may also occur if a stallion covers a mare with a Caslick's prior to episiotomy, or if the mare suddenly lunges to one side while being covered. The long-term effect of such trauma will not only depend on the physical recovery of the penis but also on the stallion's psychological recovery. Such trauma can make a stallion very reluctant to cover a mare again, especially if he is young and only just beginning to learn the job. Damage to the urethra within the penis is evident as blood contamination of any

semen produced, termed haemospermia. Haemospermia may also be caused by bacterial or viral infections of the urethra or by penis damage from sutured mares. Contamination of semen with blood not only indicates the risk of possible infection transfer but is also associated with low fertility rates. Infective agents causing haemospermia include *Habronema* larvae (summer sores), *Strongylus edentatus* larvae, *Trypanosama equiperdum* (dourine), herpesvirus (coital exanthema or genital horse pox) and herpesvirus I (penile paralysis).

Tumours of the penis are not often malignant, but lesions due to melanomas, sarcoids, carcinomas and herpesvirus often burst and haemorrhage at covering and cause the stallion considerable pain. Blockage of the urethra or vas deferens has been reported, characterised by a normal libido but aspermatic semen (containing no sperm), even though testicular function is normal.

Not only may the penis itself be infected, but it is also the major means by which venereal infection can be passed from the stallion to the mare and vice versa. The penis has a naturally balanced microflora that causes no problem to him or any mares that he covers. However, if this balance is disturbed due to systemic infection or disease, general ill health or impaired normal disease resistance, serious consequences can result. Similarly, if he comes into contact with a contaminated mare his natural microflora balance may be breached and allow the invasion of foreign infective agents. The prepuce area protecting the penis then provides an ideal environment in which such organisms can multiply. Contamination of the stallion's penis as mentioned may be due to contact with infected mares or alternatively poor hygiene especially at covering. Therefore, regular swabbing of stallions and all mares to be covered, especially for contagious equine metritis, and the washing of the stallion and mare during the preparation for covering go a long way to preventing venereal disease transfer.

The stallion often does not show any clinical signs of contamination, but infection is traced back through symptoms shown by mares he has covered. Symptoms of infections within the mare's tract are discussed in the following sections dealing specifically with mare infertility. Isolation and treatment of the stallion are the only course of action when such infections are suspected, as transfer from mare to mare via a stallion in a busy season is very easy and can have disastrous consequences on fertility rates.

A number of venereal diseases are transferred via the stallion's penis. Coital exanthema may be characterised by lesions, though the stallion can be infected even if no clinical symptoms are evident. Dourine, thankfully not evident in the United Kingdom, but of importance in the warmer climates of Asia, Africa, South

America and Southeast Europe, is also a sexually transmitted disease. It is now a notifiable disease in the United Kingdom, and horses require a negative dourine certificate before being imported into the country.

Truly sexually transmissible bacteria, that is those transferred into the uterus of the mare at covering, include *Streptococcus faecalis, Staphlococcus albus, Sreptococcus zooepidemicus, Klebsiella aeruginosa, E.coli, Staphlococcus aureus, Pseudomonas aeruginosa* and *Haemophilus equigenitalis*. The presence of these infections in the mare causes endometritis, inflammation and infection of the uterine endometrium, which will be discussed in some detail in the following section on mare infertility.

Complete prevention of venereal disease is difficult but is aided by full hygienic precautions prior to covering, though complete disinfection of the stallion's penis is impossible. Regular swabbing of the penile urethra, urethra fossa and prepupuce will ensure that any bacteria present are identified so that appropriate treatment can be given. Treatment itself, however, can cause problems as systemic antibiotics may affect the natural microfloral balance which will take time to restore. Topical application to such a sensitive area may also cause dryness and cracking, causing pain at covering.

Semen Abnormalities
Semen abnormalities will be discussed in full in Chapter 20, along with the evaluation of semen. In summary most infections and trauma of the male reproductive tract result in adverse effects on sperm production and, hence, fertility. The normal parameters expected of a semen sample are given in Table 19.2.

Table 19.2 The normal parameters of a semen sample.

Volume	30–300 ml
Concentration	30–600 million/ml
Morphology	minimum 65% physiologically normal
Live : dead ratio	6.5 : 3.5
Motility	minimum 40% progressively motile
Longevity at room temperature	45% alive after 3 hours
	10% alive after 8 hours
pH	6.9–7.8
White blood cells	< 1500/ml
Red blood cells	< 500/ml

Infection and/or abnormalities of the reproductive tract may affect any of these parameters but usually cause a reduction in sperm concentration, inadequate motility and poor longevity. Infection, rather than trauma, is often characterised by high leucocyte counts. If a semen sample does not meet the above parameters, the cause of the problem should be identified before the stallion is used for his own protection and that of any mare.

Mare Infertility

On average 50–80% of services result in the birth of a live foal. This figure, however, varies with the population of mares (Osbourne, 1975; Sullivan, Turner, Self, Gutteridge and Bartell, 1975). Age has a detrimental effect on fertility, rates of 75% being given for 4-year-olds compared to 50% for mares over 20 years old (Day, 1939). Man's interference in the reproduction of the horse also has had a significant effect on fertility rates. Fertility rates for wild horses and ponies are in the region of 95% compared with 60% for hand bred mares. Embryo mortality is held responsible for a significant amount of the apparent infertility. Rates of 5–14% are reported for in hand mating compared to 1–2.5% for free mating. Most embryo mortality occurs prior to Day 40 (Ginther, 1985); therefore most measurements of infertility will also include this early embryo mortality.

Fertility rates have suffered considerably from man's interference, not only directly in the mating process, but in man's selection of breeding stock. The horse today, especially within the Thoroughbred industry, is selected for performance with little emphasis on reproductive competence, and the two often, unfortunately, do not go together. Reproductive performance has largely been ignored in the improvement of the equid. This is in contrast to other farm livestock, where reproductive efficiency is of utmost importance. It is evident, therefore, that barrenness in a mare at the end of the season is not necessarily due to pathological infertility but is also a consequence of other environmental and managerial influences and the natural tendency in many mares to take a season off breeding. Some mares are consistently barren in early life but successfully breed later on in life. Failure to produce an offspring in a particular year is, therefore, a complicated problem and is not to be confused with infertility and embryo mortality. Before barrenness is discussed in any detail a few definitions should be given.

Fertile   able to produce a live foal
Infertility   temporary inability to reproduce
Barrenness   lack of a pregnancy at the end of the season, but

perfectly capable of producing a foal as demonstrated in previous years

Subfertility   inability, either temporary or permanent, to reproduce at full potential

Sterility   permanent inability to reproduce

The causes of barrenness in a mare at the end of a season are numerous and, as in the case of the stallion, can be divided into extrinsic and intrinsic factors. These will be discussed in turn.

Extrinsic Factors Affecting Reproductive Performance in the Mare

Extrinsic factors affecting fertility in the mare are those external factors that affect her reproduction and may be divided into lack of use, subfertile or infertile stallion, poor stallion management, poor mare management and artificially imposed breeding season.

Lack of Use

A mare may not be covered in a particular season due to design or due to unavoidable circumstances. If that year's foal was born late she may not be covered that season by design in order to allow her to return to foaling earlier in following years. Disease may also preclude a mare from use in a particular year. Such infections include *Haemophilus equigenitalis, Pseudomonas aeruginosa, Klebsiella aerogenes*, equine coital exanthema, equine herpesvirus, neurological diseases, equine flu, strangles, *Salmonella*, etc. All of these will preclude a mare from being covered in a particular season, either because she herself is not in a fit condition to carry a foal successfully and/or there is a danger of disease transfer to the stallion either directly in the case of venereal diseases or to other stock on the stud from general infectious diseases. The diseases listed above will be discussed in further detail in relation to the specific areas of the mare's reproductive tract that they infect.

Finally in the case of a performance horse she may not be bred in a particular year due to work commitments, though with the advances in embryo transfer such mares may in fact have a foal via a recipient mare, hence avoiding being pregnant herself.

Subfertile or Infertile Stallion

Half the responsibility for the success or failure of a covering lies with the stallion and half with the mare. If a mare, therefore, is covered by an infertile or subfertile stallion, her chances of producing a foal are significantly reduced through no fault of her own. The causes of infertility in the stallion have already been discussed in the previous section. It is good practice as a safeguard to evaluate a stallion's semen at the beginning of each season and at the first indication of fertility problems. He can then be taken out of work and treated accordingly if a problem is identified. Mares can only be expected to perform to their full reproductive potential if they are mated to a stallion whose semen meets the minimum parameters. In addition, a stallion must be physically capable of covering a mare effectively as a good semen evaluation in the absence of the ability or willingness to mate is of no use in the natural service of a mare. To a certain extent this problem may be bypassed by the use of artificial insemination, but in such cases it must be certain that the stallion's lack of libido or ability is not due to a potentially heritable fault.

Poor Stallion Management
Stallion management has been detailed in Chapters 12, 13 and 19. If any aspect of a stallion's management is not correct then there is the potential for his fertility to be affected, especially in his management associated with the act of covering a mare. It is not only the management of the stallion immediately around the time of covering or during that particular season, but also his management in the earlier formative years of his life that will affect his willingness and ability to cover mares and, hence, affect apparent mare fertility rates.

The mere fact that man is today so closely involved in the vast majority of equine coverings has a potential detrimental effect on fertility rates, especially if those involved are not experienced. In the natural system, and in those systems where the stallion is allowed to run out free with his mares, he will cover his mares at least 4–5 times per oestrous period. Man's interference now dictates a maximum, in most cases, of two coverings per oestrus, at a time largely dictated by man and not the horse. Detection of oestrus which is needed in such systems is often inaccurate, resulting in the mare not being covered at the most opportune time. Incorrect detection of successful ejaculation is a common cause of apparent infertility. In some stallions it is very easy to detect by the flagging of their tails; in others it is less obvious and diagnosis has to be by means of feeling for the ejaculation pulsations through the urethra along the underside of the penis. However, the actual act of feeling for ejaculation through the urethra can put the stallion off his job and lead to ejaculation failure. Man's additional interference in trying to push the breeding season forward in line with the Thoroughbred industry's registration date of 1 January for all foals causes even more problems with apparent infertility. Even with the use of artificial lights and hormone treatments a stallion's performance during the months of January and February is lower than during his true breeding season. Fertility rates for mares covered within this period cannot be expected to be up to normal expectations.

Behavioural abnormalities can arise from inappropriate stallion management, especially during his early formative years, and may lead to such problems as failure to obtain or maintain an erection, incomplete intromission and ejaculation failure (Pickett and Voss, 1975; Kenney, 1975). These problems are especially evident in ex-performance horses, whose management for a significant period of their life has been geared towards repressing their sexual urges and, therefore, they are often inhibited when turned out to stud. Once established, these habits are hard to correct and make a considerable contribution to the apparently high infertility rates in mares through no fault of their own.

Poor Mare Management
Mare management has been discussed in detail in Chapter 12, 'Preparation of the Mare and Stallion for Covering'; Chapter 13, 'Mating Management'; and Chapter 14, 'Management of the Pregnant Mare'. Any deficiences or inadequacies in mare management during any of these periods can lead to poor fertility rates. Of special significance is mare management at mating, especially inaccurate oestrous detection. This can be largely alleviated by better, more experienced management and the use of veterinary diagnosis. When such measures are introduced mare fertility rates are seen to improve significantly.

Imposed Breeding Season
As mentioned in the discussion on stallion management there is considerable pressure these days for foals to be born as soon as possible after 1 January. The methods by which the mare's breeding season may be advanced in order to achieve this are discussed in Chapter 12. Whatever the treatment used, conception rates out of the natural breeding season are never as high as normal expectations within the season and hence the imposition of such an arbitrary breeding season puts unfair constraints on a mare's potential fertility rate.

Intrinsic Factors Affecting Reproductive Performance in the Mare

Intrinsic factors affecting reproduction in the mare are internal faults or problems including age and chromosomal, hormonal, pituitary, ovarian, fallopian tube, uterine, cervical, vaginal and vulval abnormalities. All these will be discussed in turn; many, however, are closely interrelated and causal.

Age

Age does have a bearing on fertility. Fertility rates in young mares less than 8 years of age tend to be lower than in mares between 8 and 15 years of age. Day (1939) reported fertilisation rates of 75% for 4-year-old mares compared to 50% for mares 20 years old or over. After 15 years of age fertility rates tend to tail off though, providing a mare is in good physical condition, she may breed successfully well into her twenties. It should, however, be questioned whether breeding at such an advanced age is good for the mare. There is now an alternative to putting an older mare through the stresses of carrying a pregnancy to term, and that is the use of embryo transfer. Success rates are similar to those using ova from younger mares, though there is some evidence that ova from older mares may be less viable.

Chromosomal Abnormalities

The normal chromosomal complement for the equid is 64 (32 pairs), the female complement being denoted as 64XX. Various variations reported on the normal complement include 63XO (a female with a single X chromosome). The condition is Turners syndrome, and it is one of the more common chromosomal abnormalities. Such mares are characterised by small rudimentary ovaries, flaccid uterus, no ovarian activity and, therefore, permanent anoestrus. They also tend to be short in stature. Some mares show a mosaic chromosomal configuration, being 64XX in some cells, 63XO in others. Such mares show erratic oestrous cycles with no ovulation (Chandley, Fletcher, Rossdale, Peace, Ricketts, McEnery, Thorne, Short and Allen, 1975). Very rarely a chromosomal complement of 64XX/65XXY is found and such horses are termed intersex. They may show penis-like clitorises and degenerated, underdeveloped testes. Though such animals may show oestrous cycles because of the presence of ovaries, they show no ovulation.

Positive diagnosis of chromosomal abnormalities is only possible by genetic mapping, though they may be indirectly indicated by inactive ovaries, abnormal external genitalia, underdeveloped uterine endometrium and lack of oestrous behaviour (Ricketts, 1975, 1978; Halnan, 1985).

Hormonal Abnormalities

As discussed in Chapter 3, the control of the mare's reproduction is a finely balanced mechanism controlled by the hypothalamic-pituitary-ovarian axis. Abnormalities/inefficiencies of any of these centres can cause imbalances throughout the whole axis. The majority of hormonal deficiencies are associated with pituitary abnormalities. Complete failure, or neoplasia, of the pituitary is relatively rare in the horse, but temporary malfunction may occur, especially in association with the transitional period at the beginning or end of the breeding season. During this transitional period mares tend to suffer from delayed oestrus, prolonged oestrus and dioestrus, silent ovulations, split oestrus (oestrus over a period of up to 3 weeks with possibly a quiescent period in the middle), etc. It seems evident that a period of time is required before the mare's system settles down to the regular 20–22 day cycle seen throughout the remainder of the season until the transition period into anoestrus.

Diagnosis of such conditions is firstly via a mare's behaviour and the seeming inability to detect oestrus or conversely apparent continual oestrus in such mares. Diagnosis of the cause is helped via rectal palpation by which ovarian activity can be assessed. The incidence of these hormonal deficiencies or abnormalities is particularly evident today, as man continually attempts to breed mares earlier and earlier in the season. The use of exogenous hormonal treatments and/or light treatment does successfully bring forward oestrous behaviour, but does not eliminate the transition period which may still be associated with the problems detailed above.

Pituitary or hypothalamic tumours are rare in mares, but if they do occur they are associated with muscle wasting, hypoglycaemia, docility, alopecia, blindness and uncoordinated movement in addition to prolonged anoestrus.

Ovarian Abnormalities
Occasionally ovaries may be absent because of surgical intervention by a mare's previous owner, or alternatively due to chromosomal abnormalities. Inactive ovaries and ovulation failure are seen in mares and are exacerbated by the imposition of an arbitrary breeding season (Kenney, Coon, Ganjam, and Channing, 1979). Follicular atresia is responsible for some incidences of ovulation failure. In such cases some follicles develop normally until the 3 cm stage, but there is a failure of any follicle to become dominant and develop beyond this critical 3 cm stage. Conditions such as ovarian hypoplasia, granulosa cell tumours, ovarian cysts, uterine infections and malnutrition have all been implicated in follicular atresia (Bosu, 1982; Pugh, 1985). The best cure, however, seems to be time, especially in mares encountering problems during the transitional stage of the breeding season. Often succeeding cycles will not show such problems.

Corpora lutea (CL) persistence and conversely failure are also causes of infertility in the mare, manifesting themselves as long or short oestrous cycles respectively. Failure of the CL is less evident than problems with persistent CL. However, failure of the CL is implicated in experiments using progesterone supplementation with some success to prevent abortion (Ganjam, Kenney and Flickinger, 1975). The presence of a persistent CL is more common in mares and is an important cause of anoestrus. The normal lifespan of a CL is 14 days after which, in the absence of a pregnancy, PGF2α is secreted by the uterine endometrium. A persistent CL is presumably, therefore, a failure in the release of PGF2α or a failure of the CL to react. The presence of such conditions is implicated by the lack of oestrous behaviour and is confirmed by rectal palpation. The failure of the luteolytic message may be linked to uterine infection rendering the uterine endometrium unable to produce PGF2α. Embryo mortality is another potential cause of a persistent CL. Treatment with PGF2α injection is normally successful.

Ovarian cysts can also be a cause of anoestrus. Most cysts are associated with follicles though their presence must not be confused with large, perfectly normal follicles that have been reported to reach up to 20 cm in diameter. Ova fossa cysts are also reported, especially in older mares. They seem to be associated with the epithelium of the fimbrae and cause blockage of the ova fossa and, therefore, fertilisation failure. They are often evident as a bundle of cysts similar to a bunch of grapes near the ova fossa. In the extreme they may also interfere with the blood supply to the rectum (Prickett, 1966; Stabenfeldt, 1979).

Granulosa cell cysts are the most common tumour within the equine ovary and an important cause of anoestrus. They normally affect mares between the ages of 5 and 7 and are usually associated with a single ovary. The cysts are usually large and fluid filled and are reported to cause ovaries to weigh up to 8 kg (Norris, Taylor and Garner, 1968). The symptoms shown by such mares depends on the hormones secreted by the tumours: oestrogen, testosterone or progesterone. Oestrogen production is the most common and may result in nymphomaniac behaviour (prolonged oestrus) in the absence of ovulation. Testosterone producing cysts result in stallion-like behaviour and progesterone producing cysts in continual anoestrus. Removal of the ovary responsible allows the resumption of normal reproductive activity by the remaining ovary (Meager, 1978).

Ovarian teratomas containing hair and teeth have been reported but very rarely occur (Rossdale and Ricketts, 1980) and, unlike teratomas seen in stallions, prove to be benign. Parovarian or paroophoron cysts are also common, but rarely cause problems. They develop from regressed woolfian ducts. Another cause of anoestrus is hypoplasia of the ovary, in which mares show immature ovaries as far as size is concerned even at mature body size. No ovarian activity and, therefore, increase in ovarian size is detected during the breeding season.

As far as infection or disease is concerned, the ovary is essentially unaffected and many ovarian abnormalities and, hence, ovarian infertility occur as a result of deformities, tumours, cysts, etc, not pathogenic agents.

Multiple ovulations should be mentioned, as they are largely undesirable as far as equine reproduction is concerned. The mare's reproductive tract is not really capable of maintaining more than one foetus satisfactorily. Twinning causes crowding, often leading to the failure of one foetus to survive due to inadequate nutritional intake from a placenta of limited size. Such restriction results in either spontaneous abortion of one or both foetuses or the mummification of the smaller twin, which is passed out at the parturition of the surviving twin. Due to the fact that the placenta is a finite size as far as the surface area of the uterus is concerned, by Day 120 the death of the smaller foal does not allow the placenta of the remaining foal to take over the uterine surface vacated by the dead foetus. Thus the surviving foal will be born smaller than expected for a single birth and will often seem illthrifty. The affects of such an adverse uterine environment may be long term, the foal never being able to fulfil its genetic potential. If twins do survive to term, they too will be of a small birth weight and illthrifty (Figure 19.6) (McDowell, 1985).

Twins are, therefore, not desirable and management in general is geared towards avoiding them. Their incidence can be linked to infertility in the mare, as, if twins are detected by rectal palpation in early pregnancy, they are often aborted by the injection of PGF2α and the mare is in essence infertile to that covering. The exact incidence of twinning is unknown, but it is suggested that up to 25% of oestrous cycles are characterised by more than one follicle developing beyond the 3 cm stage and are, therefore, capable of ovulating. However, it is known that twinning is evident in less than 25% of coverings and hence some of these larger follicles must regress, or ovulate at an inappropriate time in relation to oestrus to result in fertilisation. Twinning is an inherited trait and can, therefore, be avoided or an awareness of its possibility gained by studying mare's breeding history and any previous incidences of twinning (Jeffcote and Whitwell, 1973; Ginther, 1982; Miller and Woods, 1988).

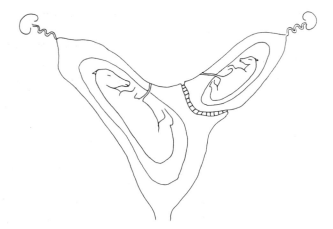

Figure 19.6   The division of the uterus between twins varies: normally one will dominate at the expense of the other (above); occasionally the division will be equal and both twins may survive (below).

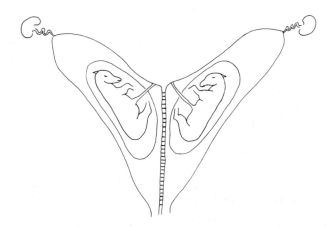

Follicular Tube Abnormalities

The extent of follicular tube abnormality is disputed. Arthur (1958), using slaughter house material, demonstrated that the incidence of such abnormalities in the United Kingdom was very low. However, work in Belgium surveying 200 mares demonstrated a 40% incidence of abnormalities normally involving adhesions of the infundibulum to other parts of the reproductive tract (Vandeplashe and Henry, 1977). The incidence of such abnormalities was also demonstrated to be greater on the right-hand side of the tract rather than the left-.

Inflammation of the fallopian tubes, termed salpingitis, is relatively rare, but if a mare is suffering or has suffered from endometritis (inflammation/infection of the uterine endometrium), salpingitis often results. Bacteria causing salpingitis may also come from the body cavity or from the blood supply. Complete blockage of the fallopian tubes is rare, but inflammation can interrupt the passage of ova towards the utero-tubular junction. Infertility or subfertility may be seen as a result. Occasionally infection may cause inflammation of the valve at the utero-tubular junction, affecting the passage of sperm and/or fertilised ova. This has been suggested to be the cause of high early embryo mortality rates in mares due to delays in fertilised ova release into the uterus, toxic effects of inflammation and possible loss into the abdominal cavity. Rarely an ovarian cyst may be seen to block the entry to the fallopian tube at the isthmus end. Tumours of the fallopian tube are extremely rare (Allen, 1979).

Uterine Abnormalities
Abnormalities of the uterus due to congenital problems or infections are relatively well understood in the mare, compared to abnormalities of the remainder of the tract. Due to the development of such techniques as uterine biopsy and the use of the endoscope, great advances have been made in understanding changes within the uterus.

Hypoplasia Hypoplasia is the term given to immaturity of the uterus as a result of a failure to develop to a stage capable of maintaining a pregnancy. The endometrial glands are those that seem to suffer the most; they tend to be very small and so incapable of supporting a pregnancy. As a result even if fertilisation does occur, embryo mortality is high. It is known that mating too close to puberty results in high embryonic mortality rates and this may well be due to hypoplasia, simply because the uterus has not had the time to develop. The actual age at which the uterus is fully mature depends very largely on the individual mare. Anywhere between 18 months and 4 years is considered acceptable. Hypoplasia at 4 years old or over is indicative of permanent problems and may be associated with chromosomal abnormalities.

Hypoplasia may also be a result of delayed involution post partum. If the uterine endometrium fails to return to normal post partum, the mare is unable to carry another pregnancy. Many studs carry out a routine internal examination 48–96 hours post partum to assess, among other things, that uterine involution is occurring as expected. If complete uterine involution has not occurred by the foal heat, mating can increase the chances of infection and the likelihood of a successful pregnancy is reduced. Uterine infection also tends to perpetuate the problem of involution. Occasionally a mare may suffer from uterine asymmetry, in which one horn has failed to involute properly; it is suggested that this affects sperm motility and, therefore, fertility rates (Kenney, 1978; Ricketts, 1975; McKinnon, 1987 and 1988).

Uterine Atrophy   Uterine atrophy or senility is caused by a decrease in the number of endometrial glands due to atrophy or an inability to regenerate themselves. Atrophy is associated with ovarian incompetence and progressive wear and tear through numerous pregnancies. It also seems to be more apparent later in the season and, as such, seems to be associated with a decrease in oestrous cycles and ovarian activity. As a rule such atrophy during the late season is of little concern, but evidence of it in early season is worrying as this may be indicative of permanent problems with fertility.

Immunological Incompetence   Immunological incompetence or lymphstasis is associated with an overproduction of antibodies as a result of endometritis. If the condition develops further, immunological rejection of the foetus may result (Asbury, 1984, 1987).

Luminal Cysts   Uterine luminal cysts are generally thin walled and greater than 3 cm in diameter, filled with lymph (Colour plate 45). They are particularly evident in mares 10 years old or over, and may disrupt implantation. If they are present in any number, they can also significantly reduce the surface area of the uterus available for placental attachment and lead to embryo mortality. Treatment is normally by means of puncturing the cysts via curettage or endoscopic manipulation (Ricketts, 1978; Mather, Refsal, Gustatsson, Seguin and Whitmore, 1979; Wilson, 1985).

Uterine curettage is a mechanical or chemical irritation of the surface of the uterus carried out by a veterinary surgeon. The reasoning behind such treatment is that the irritation stimulates nerve endings and initiates a cleansing and regeneration process within the uterine endometrium. The most popular curette is the Fort Dodge, which consists of a curved plate with a cutting edge on the end of a long shaft. This is introduced into the uterus, carefully avoiding damage to the vagina and cervix en route. The curette is then pushed and pulled along to scrape the entire surface of the uterus. Curettage was once very popular but has been discredited recently as infective and also as the potential cause of excess scar tissue and uterine adhesions. An alternative is the infusion of kerosene as a chemical irritant, which is reported to have a similar cleansing effect to curettage (Bracher, 1992).

Ventral Uterine Dilation   Ventral uterine dilation is caused by muscle atrophy in one uterine horn, resulting in outfoldings of one side of the uterus. This is again more common in older multiparous mares due to a weakening of the uterine myometrial layers. It often occurs at the implantation site, especially after several pregnancies, and the dilation is caused by a gradual weakening of the wall at an area of repeated excessive stretching. Treatment is, unfortunately, relatively unsuccessful but some beneficial results have been reported after treatment with oxytocin (Kenney and Ganjam, 1975).

Uterine Neoplasia   Neoplasia or tumours within the uterus are rare, but when present can cause persistent haemorrhages. Treatment may be attempted by surgery with some success (Kenney, 1978; Bostock and Owen, 1975).

Uterine Infections   One of the major causes of infertility in the mare is endometritis or infection of the uterine endometrium (Bennett, 1987). There are several factors that predispose the mare's tract to such infections and these will be discussed in the following sections on cervical and vulval abnormalities. There is also evidence that immunological or endocrine deficiencies may also predispose the mare to uterine infection and these may be inherited, leading to a heritable predisposition to endometritis.

Unfortunately the mare's reproductive tract is not well designed for the easy natural removal of infective organisms or their resulting pus or exudate. Established infections are difficult, therefore, for the mare's system to eliminate naturally and can easily develop into chronic infections. Chronic infections, not identified, can cause serious problems if not noticed and treatment administered immediately. Temporary infertility nearly always results, and, if the infection damage is too great, permanent infertility will result. Bacterial infection is nearly always introduced either by mating or by inadequate hygiene precautions at internal examination of the mare's tract. If all the hygiene precautions outlined in Chapter 13 are adhered to, the risk of infection is significantly reduced.

One of the major problems with uterine infections is that they can remain undetected for prolonged periods of time, thus not only reducing the mare's fertility rates but also risking transfer to any stallions used to cover her and, hence, on to other mares. Regular swabbing is not only compulsory in some studs, but is a good practice in order to ensure that all chronic endometritis infections are identified and treated immediately.

Endometritis is characterised by excess mucus often seen exuding from the vulva, high leucocyte counts and increased uterine blood supply. If oedema (fluid accumulation) results the uterus can be felt, via rectal palpation, as large and flaccid. The mare may also show shortened oestrous cycles due to the irritation of the uterine wall resulting in premature CL regression.

There are three main types of uterine infections: acute, chronic and septicaemic. Acute infections develop rapidly, giving immediate symptoms, and are characterised by irregular oestrous cycles and evidence of exudate or pus, milky or creamy white in colour on the mare's tail. Internally, the infection causes deep haemorrhage and degeneration of luminal epithelial cells and in severe cases also degeneration of the deeper stroma cells leading to areas of missing endometrium. This can lead to hypertrophy and abscessed uterine glands (Rooney, 1970). Acute endometritis is a major cause of infertility in the mare, providing a hostile environment for both sperm and developing ova. Elevated leucocyte levels ingest any ova or sperm present.

Infection is normally introduced directly in through the cervix at mating by the stallion's penis. There is evidence that acute endometritis is a common occurrence as a result of mating, but the mare's system can normally produce enough leucocytes to cope with the infection and the uterus is back to normal within 72 hours, ready to receive a fertilised ovum. Occasionally the endometritis persists and becomes an acute or potentially chronic infection. The bacteria responsible for such infections can normally be isolated by cervical swabbing. The circumstances under which post mating endometritis is not cleared from the body include general stress, a decline in the mare's general wellbeing and cervical, vaginal or vulval abnormalities leading to inadequate protection of the reproductive tract. Even if no physical abnormalities are evident, acute endometritis may result, as some mares are seemingly more susceptible than others, possibly due to leucocyte incompetence.

Treatment for acute endometritis begins with an attempt to correct any physical abnormalities that may be predisposing the mare to infection. The existing infection is then normally treated with uterine infusion using a broad spectrum antibiotic or antiseptic, in 400–500 ml of sterile saline as a carrier base (Threlfall, 1979). Infusion is by means of an indwelling catheter passed through the cervix and into the uterus. The end of the catheter is looped into two ram's horn shapes and this helps keep the catheter in place and allows repeated infusions without the need to change the catheter (Figure 19.7).

The catheter is most easily introduced through a relaxed cervix, normally associated with oestrus. However, this is not always possible as endometritis often leads to anoestrus and, hence, a tight cervix. Treatment with PGF2α is sometimes used to bring the

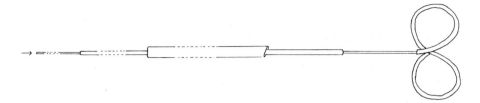

Figure 19.7   An infusion catheter for the treatment of mares with antibiotic solution in cases of endometritis.

mare into oestrus sooner, reducing the delay until treatment can be commenced and also as a series of injections to allow a series of short cycles and, therefore, speed up a series of treatments. Infusion times depend upon the severity of the infection and may vary from daily infusions for 3–5 days up to 15 days for infections that have not responded to shorter periods. Such antibiotic treatment must be used with care, as some antibiotics may cause necrosis or erosion of the endometrium. Bacterial resistance may also be a problem and excessive use may allow fungal infections to set up within the uterus that will themselves require treatment.

It is possible to use systemic antibiotics but evidence for their success is inconclusive. Acute endometritis, due to an inherent incompetence in the mare's defence system, can also be reduced by prophylactic measures at mating. In such cases the use of artificial insemination with antibiotic semen extender significantly improves fertility rates, by reducing the challenge presented to the mare's defence mechanism. In the Thoroughbred industry where mares seem to be particularly susceptible to endometritis, AI is not applicable, as offspring conceived in this way are not eligible for registration. In these mares pre and post mating infusions can help fertility rates. Infusion should be carried out using 100–300 ml of antibiotic semen extender immediately prior to covering. Covering should also be limited to one per oestrus to minimise the challenge to the mare's own defence system (Kenney, Bergman, Cooper and Morse, 1975; Kenney and Ganjam, 1975). Post covering infusion using a similar fluid may also be used but it runs the risk of washing out semen still within the vagina (Bennett, 1986).

(a) Potential Acute Endometritis Organisms

There are several bacteria that potentially can cause acute endometritis and include the following:

Haemophilus equigenitalis is an extremely contagious bacterium causing contagious equine metritis (CEM) (Tainturier, 1981). It was first isolated in the United Kingdom by Crowhurst (1977) in Newmarket, where it spread rapidly and widely due to the reluctance of infected mare owners not to present their mares for service. The bacterium involved is a rod bacterium, changing in shape to a sphere with age. The stallion is seemingly not affected by the bacterium but is the prime means by which it is spread from mare to mare. In the mare the typical symptoms of acute endometritis are seen, characterised by uterine, cervix and vaginal inflammation along with copious grey discharge within 2–5 days of infection. In rarer instances the mare may not show any clinical symptoms but still be a carrier capable of infecting a stallion. At the other extreme the infection may develop on to give chronic endometritis.

Due to the bacterium's highly contagious nature strict measures have been introduced in many countries to help reduce its incidence. In the United Kingdom the Horse Race and Levy Betting Board in 1978 set up a scientific committee of inquiry after the disaster of the previous breeding season. As a result the committee recommended a series of up to three swabs should be taken from all mares prior to service to detect the presence of the bacteria. In Thoroughbred studs today no mares are accepted unless they have a negative CEM certificate on arrival, and they are swabbed at least once again at the first signs of heat at the oestrus when they are to be covered. This swab must also be negative for covering to go ahead. The bacteria are harboured in the clitoral sinuses, clitoral fossa and urethral opening, and it is these areas that are routinely swabbed. If an infection is detected, treatment can be via penicillin both locally applied or systemic. Surgical removal of the clitoral sinuses has also been used to prevent a reoccurrence. There is also a serological (blood) test that can help in the detection of carrier mares. The test, however, is not 100% accurate (Benson, 1978; Hughes, 1979).

The code of practice produced by the Thoroughbred Breeders Association classifies mares as follows:

High risk mares (those in which CEM has been previously isolated) require 3 negative swabs

- one prior to arrival
- one immediately upon arrival
- one at oestrus of covering

Low risk mares (those in which CEM has not been previously isolated) require 2 negative swabs

- one prior to arrival
- one at oestrus of covering

Swabs should be taken from the clitoral sinuses, clitoral fossa and urethral opening plus an endometrial swab (Plate 19.1). Codes of practice are also laid down for stallion swabbing. All stallions and teasers should have swabs taken from the genital organs after 1 January each year and before the covering season. No infected animals should be used. CEM is now eradicated from Britain and is a notifiable disease. Codes of practice are also laid down for exportation, importation and cases of suspected CEM abortion (Rossdale, Hunt, Peace, Hope, Ricketts and Wingfield-Digby, 1979; Platt, Atherton and Simpson, 1978; Crowhurst, Simpson, Greenwood and Ellis, 1979).

Plate 19.1

Many studs require all mares to be swabbed prior to service. The swabs are taken from clitoral sinuses and fossa along with, as in this plate, the urethral opening and endometrium.

Klebsiella aeruginosa are encapsulated rod shaped bacteria and often cause the development on from acute to chronic endometritis. The symptoms include a vulval discharge 2–7 days post coitum and the infection is usually passed at service to a carrier stallion if adequate hygiene precautions have not been taken at covering. It is also possible for transfer to occur at teasing or at an internal veterinary examination without adequate hygiene precautions. The disease is endemic and widespread, but diagnosis is reasonably accurate via cervico-uterine swabbing. Unfortunately, the bacteria are relatively insensitive to antibiotics and antiseptic washing agents. There is in fact evidence to suggest that *Klebsiella aeruginosa* infection is increased by indiscriminate use of antibiotics and antiseptics, which kill off the naturally existing microflora and allow invasion by opportunist bacteria such as *Klebsiella aeruginosa*.

Pseudomonas aeruginosa, a slender rod bacterium with rounded ends and flagella, is found widely within the environment. As far as reproductive tract infection is concerned it is an opportunistic bacterium, taking advantage of mares with lowered resistance. It is usually introduced into the mare via a carrier stallion. *Pseudomonas aeruginosa* may be isolated in a stallion's semen or in swabs taken from the urethral fossa, but he rarely shows clinical symptoms. In the mare *Pseudomonas aeruginosa* causes a greenish blue or yellowish green exudate, and it seems to be more prevalent in older mares. As with *Klebsiella aeruginosa* its presence is also associated with excessive use of antibiotics and/or antiseptics removing the natural microflora and allowing infections to take over. In addition it is itself relatively resistant to antibiotics and antiseptics. Early diagnosis and cessation of natural cover is the best course of action (Hughes, Loy, Atwood, Astbury and Burd, 1966; Hughes and Loy, 1975).

Streptococcus zooepidemicus is implicated in 75% of acute endometritis cases. They are spherical bacteria found normally in chain formation. They are found in the general environment, including the intestine and mucus membranes. Streptococci are classified into subgroups alpha and beta, *Streptococcus zooepidemicus* is a beta streptococcus and, as such, causes the destruction of red blood cells and has a major role in initiating infection of the mare's cervix and uterus (Hughes and Loy, 1969).

Staphlococcus aureus are spherical or oval bacteria normally evident in clusters. It is again an opportunistic bacterium normally found associated with skin and mucous membranes and under suitable conditions, such as the disruption of the natural microflora, ill health or stress, will invade the reproductive tract of the mare, causing destruction of endometrial tissue.

Hemolytic E. coli is a rod shaped bacterium that is found either alone or in short chains. It is the second most common cause of uterine infection. It is naturally found in the intestine and is associated in particular with faecal contamination. It can cause not only acute endometritis but severe systemic infection which can prove fatal.

Proteus are multi-form bacterial rods covered with flagella and associated with the intestine. They are also opportunistic bacteria taking advantage of situations such as stress, ill health or a disruption in the natural microflora, in order to invade the mare's uterus.

(b) Chronic Uterine Infections
Chronic uterine infections or pyrometra, which is relatively rare, is normally a development from an acute uterine infection, due to a mare's inability to fight the initial infection. Such infection is often long-term and can be extremely damaging to the uterine tissue and cause permanent infertility. Pyrometra uteruses become large and pendulous with flaccid walls with accumulated pus within. In time the uterine walls may be leathery, tough and fibrous, due to continual infection. Such mares often appear healthy in themselves but often do not show oestrus cycles due to the inability of the uterus to produce PGF2α. Pyrometra may be associated with a blockage of the uterus, resulting in a build-up of exudate and pus within the uterus with no normal drainage. Treatment is normally by antibiotic infusion or injection but the prognosis is poor. If left untreated, pyrometra may develop into septicaemia (Hughes, Stabenfeldt, Kindall, Kennedy, Edquist, Nealy and Schalm, 1979; Ricketts, 1978).

Chronic endometritis may be caused by wear and tear rather than by infection, and such endometritis is classified as either infiltrative or degenerative. Infiltrative endometritis may be a result of changes within the uterus due to a busy breeding career, and it is associated with a natural increase in leucocytes in response to the normal bacterial challenge post coitum (Ricketts, 1978; Van Furth, Cohn, Hirsch, Humphrey, Spector and Langevoot, 1972).

Degenerative endometritis is associated with infertility and is caused again by wear and tear due to repeated gestations, especially if there has been a history of infections. Degeneration of the endometrial glands results in a failure to return to normal post parturition, leaving lesions containing lymph fluid. Treatment is by the stimulation of the growth of new healthy endometrium. This was traditionally achieved by curettage, as discussed earlier in this chapter, or alternatively by chemical irritation of the uterus by the infusion of kerosene, followed by the infusion of antibiotics to help fight/prevent infection (Bergman and Kenney, 1975; Kenney and Ganjam 1975; Witherspoon, 1972).

(c) Septic Metritis

Septic metritis is the most serious uterine infection usually associated with the retention of placental material or foetal tissue at parturition. Decomposition of this retained tissue, along with infection introduced through the cervix via air sucked in immediately post partum, or on hands used to aid parturition, is the cause. Spread of infection is rapid and the absorption of toxins results in severe, potentially fatal systemic infection. Prevention is infinitely better than cure, and absolute hygiene at parturition, plus ensuring that complete expulsion of all placental tissue has taken place post partum, is essential.

Cervical Abnormalities

Cervical tumours are extremely rare, but damage to the cervix as a result of a difficult parturition is reasonably common. Lacerations or injuries to the cervix often do not heal properly, either causing adhesions that may block the entrance to the uterus through the cervix or cause cervical incompetence and, therefore, allow infection to enter the uterus. As discussed in some detail in Chapter 1, the cervix naturally forms the final seal protecting the upper reproductive tract from infection. If this seal is incompetent, especially if the vestibular or vaginal seal and vulval seal are also incompetent, the uterus is easily infected by bacteria sucked in through the vagina. The prognosis in most cases is poor. However, minor adhesions may be treated surgically by cutting the scar tissue and inserting a plastic tube to prevent their reoccurrence. Inherited cervical incompetence is not normally seen but has been reported in pony mares (Lieux, 1972; Brown, 1984).

Infection of the cervix, termed cervicitis, is usually associated with and is often the initial cause of endometritis. Infection causes inflammation and pus accumulation.

Vaginal Abnormalities

As discussed in Chapter 1, pneumovagina (sucking of air and hence bacteria into the vagina) is a common cause of infertility, especially in Thoroughbred mares, and is due to incompetence of the vestibular and vulval seals associated with poor perineal conformation. This can be cured quite successfully by a Caslick's operation, in which the sides of the lower part of the vulva are cut and sutured together to form a seal and, hence, prevent air and any associated faeces being sucked into the vagina causing contamination. Full details of vulval incompetence and its treatment are given in Chapter 1.

Dourine, a sexually transmitted protozoan, now eradicated from the United Kingdom but still prevalent in many temperate countries, causes vaginal and vulval infection and inflammation along with discharge. If left untreated, it will develop on to cause raised rings of the mare's coat all over her body followed by depigmentation of the genitals along with fever and death in 50–75% of cases.

Urine pooling within the vagina is also associated with infection, which can easily spread to the rest of the reproductive tract and, hence, adversely affect fertility. Normally such a problem is only seen in older multiparous mares, with pendulous reproductive tracts due to the continual stretching and weakening with successive pregnancies. Occasionally, it may also be seen as a temporary problem at foal heat, but in most circumstances it will have rectified itself by the second heat post partum. If it is evident it is essential that such a mare is not covered, as there is an increased chance of uterine infection. Treatment using oxytocin has proved reasonably successful (Monin, 1972).

Damage at parturition may also occur, associated with abnormal foal positions. Superficial damage will correct and heal itself, but there is always the risk of adhesions. Severe adhesions may cause the mare pain at subsequent coverings. In the extreme, rectal vaginal fissures may be caused by the foal's foot ripping into the rectum at parturition. In such cases immediate action should be taken to prevent urine, air or faeces entering the abdominal cavity and causing peritonitis and death.

Occasionally a persistent hymen may be evident, dividing the vagina into anterior and posterior sections. If the hymen is not broken manually prior to the first service, then it may tear causing pain to the mare and the development of scar tissue.

Vulval Abnormalities
Vulval abnormalities most commonly involve incompetent perineal conformation as discussed in some detail in Chapter 1, leading to vulval seal incompetence and an increase in the chance of infection entering the upper reproductive tract. Lacerations of the vulva at parturition can again cause problems associated with adhesions and the development of vulval seal incompetence due to incorrect healing of the vulva. Failure to cut a Caslick's at mating or parturition will also cause tearing of the vulva that may not heal without the development of adhesions (Aanes, 1969, 1973).

Haemorrhage of the vulval lips may also be evident due to the bursting of varicose veins. This has little effect on fertility but may cause discomfort at breeding. Neoplasms of the vulva may also be evident. Most commonly these are melanomas originating in the pigment-producing cells of the skin and are especially prevalent in grey mares. These tumours can spread from the perineal area to under the tail and eventually throughout the rest of the body. Squamous cell carcinoma, normally associated with the penis, may also be seen on the vulval lips.

As mentioned previously enlarged clitorises, sometimes in the form of a vestigial penis, are associated with chromosomal abnormalities and such animals are sterile.

Finally, equine coital exanthema, genital horse pox, caused by a herpesvirus infection, may also lead to pain at covering and can, therefore, affect fertility. Such infection causes vesiculation and ulceration of both the vulva and the penis. It is a sexually transmitted disease and symptomless carriers are possible. There is no direct effect on fertility but in order to prevent transfer natural covering must cease. Treatment with antibacterial creams or powders prevents secondary infections and helps the natural healing of the lesions (Pascoe, Spradbrow and Bagust, 1968, 1969; Gibbs, Roberts and Morris, 1972).

Conclusion

The causes of infertility, or the failure to produce an offspring, whether on a temporary or permanent basis, are numerous. Some are treatable but many may cause the mare or stallion to be useless for breeding. It is essential, therefore, that all potential breeding stock is given a thorough examination prior to purchase in order to ensure that it is capable of fulfilling its reproductive potential.

References and Further Reading

Aanes, W.A. (1969) Surgical repair of third degree perineal laceration and rectovaginal fistula in the mare. J.Am.Vet.Med.Ass. 144, 485.

Aanes, W.A. (1973) Progress in recto-vaginal surgery. Proc.19th Ann.Conv. Am.Ass.Equine Practnrs., p. 225.

Allen, W.E. (1979) Evaluation of uterine tube function in pony mares. Vet.Rec., 105, 364.

Arthur, E.H. (1958) An analysis of the reproductive function of mares based on post mortem examination. Vet.Rec., 70, 682.

Asbury, A.C. (1984) Uterine defense mechanisms in the mare. The use of intrauterine plasma in the management of endometritis. Theriogenology, 21, 387.

Asbury, A.C. (1987) Infectious and immunological considerations in mare infertility. Compend.Cont.Ed.Pract.Vet., 9, 585.

Asbury, A.C, and Lyle, S.K. (1993) Infectious causes of Infertility. In: Equine

Reproduction. Ed. A.O. McKinnon and J.L. Voss. Lea and Feibiger, Philadelphia & London, p. 381.

Bennett, D.G. (1986) Therapy of endometritis in mares. J.Am.Vet.Med. Ass., 188, 1390.

Bennett, D.G. (1987) Diagnosis and treatment of equine bacterial endometritis. J.Equine Vet.Sci., 7, 345.

Benson, J.A. (1978) Serological response in mares affected by contagious equine metritis. Vet.Rec., 102, 277.

Berndstone, W.R., Hoyer, J.M., Squires, E.L. and Pickett, B.W. (1979) The influence of exogenous testosterone on sperm production, seminal quality and libido of stallions. J.Reprod.Fert.Suppl., 27, 19.

Bishop, M.W.H., David, J.S.E. and Messervy, A. (1966) Cryptorchidism in the stallion. Proc.Royal Soc.Med., 59, 769.

Blanchard, T.L. and Varner, D.D. (1993). Testicular Degeneration. In: Equine Reproduction. Ed. A.O. McKinnon and J.L. Voss. Lea and Feibiger, Philadelphia & London, p. 855.

Blue, B.J., Pickett, B.W., Squires, E.L., McKinnon, A.O., Nett, T.M., Amann, R.P, and Shiner, K.A. (1991) Effect of pulsatile or continuous administration of GnRH on the reproductive function of the stallion. J.Reprod.Fert.Suppl., 44, 145.

Bostock, D.E, and Owen, L.N. (1975) Neoplasia in the cat, dog and horse. Wolfe Medical Publications Book.

Bosu, W.T.K. (1982) Ovarian disorders: Clinical and morphological observations in 30 mares. Can.Vet.J., 23, 6.

Bosu, W.T.K. and Smith, C.A. (1993) Ovarian Abnormalities. In. Equine Reproduction. Ed. A.O. McKinnon and J.L. Voss. Lea and Feibiger, Philadelphia & London, p. 397.

Bowling, A.T. and Hughes, J.P. (1993) Cytogenetic Abnormalities. In: Equine Reproduction. Ed. A.O. McKinnon and J.L. Voss. Lea and Feibiger, Philadelphia & London, p. 258.

Boyle, M.S., Skidmore, J., Zhang, J. and Cox, J.E. (1991) The effects of continuous treatment of stallions with high levels of a potent GnRH analogue. J.Reprod.Fert.Suppl., 44, 169.

Bracher, V. (1992) Equine Endometritis. PhD thesis. University of Cambridge.

Brook, D. (1993) Uterine cytology. In: Equine Reproduction. Ed. A.O. McKinnon and J.L. Voss. Lea and Feibiger, Philadelphia & London, p. 246.

Brook, D. and Franket, K. (1987) Electrocoagulative removal of endometrial cysts in the mare. J.Equine Vet.Sci., 7, 77.

Brown, J.S. (1984) Surgical repair of the lacerated cervix in the mare. Theriogenology, 22, 351.

Burgman, R.V. and Kenney, R.M. (1975) Representiveness of a uterine biopsy in the mare. Proc. 21st Ann.Conv.Am.Ass.Equine Practnrs., 355.

Casslick, E.A. (1937) The vulva and vulvo-vaginal orifice and its relation to genital health of the Thoroughbred mare. Cornell Vet., 27, 178.

Chandley, A.N., Fletcher, J., Rossdale, P.D., Peace, C.K., Ricketts, S.W., McEnery, R.J., Thorne, J.P., Short, R.V. and Allen W.R. (1975) Chromosome abnormalities as a cause of infertility in mares. J.Reprod.Fert.Suppl., 23, 377.

Couto, M.A. and Hughes, J.P. (1993) Sexually transmitted (venereal) diseases of horses. In: Equine Reproduction. Ed. A.O.McKinnon and J.L. Voss. Lea and Feibiger, Philadelphia & London, p. 845.

Cox, J.E. (1993) Developmental abnormalities of the male reproductive tract. In: Equine Reproduction. Ed. A.O. McKinnon and J.L. Voss. Lea and Feibiger, Philadelphia & London, p. 895.

Crouch, J.R.F., Atherton, J.G, and Platt, H. (1972) Venereal transmission of Klebsiella aerogenes in a thoroughbred stud farm from a persistently infected stallion. Vet.Rec., 90, 21.

Crowhurst, R.C. (1977) Genital infection in mares. Vet.Rec., 100, 476.

Crowhurst, R.C., Simpson, D.Y., Greenwood, R.E.S. and Ellis, D.R. (1979) Contagious equine metritis. Vet.Rec., 104, 465.

Day, F.T. (1939) Some observations on the causes of infertility in horse breeding. Vet.Rec., 51, 581.

DeGannes, R.V.G. (1982) Old and new treatment for broodmares. Proc. 28th Ann.Conv.Am.Ass.Equine Practnrs., p. 405.

De Vries, P.J. (1993) Diseases of the testes, penis and related structures. In: Equine Reproduction Ed: A.O. McKinnon and J.L. Voss. Lea and Feibiger, Philadelphia & London, p. 878.

Doig, P.A., and Waelchi, R.O. (1993) Endometrial biopsy. In Equine Reproduction. Ed. A.O. McKinnon and J.L. Voss. Lea and Feibiger, Philadelphia & London, p. 225.

Farrely, B.T, and Mullaney, P.E. (1964) Cervical and uterine infection in thoroughbred mares. Ir.Vet.J., 18, 201.

Ganjam, V.K., Kenney, R.M. and Flickinger, G. (1975) Plasma progesterone in cyclic, pregnant and post-partum mares. J.Reprod.Fert., 23, 441.

Gibbs, E.P.Y., Roberts, M.C. and Morris, J.M. (1972). Equine coital exanthema in the U.K. Equine Vet.J., 4, 74.

Ginther, O.J. (1982) Twinning in mares: A review of recent studies. J.Equine Vet.Sci., 2, 127.

Ginther, O.J. and Pierson R.A. (1984) Ultrasonic anatomy and pathology of the equine uterus. Theriogenology, 21, 505.

Ginther O.J. (1985) Embryonic loss in mares. Incidence and ultrasonic morphology. Theriogenology, 24, 73.

Greenhof, G.R. and Kenney, R.M. (1975) Evaluation of the reproductive status of non-pregnant mares. J.Am.Vet.Med.Ass., 167, 449.

Halnan, C.R.E. (1985) Sex chromosome mosaicism and infertility in mares. Vet.Rec., 116, 542.

Hughes, J.P. (1979) Contagious equine metritis. A review. Theriogenology, 11, 209.

Hughes, J.P. (1993) Developmental anomalies of the female reproductive tract. In: Equine Reproduction. Ed. A.O. McKinnon and J.L. Voss. Lea and Feibiger, Philadelphia & London, p. 408.

Hughes, J.P, and Loy, R.G. (1969) Investigation on the effect of intrauterine inoculation of Streptococcus zooepidemicus in the mare. Proc.15th.Ann. Conv.Am.Ass.Equine Practnrs., p. 289.

Hughes, J.P. and Loy, R.G. (1975) The relation of infection to infertility in the mare and the stallion. Equine Vet.J., 7, 155.

Hughes, J.P., Loy, R.G., Atwood, C., Astbury, A.C, and Burd, H.E. (1966)

The occurrence of pseudomonas in the reproductive tract of mares and its effect on fertility. Cornell Vet. 56, 595.

Hughes, J.P., Stabenfeldt, G.H., Kindhal, H., Kennedy, P.C., Edquist, L.E., Nealy, D.P. and Schalm, O.W. (1979) Pyrometra in the mare. J.Reprod.Fert.Suppl., 27, 321.

Jeffcote, L.B. and Whitwell, K.E. (1973) Twinning as a cause of foetal and neonatal loss in the thoroughbred mare. J.Comp.Pathol. 83, 91.

Johnson, L., Varner, D.D. and Thompson, D.L. (1991) Effect of age and season on the establishment of spermatogenesis in the horse. J.Reprod. Fert.Suppl., 44, 87.

Kenney, R.M. (1975) Clinical fertility evaluation of the stallion. Proc. 21st.Ann.Conv.Am.Ass.Equine Practnrs. 336.

Kenney, R.M. (1975) Prognostic value of endometrial biopsy of the mare. J.Reprod.Fert.Suppl., 23, 347.

Kenney, R.M. (1978) Cyclic and pathological changes of the mare endometrium as detected by biopsy, with a note on early embryonic death. J.Am.Vet.Med.Ass. 172, 241.

Kenney, R.M. (1990) Estimation of stallion fertility: the use of sperm chromatin structure assay, DNA Index and karyotype as adjuncts to traditional tests. Proc. 5th.Int.Symp. Equine Reprod., 42–43.

Kenney, R.M., Bergman, R.V., Cooper, W.L. and Morse, G.W. (1975) Minimal contamination techniques for breeding mares: technique and preliminary findings. Proc.21st.Ann.Conv.Am.Ass.Equine Practnrs. 327.

Kenney, R.M., Condon, W.A., Ganjam, J.K. and Channing, C. (1979) Morphological and biochemical correlates of equine ovarian follicles as a function of their state of viability or atresia. J.Reprod.Fert.Suppl., 27, 163.

Kenney, R.M. and Ganjam, V.K. (1975) Selected pathological changes of the mare's uterus and ovary. J.Reprod.Fert.Suppl., 23, 335.

Kenney, R.M., Ganjam, V.K., Cooper, W.R. and Lauderdale, J.W. (1975) The use of PGF2α – THAM salt in mares in clinical anoestrus. J.Reprod.Fert. Suppl., 23, 247.

Kenney, R.M., Kent, M.G., Garcia, M.C. and Hurtgen, J.P. (1991) The use of DNA Index and karyotype analysis as adjuncts to the estimation of fertility in stallions. J.Reprod.Fert.Suppl., 44, 69.

LeBlanc, M.M. (1993) Vaginal examination In: Equine Reproduction. Ed. A.O. McKinnon and J.L. Voss. Lea and Feibiger, Philadelphia & London, p. 221.

LeBlanc, M.M. (1993) Endoscopy. In. Equine Reproduction. Ed. A.O. McKinnon and J.L. Voss. Lea and Feibiger, Philadelphia & London, p. 255.

Lieux, P. (1972) Reproductive and genital disease. In: Equine Medicine and Surgery. Ed. E.J. Catcott and J.F. Smithcors. 2nd Ed., p.567. Wheaton 111, American Veterinary Publications Book.

Love, C.C., Garcia, M.C., Riera, F.R. and Kenney, R.M. (1991) Evolution of measures taken by ultrasonography and calipers to estimate testicular volume and predict daily sperm output in the stallion. J.Reprod.Fert. Suppl., 44, 99.

Mather, E.C. (1979) The use of fibreoptic techniques in clinical diagnosis and visual assessment of experimental intrauterine therapy in mares. J.Reprod.Fert.Suppl., 27, 293.

Mather, E.L., Refsal, K.R., Gustatsson, B.K., Seguin, B.E. and Whitmaore, H.C. (1979) The use of fibreoptic techniques in clinical diagnosis and assessment of experimental intrauterine therapy in mares. J.Reprod.Fert. Suppl., 27, 293.

Meager, D.M. (1978) Granulosa cell tumours in the mare. A review of 78 cases. Proc.23rd.Ann.Conv.Am.Ass.Equine Practnrs., p. 133.

McDowell, K.J. (1985) Effect of restricted conceptus mobility on maternal recognition of pregnancy in mares. Equine Vet.J., 3, 23.

McKinnon, A.O. (1987) Diagnostic ultrasonography of uterine pathology in the mare. Proc. 33rd. Ann.Conv.Am.Ass.Equine Practnrs.

McKinnon, A.O. (1988) Ultrasonic studies on the reproductive tract of mares after parturition: Effect of involution and uterine fluid on pregnancy rates in mares with normal and delayed first postpartum ovulatory cycles. J.Am.Vet.Med.Ass. 192, 350.

McKinnon, A.O. and Carnevale, E.M. (1993) Ultrasonography. In: Equine Reproduction. Ed. A.O. McKinnon and J.L. Voss. Lea and Feibiger, Philadelphia & London, p. 211.

McKinnon, A.O. and Beldon, J.O. (1988). A urethral extension technique to correct urine pooling (vesico-vaginal reflux) in mares. J.Am.Vet.Med.Ass. 192, 647.

McKinnon, A.O., Voss, J.L., Squires, E.L. and Carnevale. E.M. (1993) Diagnostic ultrasonography. In: Equine Reproduction. Ed. A.O. McKinnon and J.L. Voss. Lea and Feibiger, Philadelphia & London, p. 266.

McKinnon, A.O. and Voss, J.L. (1993) Breeding the problem mare. In: Equine Reproduction. Ed. A.O. McKinnon and J.L. Voss. Lea and Feibiger, Philadelphia & London, p. 368.

Miller, A. and Woods, G.L. (1988) Diagnosis and correction of twin pregnancy in the mare. Vet.Clin.No.Am. (Equine Pract) 4, 215.

Monin, T. (1972) Vaginoplasty: a surgical treatment for urine pooling in the mare. Proc.18th.Ann.Conv.Am.Ass. Equine Practnrs. 99.

Norris, H.Y., Taylor, W.B. and Garner, F.M. (1968) Equine ovarian granular tumours. Vet.Rec., 82, 419.

Osbourne, V.E. (1975) Factors influencing foaling percentages in Australian mares. J.Reprod.Fert. Suppl., 23, 477.

Pascoe, R.R. (1979) Observations on the length and angle of declination of the vulva and its relationship to fertility in the mare. J.Reprod.Fert. Suppl., 27, 299.

Pascoe, R.R., Spradbrow, P.B. and Bagust, T.Y. (1968) Equine coital exanthema. Aust.Vet.J., 44, 485.

Pascoe, R.R., Spradbrow, P.B, and Bagust, T.Y. (1969) An equine genital infection resembling coital exanthema associated with a virus. Aust.Vet.J., 45, 166.

Peterson, F.B., Mcfeey, R.A, and David, J.S.E. (1969) Studies on the pathogenesis of endometritis in the mare. Proc.15th.Ann.Conv.Am.Ass. Equine Practnrs. 279.

Pickett, B.W. and Voss, J.L. (1972) Reproductive management of stallions. Bull.Colo.St.Univ.Agric.Exp.Stat.Anim.Reprod.Lab.Gen. Ser. 934.

Pickett, B.W. and Voss, J.L. (1975) Abnormalities of mating behaviour in domestic stallions. J.Reprod.Fert. Suppl., 23, 129.

Pickett, B.W. and Voss, J.C. (1975) The effect of semen extenders on mare fertility. J.Reprod.Fert.Suppl., 23, 95.

Platt, H., Atherton, J.G. and Simpson, D.Y. (1978) The experimental infection of ponies with contagious equine metritis. Equine Vet.J., 10, 153.

Pouret, E.J.M. (1982) Surgical technique for the correction of pneumo- and urovagina. Equine Vet.J., 14, 249.

Prickett, M.E. (1966) Pathology of the Equine ovary. Proc. 12th Ann.Conv. Am.Ass.Equine Practnrs., 145.

Probett, B.W. and Voss, J.L. (1972) Reproductive management of stallions. Proc. 18th.Ann.Conv.Am.Ass.Equine Practnrs, 501.

Pugh, D.G.(1985) Equine ovarian tumours. Compend.Cont.Ed. Pract.Vet. 7, 710.

Ricketts, S.W. (1975) Endometrial biopsy as a guide to diagnosis of endometrial pathology in the mare. J.Reprod.Fert.Suppl., 23, 341.

Ricketts, S.W. (1978) Histological and histopathological studies of the endometrium of the mare. Fellowship thesis, Royal College of Vet. Surgeons.

Ricketts, S.W. and Rossdale, P.D. (1979) Endometrial biopsy findings in mares with contagious equine metritis. J.Reprod.Fert.Suppl., 27, 355.

Ricketts, S.W., Young, A. and Medici, E.B. (1993) Uterine and clitoral cultures. In: Equine Reproduction. Ed. A.O. McKinnon and J.L. Voss. Lea and Feibiger, Philadelphia & London, p.234.

Roberts, S.Y. (1971) Veterinary Obstetrics and Genital Disease. Edwards, Ann Arbor,Michigan.

Roger, J.F. and Hughes, J.P. (1991) Prolonged pulsatile administration of gonadotrophin releasing hormone (GnRH) to fertile stallions. J.Reprod.Fert. Suppl., 44, 155.

Rooney, J.R. (1970) Autopsy of the horse, Technique and interpretation. Baltimore, Williams and Wilkins.

Rossdale, P.D., Hunt, M.D.N., Peace, C.K., Hopes, R., Ricketts, S.W. and Wingfield-Digby, N.Y. (1979) C.E.M.: The case of A.I. Vet.Rec., 104, 536.

Rossdale, P.D. and Ricketts, S.W. (1980) Equine Stud Farm Medicine, 2nd Ed. Baillière Tindall, London.

Schmacher, J. and Varner, D.D. (1993). Neoplasia of the stallion's reproductive tract. In: Equine Reproduction. Ed. A.O. McKinnon and J.L. Voss. Lea and Feibiger, Philadelphia & London, p. 871.

Sertich, P.L. (1993) Cervical problems in the mare. In: Equine Reproduction. Ed. A.O. McKinnon and J.L. Voss. Lea and Feibiger, Philadelphia & London, p. 404.

Shideler, R.K. (1993) External examination. In: Equine Reproduction. Ed. A.O. McKinnon and J.L. Voss. Lea and Feibiger, Philadelphia & London, p. 199.

Shideler, R.K. (1993) Rectal palpation. In: Equine Reproduction. Ed. A.O. McKinnon and J.L. Voss. Lea and Feibiger, Philadelphia & London, p. 204.

Simpson, D.J.S. and Eaton-Evans, W.E. (1978) Isolation of the CEM organism from the clitoris of the mare.

Stabenfeldt, G.H. (1979) Clinical findings, pathological changes and endocrinological secretory patterns in mares with ovarian tumours. J.Reprod.Fert.Suppl., 27, 277.

Stashak, T.S. (1993) Inguinal hernia. In: Equine Reproduction, Ed. A.O. McKinnon and J.L. Voss. Lea and Feibiger, Philadelphia & London, p. 925.

Sullivan, J.J., Turner, P.C., Self, L.C., Gutteridge, H.B. and Bartell, D.E. (1975) Survey of reproductive efficiency in the Quarter-horse and Thoroughbred. J.Reprod.Fert. Suppl., 23, 315.

Tainturier, D.J. (1981) Bacteriology of Hemophilus equigenitalis. J.Clin. Microbiol. 14, 355.

Threlfall, W.R. (1979) Antibiotic infusion of the uterus of the mare. Proc. Soc.Theriogenol, p. 45.

Van Camp, S.D. (1993) Uterine abnormalities. In: Equine Reproduction. Ed. A.O. McKinnon and J.L. Voss. Lea and Feibiger, Philadelphia & London, p. 392.

Van Furth, R., Cohn, Z.A., Hirsch, J.G., Humphrey, J.H., Spector, W.G. and Langevoot, H.L. (1972) The mononuclear phagocycte system: a new classification of macrophages, monocytes and their precursor cells. Bull.Wld.Hlth.Org. 46, 845.

Vandeplassche, M. and Henry, M. (1977) Salpingitis in the mare. Proc. 23rd. Ann.Conv.Am.Ass.Equine Practnrs., 123.

Varner, D.D., Taylor, T.S, and Blanchard, T.L. (1993) Seminal vesiculitis. In: Equine Reproduction Ed. A.O. McKinnon and J.L. Voss. Lea and Feibiger, Philadelphia & London, p. 861.

Vaughan, J.T. (1993) Penis and prepuce. In: Equine Reproduction. Ed. A.O. McKinnon and J.L. Voss. Lea and Feibiger, Philadelphia & London, p. 885.

Voss, J.L. and McKinnon, A.O. (1993) Hemospermia and urospermia. In: Equine Reproduction. Ed. A.O. McKinnon and J.L. Voss. Feibiger, Philadelphia & London, p. 864.

Wilson, G.L. (1985) Diagnostic and therapeutic hysteroscopy for endometrial cysts in mares. Vet.Med. 8, 59.

Witherspoon, D.M. (1972) Technique and evaluation of uterine curettage. Proc.18th.Ann.Conv.Am.Ass.Equine Practnrs. 51.

Zafracas, A.M. (1975) Candida infection of the genital tract in thoroughbred mares. J.Reprod.Fert.Suppl., 23, 349.

CHAPTER 20

Artificial Insemination

RESEARCH into artificial insemination (AI) in equids has been limited, as a number of breed societies will not accept progeny conceived in this way for registration. The most noteworthy of these in Britain is the Thoroughbred Breeders Society, which in turn provides a large amount of funding both directly and indirectly for equine research, especially that related to reproduction. Most of the other breed societies within Britain do accept the progeny of AI but some limit the number that can be registered per stallion per year. In the United States there is still some opposition to its use, though its popularity is now taking off. This opposition to its use has limited the work done and hence its development. It has, therefore, some considerable way to go before it reaches the sophistication of AI in cattle.

Semen Collection

Semen collection can be carried out using one of several methods. The easiest method is the collection of dismount samples. The drips of semen are collected from the stallion after withdrawal from the mare into a sterile jar. This method can be unreliable and the quality of sample is very variable and the majority of the sample is left within the mare. Samples often contain high levels of secretions from the accessory glands, and hence low sperm concentrations. The semen is also likely to contain pathogenic organisms originating from either the mare or the stallion.

Semen can also be collected from the anterior vagina immediately post mating. However, the semen at collection has already come into contact with the vaginal secretions. The vagina tends to be acid and as such is known have detrimental effects on sperm survival. Assessment of sperm viability, motility and live/dead ratios may well give inaccurate results. The sperm may also have become contaminated by pathogenic organisms originating from the mare. Identification of the origin of any pathogenic organisms found in

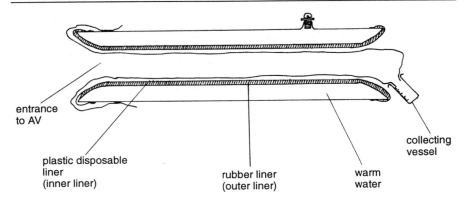

entrance
to AV

plastic disposable
liner
(inner liner)

rubber liner
(outer liner)

warm
water

collecting
vessel

Figure 20.1   A diagrammatic representation of an equine artificial vagina.

such a sample is very difficult. Both these methods of collection do not allow the total volume of semen produced to be assessed.

Condoms have been developed and used in horses for the collection of semen. They work very well, but do have a tendency to burst or come off the penis resulting in the whole sample being lost.

Finally, the best and now most commonly used method of semen collection is by artificial vagina (AV). The first AV was developed for use in horses in Russia at the beginning of this century. Subsequent development of the AV was for use in cattle rather than horses. Various models are available but they are all based on the same principles. They provide a warm jacket under some pressure with a collecting vessel at the end, in an attempt to mimic the natural vagina. Figure 20.1 illustrates diagrammatically the version of AV shown in Plate 20.1.

Most AVs consist of an outer solid casing, possibly of aluminium to minimise the weight, with two rubber linings, an outer and an inner. The outer lining and the casing form a water jacket, into which warm water is passed by means of a tap. The temperature of the water used is usually in the order of 50–60°C; it must be warm enough to ensure that, at use, the inside of the AV is only slightly above body temperature at 40–44°C. The amount of water used must be adequate to ensure that the pressure within the lumen of the AV mimics, as closely as possible, the pressure exerted against the insertion of the penis into the natural vagina. The inner lining of the AV provides a sterile inner environment and is connected to the collecting vessel. Before use this inner lining is lubricated with sterile obstetric lubricant to aid the stallion. The collecting vessel, as with the lumen of the AV, must be at 40–44°C during collection.

Plate 20.1   *An artificial vagina (AV) for use in horses.*
There are many designs but basically an AV consists of an outer and inner rubber lining, between which warm water is passed via the inlet on the top. The AV is lined with a disposable liner prior to collection, shown here as clear plastic. The semen is then collected either in the bottom of this bag or through a filter in the bottom of the liner into a collecting vessel.

An end-on view of the above AV through the end into which the stallion ejaculates. This end has a smaller lumen in order to mimic the mare's vagina.

To aid in the maintenance of this warm temperature and to protect the sperm from ultraviolet light the whole AV and collecting vessel may be enclosed within a protective jacket. The fully assembled AV prior to use is in the order of 50 cm long and can weigh up to 10 kg (Plate 20.2).

Plate 20.2   *To protect the AV from ultraviolet light and to help maintain the required temperature, the whole thing can be enclosed within a protective jacket. (Thoroughbred Breeders Equine Fertility Unit, Newmarket)*

As discussed, sperm are susceptible to sunlight and temperature change. If the temperature of the AV or collecting vessel is too cold, sperm will suffer from cold shock and easily die. If the temperature of the AV is too hot, there is a similar effect on the sperm and you run the additional risk of putting the stallion off using an AV, and also possibly natural service, for life. The stallion seems less sensitive to temperature than other farm livestock but temperatures above 50°C must be avoided.

Stallions can be trained relatively easily to use an AV. Initial training uses an oestrous mare, prepared as for normal covering, to encourage mating and ejaculation into the AV. Use of an oestrous mare does, however, involve risks, not least of all her accidental mating. Many stallions are, therefore, trained to mount a dummy mare (Plate 20.3).

Plate 20.3 *Many stallions are trained to use a dummy mare. This eliminates the chance of accidental mating of a jump mare. (Thoroughbred Breeders Equine Fertility Unit, Newmarket)*

Many stallions are quite happy to use such a dummy, especially if an oestrous mare is in the vicinity. Some stallions are not keen to mount a dummy usually as a result of low libido, a possible consequence of coming into stallion work late in life after time as a performance horse, or due to incorrect AI management in the past. These stallions may well require the extra stimulus of a jump mare. This mare may be a naturally occurring nymphomaniac, that is in continual state of oestrus due to hormonal imbalance, or she may have been treated with oestradiol 17β to induce heat without ovulation. Not all nymphomaniac mares, however, are suitable, as this condition can cause them to show unpredictable behaviour.

A stallion is prepared for semen collection with an AV in the same manner as he would be for natural mating (see Chapter 13). His covering bridle is used and semen is collected in the normal covering yard. If a jump mare is to be used, she is also prepared as for natural service and such a mare must be of a quiet and calm disposition. This is especially important during the training of a young stallion (Plate 20.4).

It is essential that all equipment is present and in satisfactory order and at the right temperature for both collection and subsequent

Plate 20.4   *It essential that a jump mare used for the collection of semen is of a quiet disposition, especially when training young stallions. (Thoroughbred Breeders Equine Fertility Unit, Newmarket)*

handling before the stallion is brought in. Everyone involved must know their job and what is expected of them. Collection of semen always carries a risk due to the unpredictable nature of stallions, especially when covering. As is regular with in hand covering, all handlers should wear hard hats, especially the semen collector. Up to three handlers will be needed if a jump mare is to be used, one to hold the stallion, one to collect the semen and one to hold the jump mare. All handlers should stand on the same side of the stallion. If the collector is right-handed, the right-hand side of the stallion is the best. This ensures that in the event of accidents the stallion can be pulled away by his handler to the right and minimise the chances of the semen collector getting kicked. The side used does not affect the sample collected but once a stallion has got used to semen being collected from one side it is best to try to stick to that side in future to avoid putting him off.

The stallion is allowed to mount and the collector diverts the penis, when erect, towards the AV. The stallion should be allowed to gain intromission and enter the AV of his own free will and not have the AV forced upon him. The AV can be stabilised by being held against the hindquaters of the mare if present or back of the dummy (Plate 20.5). The occurrence of ejaculation is noted as in natural covering by the flagging of the tail or by feeling for the contractions of the urethra and the passage of the seminal plasma along the base of the penis.

After collection the collecting vessel must be carefully removed from the AV and the semen evaluated as soon as possible. If it is not possible to carry out assessment immediately, it can be stored at 40–44°C for up to 24 hours without appreciable reduction in its viability, and a reasonably accurate evaluation of the semen can still be obtained.

Prior to evaluation the gel fraction of semen has to be removed. This gel fraction is the later secretion of the accessory glands, especially the vesicular glands or seminal vessicles, and has a very low concentration of semen. The production of this gel is nature's way of helping to minimise the loss of semen out of the mare's tract post mating. Different species produce differing amounts, the extreme being the mouse whose semen produces a clot within the vagina. The gel fraction of semen can be removed by several methods: careful aspiration with a sterile syringe; filtration through a gauze or by an in line filtration system incorporated into the AV; or careful decanting off of the gel. Filtration is the most popular method used today.

Plate 20.5   The stallion should be allowed to gain intromission of his own free will. The AV may be stabilised by being held against the hindquarters of the mare. (Thoroughbred Breeders Equine Fertility Unit, Newmarket)

Semen Evaluation

It must be remembered that at all times, including during handling and evaluation, semen must be kept warm at 38–44°C (slightly lower than the temperature actually at collection). Sperm are very susceptible to cold shock and if any instruments, slides, microscope stages, etc are not pre warmed then results obtained will be misleading. Semen can be evaluated under several categories and not all evaluation involves all the assessments detailed below. It will vary according to what it is hoped to achieve by the assessment and the reasons for it being carried out. It is advisable for all the assessments to be carried out before a stallion enters an AI programme for the first time, and it is as well to repeat this full assessment at the beginning of every season. Appearance, motility, concentration and possibly morphology should be ideally assessed in each sample collected.

Appearance

Stallion semen is normally milky white in colour with a thickness equivalent to single cream. It should contain no evidence of blood staining or clots. If these are present the sample should be discarded and the stallion examined for possible damage or internal haemorrhage. The normal volume of semen produced by a stallion at each ejaculate varies considerably, from 30 to 250 ml, but on average most stallions produce about 100 ml, and of this the gel fraction takes up normally 20–40 ml. However, considerable variation is evident both between different stallions and between different collections throughout the year. As discussed in Chapter 2, semen quality and quantity vary within the breeding season.

Motility

Sperm motility is graded on a scale of 0–5, 0 being classified as very poor and 5 as excellent. Motility must be assessed immediately on a pre warmed slide examined under a microscope in order to get an accurate measurement. Semen can be examined undiluted when the wave movements of the sample are graded, or it can be assessed diluted when the individual sperm can be graded, although dilution itself may affect motility. Assessment, therefore, tends to be subjective and dependent on the examiner's previous experience. As a rough guide the following description of the grades 0–5 is often used.

Motility Evaluation

Grade	Description
0	immotile
1	stationary or weak rotatory movements
2	backward and forward movement or rotatory movement, but fewer than 50% of cells are progressively motile and there are no waves or currents
3	progressively rapid movement of sperm with slow currents, indicating that about 50–80% of the sperm are progressively motile
4	vigorous, progressive movement with rapid waves, indicating that about 90% of the sperm are progressively motile
5	very vigorous forward motion with strong, rapid currents, indicating that about 100% of the sperm are progressively motile

Because of the subjectivity of microscopic assessment, other methods are being developed including a dark field photographic method (Elliott, Sherman, Elliott and Sullivan, 1973) and cinematographic techniques (Plewinska-Wierzbowski and Bielanski, 1970). However, they tend to be complicated and take longer than a quick assessment by an experienced eye.

Whatever assessment method is used it is necessary to assess the movement of the sperm within the sample. The kind of movement required is progressive and/or oscillatory (movement on the spot); movement in very tight circles is classified as abnormal. If individual sperm are not assessed then the wave patterns of undiluted sperm can be used to give an indication of individual sperm movement. Some evaluators classify sperm motility as a ratio of those showing progressive movement:oscillatory movement.

As mentioned, semen may be extended in order to view individual sperm movement. The extender used must have a minimal effect on the sperm, while allowing accurate motility assessment, and it must be remembered that extenders may affect the accuracy of the results obtained. Extenders include the following: dried skim milk diluent, consisting of skim milk, penicillin and streptomycin; and cream gelatin diluent, consisting of gelatin, cream, penicillin, streptomycin and polymyxin.

Semen containing at least 40% actively progressively motile spermatozoa, that is grade 2/3, can be considered as adequate for AI.

Morphology

Morphological examination is a common method of trying to assess a stallion's fertility. Sperm are examined microscopically in a diluted semen sample. Individual sperm are examined to assess the number that show abnormalities, and this figure is then expressed as a percentage of the total number of sperm. Viewing of individual sperm can be enhanced by using a nigrosin-eosin stain, which stains the heads of dead sperm violet/purple. Figure 20.2 illustrates the kinds of abnormalities that may be encountered.

Morphology assessments usually involve the quality of mid-piece and tail, the only assessment of the heads is for shape and the apparent integrity of the acrosome region (Kenney, Hurtgen, Pierson, Witherspoon and Simons, 1983). Abnormalities can be due to a number of causes including the failure of complete spermatogenesis, failure of sperm maturation and sperm damage occurring after ejaculation.

Failure of spermatogenesis is usually characterised by sperm with two heads, two tails, no mid-piece, no tail, rudimentary tails or excessively coiled tails. Such abnormalities may be indicative of a long-term or even permanent problem. Maturation failure is characterised by the presence of cytoplasmic droplets on the mid-piece of the sperm. As maturation proceeds these cytoplasmic droplets progressively move down the tail and disappear. If they are

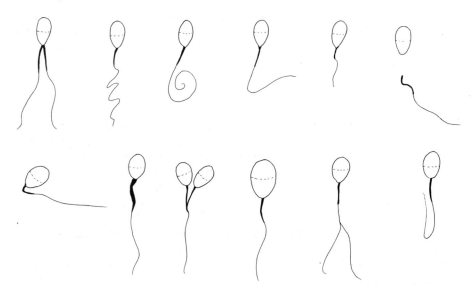

Figure 20.2   Some examples of the abnormalities that may be seen when examining sperm for morphology. From top left: double neck and tail, corkscrew tail, spiral tail, bent tail, short tail, detached head and tail. Bottom row: bent neck, thick neck, double head, large head, double tail, shoehorn tail.

present in ejaculated sperm it is an indication that the sperm are immature. This may well be only a temporary problem and possibly indicative of overuse of the stallion. A period of rest may rectify the problem. Damage occurring after ejaculation is normally characterised by a high percentage of burst sperm heads, indicating that the diluent used was not appropriate or was incorrectly made up.

Semen containing 65% or more morphologically normal sperm is considered appropriate for AI.

Live : Dead Ratio

To assess the live : dead ratio, individual sperm are examined microscopically after dilution. Staining with nigrosin-eosin in equal volume to that of semen allows the heads of dead sperm to show up as violet/purple. Dott and Foster (1972, 1975) have developed more refined methods of staining using the nigrosin-eosin stain, but allowing samples to be prepared and stored for examination over a period of three weeks.

The ratio, or percentage, of dead to live sperm is assessed in a number of samples from the collection. A live : dead ratio of 6.5 : 3.5 live sperm is acceptable for AI. However, a sample with a lower ratio may be considered usable, especially if samples from a particular stallion are rare and are particularly valuable. If such a sample has a high total sperm count it can be diluted to a lower extent still ensuring that the minimum number of live sperm for fertilisation are inseminated.

Concentration

Sperm concentration within semen samples can be assessed using either a haemocytometer or spectrophotometer. A haemocytometer consists of a counting slide with a cover slip under which a set volume of diluted semen can be trapped and viewed. The number of sperm within the sample on the haemocytometer is counted by means of the counting grid. From this figure, and the dilution rate, the concentration of sperm can be calculated. This method gives a very accurate assessment, but is rather time consuming and cannot be easily done in the field. The spectrophotometer, on the other hand, can be used more easily in the field but the results obtained can be variable. The semen is again diluted in the order of 1:30, the diluent used is 10% formalin plus 0.9% saline, and the amount of light passing through the sample is read by the spectrophotometer. The greater the concentration of sperm, the denser the sample and, therefore, the less light that will be able to pass through. Using a table of standards this density reading can be translated into a concentration reading (Pickett and Back,1973; Bielanski, 1975).

The normal range for sperm concentration is 30–600 × 10^6 sperm/ml of undiluted semen but concentration does vary with the season of collection (Pickett and Voss, 1972). A total number of 100 million sperm per inseminate is required for acceptable fertilisation rates; sperm numbers above 500 million have no beneficial effect on conception rates (Kenney, Kingston, Rajamannam and Ramburg, 1971).

pH

A standard pH meter can be used to assess the acidity/alkalinity of the semen sample. Too acid conditions are detrimental to sperm survival as seen in the mare's vagina. A pH of between 6.9 and 7.8 is acceptable, with levels of 7.3–7.7 being best (Pickett and Back, 1973).

Cytology

White Blood Cell Counts

A sample of the semen being evaluated can be mixed with an equal volume of 50% ethanol and 10% formal saline solution and stained using haematoxylin and eosin or Wright's stain and viewed under a haemocytometer (Sano, 1949; Roberts, 1971; Swerczek, 1975). A count of leucocytes greater than 1500 per ml is indicative of a problem, possibly an infection, especially if the pH of the semen is also high.

Red Blood Cell Counts

These can be counted using a similar procedure to that for leucocytes. Concentrations above 500 per ml are indicative of a problem.

Longevity

A sample of semen is assessed for longevity by the monitoring of motility over a period of time. Low motility rates after a period of time indicate a short sperm lifespan and hence reduced fertility rates. If motility is no less than 45% after 3 hours or 10% after 8 hours at room temperature, the semen may be classified as good enough for AI. It is to be noted that the longevity of the sperm in a test tube is not necessarily the same as its longevity within the mare's reproductive tract. Attempts have been made to assess longevity in the mare's tract by diluting the semen sample with 7%

glucose solution at 40°C at a ratio of 1:3 (Kenney, Bergman, Cooper and Morse, 1975).

Bacteriology

Swabs can be taken and cultured to identify the presence of infections within the semen.

A summary of the normal semen parameters is given in Table 20.1. The most regularly used to assess the potential fertilising capacity of a semen sample is motility.

Table 20.1 The acceptable range for a normal stallion's semen parameters.

Volume	30–300 ml
Concentration	30–600 million/ml
Morphology	minimum 65% physiologically normal
Live : dead ratio	6.5 : 3.5
Motility	minimum 40% progressively motile
Longevity at room temperature	45% alive after 3 hrs
	10% alive after 8 hrs
pH	6.9–7.8
White blood cells	< 1500/ml
Red blood cells	< 500/ml

When assessing the reproductive potential of a stallion, it is best to collect more than one sample for assessment. Two possible regimes used for collection for evaluation are one month of sexual rest followed by two semen collections one hour apart or by seven daily collections.

However, these take time and are really not practicable prior to stallion export. Also any delays in testing are not popular, as they delay the start of the breeding season. Most evaluation, therefore, is carried out on a single ejaculation sample.

Semen Dilution for Insemination

Once the semen has been evaluated it can be considered for insemination. Semen can be used in four ways. Firstly, used immediately undiluted to inseminate a single mare or possibly two mares, depending on the volume collected; secondly, diluted and used immediately for inseminating several mares; thirdly, diluted and refrigerated for use over the next 72–96 hours; or fourthly, diluted and frozen for use at a later date. The method used depends on the stud system and the location of the mare(s) to be inseminated. The best results are obtained using undiluted semen immediately. Such use of semen defeats one of the main objectives of AI – increasing the number of mares that can be fertilised with one ejaculate – though it may be of use as a veterinary aid.

If semen is to be extended prior to use, it should be diluted immediately. Several diluents have been developed, the three most popular being: fresh skim milk plus polymixin B; cream, gelatin and water; and skim milk plus gelatin (Kenney, Bergman, Cooper and Morse, 1975; Francel, Amann, Squires and Pickett, 1987). Two other extenders used with some success are non-fat dry skim milk (NFDSM) plus glucose, penicillin, streptomycin and deionised water; and NFDSM plus glucose, gentamicin sulphate, 8.4% $NaHCO_3$ and deionised water (Ginther, 1992). All these extenders give acceptable fertilisation rates for chilled semen. Extended semen which is not to be frozen can be stored for up to 3–4 days if extended in a 2:1 ratio of semen:extender and cooled slowly to 4°C over 4 hours and kept at this temperature until use (Allen, Bowen, Frank, Jeffcote and Rossdale, 1976; Francel, Amann, Squires and Pickett, 1987). Such treatment allows limited semen transportation as long as the temperature is kept at 4°C. Semen can only be stored for a 3–4 days at this temperature.

The extent of dilution of semen depends on the initial concentration of the sample and the motility of the sperm (Magistrini, Chanteloube and Palmer, 1987). Insemination of diluted semen containing 100 million progressively motile sperm per insemination gives good results but normally 500 million sperm per insemination is recommended in order to allow a margin for error (Pickett and Back, 1973; Cooper and Wert, 1975). The insemination volume is worked out using the following formula (PMS = progressively motile sperm):

$$\text{Insemination volume ml} = \frac{\text{Number PMS required}}{\text{Number PMS/ml}}$$

$$\text{PMS/ml} = \% \text{ PMS} \times \text{number sperm/ml}$$

Progressively motile sperm required = 100–500 million

Freezing Semen

Prolonged storage can only be achieved by freezing, but the techniques of freezing horse semen are nowhere near as refined as for cattle. Freezing horse semen has been extensively carried out in China with reported average pregnancy rates of 43% (Amann and

Pickett, 1987). Numerous extenders have been tried, based on those used for chilled insemination, but as yet results have been poor. The major problem is identifying an antifreeze or protective agent that can be used to prevent the formation of icicles within the sperm heads causing them to burst. Glycerol is used as an antifreeze in cattle semen but does not seem to have the same beneficial effect with horse semen (Tischer, 1979; Klug, Treu, Hillman and Heinze, 1975; Nishikawa, 1975; Pace and Sullivan, 1975). Great variation is also evident between stallions, with pregnancy rates of 10–47% being reported. Sophisticated systems of assessing sperm motility and, therefore, an indication of viability in thawed semen are now being developed using a computerised system based on that used in cattle, or filter systems, in order to identify frozen sperm that have maintained a high fertilisation potential (Samper, Hellander and Crabo, 1991).

Insemination Techniques

Mares are inseminated non-surgically on either Days 2 and 3 of oestrus or at 48 and 96 hours after the end of a synchronisation programme (Hyland and Bristol, 1979; Voss, Wallace, Squires Pickett and Shideler, 1979). Semen, whether diluted or undiluted, is deposited into the uterus by means of a plastic sterile pipette guided in through the cervix to the uterus by using the index finger. Alternatively, the pipette can be guided in through the cervix as per rectal palpation, the cervix can be felt through the rectal wall as a muscular tube, the pipette can then be guided through into the uterus (Figure 20.3).

Once through the cervix the insemination pipette is pushed into the uterus about 2 cm. The exact distance can be estimated by an elastic band around the tip of the pipette 4 cm from the end. Once this band is felt to have reached the cervix the insemination pipette is in place ready for the slow expulsion of semen by depressing the plunger. The tip of the pipette is flexible, allowing direction into either uterine horn. Evidence suggests that there may be an advantage as far as fertilisation rates are concerned if the semen is deposited into the horn on the same side as the ovary bearing the corpora lutea.

Uses of Artificial Insemination

AI can be used not only to increase the number of mares that can be inseminated per ejaculate but also to overcome problems

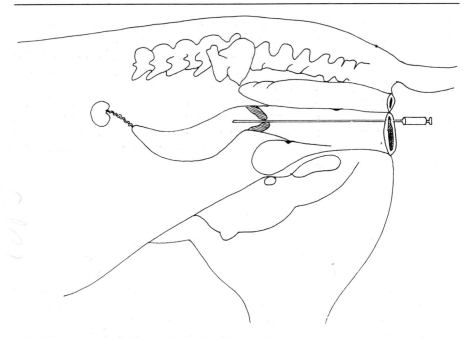

Figure 20.3   Artificial insemination in the mare.

associated with reproduction. AI provides a system of fertilisation for the mare without direct contact between the mare and stallion, thus eliminating the risk of disease transfer from the mare to the stallion and vice versa. It can be used to prevent the transfer of venereal diseases, for example, *Haemophilus equigenitalia*, equine coital exanthema, *Klebsiella aeruginosa* and *Pseudomonas aeruginosa*, and of other non-venereal diseases such as equine herpesvirus I, abortion, equine influenza, strangles and equine infectious anaemia. Semen can also be treated with extenders containing antibodies and or penicillin or by the minimal contamination technique, which involves a series of stages of centrifugation and washing with antibiotic extenders (Burns, Simpson and Snell, 1975; Kenney et al., 1975; Pickett, Sullivan, Byers, Pace and Rommenga, 1975) to minimise the natural bacterial content and reduce the number of potentially pathogenic organisms. Such semen is, therefore, useful for mares that have an increased susceptibility to uterine infection.

AI may also be used to inseminate mares with physical abnormalities, especially those caused by accident. Fertilisation of mares with genetic abnormalities is not desirable because of the risk of perpetuating the problem. Mares with behavioural problems can also be mated more safely without risk to the stallion. Mares that have tightly stitched Caslicks can be inseminated without the need for reopening and closing at each service. Stallions with behavioural problems can be trained to mount a dummy, eliminating the risk to the mare. In all these cases it is very debatable as to whether animals with abnormalities should be bred, especially if there is a chance that it may be an inheritable trait. Finally once a suitable extender for freezing horse semen can be identified, AI can be used to substantially prolong the reproductive lifetime of a stallion long after his own death.

Conclusion

The major problem facing the further development of AI in horses is the failure of the Thoroughbred industry to recognise and hence register progeny conceived this way. The Thoroughbred industry is concerned that AI will open up increased possibilities of fraud, along with the risk of exploitation of fashionable strains at the expense of others, reducing the value of bloodstock due to an excess supply. Until they can be persuaded that these problems will not necessarily materialise as a result of the use of AI, application of AI will be restricted to use in other breed societies. In America the industry is further advanced, to such an extent that 'mail order' semen is becoming increasingly popular.

References and Further Reading

Allen, W.R., Bowen, J.H., Frank, C.J., Jeffcote, L.B. and Rossdale, P.D. (1976) The current position of AI in horse breeding. Equine Vet.J., 8, 72.

Amann, R.R. and Pickett, B.W. (1987) Principles of cryopreservation and a review of cryopreservation of stallion spermatozoa. Equine Vet.Sci., 7, 145.

Bielanski, W. (1975) The evaluation of stallion's semen in aspects of fertility control and its use for artificial insemination. J.Reprod.Fert.Suppl., 23, 19.

Brinsko, S.P. and Varner, D.D. (1993) Artificial insemination. In: Equine Reproduction. Ed. A.O. McKinnon and J.L. Voss. Lea and Feibiger, Philadelphia & London, p. 798.

Burns, S.J., Simpson, R.B. and Snell, J.R. (1975) Control of microflora in stallion semen with a semen extender. J.Reprod.Fert.Suppl., 23, 139.

Cooper, W.L. and Wert, N. (1975) Wintertime breeding of mares using artificial light and insemination: six years experience. Proc.21stAnn. Conv.Am.Ass.Equine Practnrs., 245.

Dott, H.M. and Foster, G.C. (1972). A technique for studying the morphology of mammalian spermatozoa which are eosinophillic in a differential 'live/dead' stain. J.Reprod.Fert., 29, 443.

Dott, H.M. and Foster, G.C. (1975). Preservation of differential staining of spermatozoa by formal citrate. J.Reprod.Fert., 45, 57.

Elliott, F.E., Sherman, J.K., Elliott, E.J. and Sullivan, J.L. (1973) A photographic method of measuring percentage of progressing motile sperm cells using dark field microscopy. VIIIth.Int.Symp.Zootech., Milan, 160.

Francel, A.T., Amann, R.P., Squires, E.L. and Pickett, B.W. (1987). Motility and fertility of equine spermatozoa in milk extender after 12 or 24 hours at 20°C. Theriogenology, 27, 517.

Ginther, O.J. (1992) Reproductive Biology of the Mare, Basic and Applied Aspects. 2nd Ed. Equiservices 4343, Garfoot Rd., Cross Plains, Wisconsin 53528, USA.

Hyland, J.H. and Bristol, F. (1979) Synchronisation of oestrus and timed insemination of mares. J.Reprod.Fert.Suppl., 27, 251.

Kenney, R.M., Bergmann, R.V., Cooper, W.L. and Morse, G.W. (1975) Minimal contamination techniques for breeding mares: techniques and preliminary findings. Proc.21st.Ann.Conv.Am.Ass.Equine Practnrs., 327.

Kenney, R., Hurtgen, J., Pierson, R., Witherspoon, D. and Simons, J, (1983) Manual for Clinical Fertility, Evaluation of the Stallion, pp 7–54. Society of Theriogenology, Hasings, Nebraska.

Kenney, R.M., Kingston, R.S., Rajamannam, A.H. and Ramburg, C.F. (1971) Stallion semen characteristics for predicting fertility. Proc.17th Ann.Conv.Am.Ass.Equine Practnrs., 53.

Klug, E., Deegen, E., Lieske, R., Freytag, K., Martin, J.C., Gunzel, A.R. and Bader,H. (1979) The effect of vesiculectomy on seminal characteristics in the stallion.J.Reprod.Fert.Suppl., 27, 61.

Klug, E., Treu, H., Hillman, H. and Heinze, H. (1975) Results of inseminations of mares with fresh and frozen semen. J.Reprod.Fert.Suppl., 23, 107.

Magistrini, M., Chanteloube, P.H. and Palmer, E. (1987). Influence of season and frequency of ejaculation on the production of stallion semen for freezing. J.Reprod.Fert.Suppl., 35, 127.

Nishikawa, Y. (1975) Studies on the preservation of raw and frozen semen. J.Reprod.Fert.Suppl., 23, 99.

Pace, M.M. and Sullivan, J.J. (1975) Effect of timing of insemination, numbers of spermatozoa and extender components on the pregnancy rate in mares inseminated with frozen stallion semen. J.Reprod.Fert.Suppl., 23, 115.

Pickett, B.W. (1993). Collection and evaluation of stallion semen for artificial insemination. In: Equine Reproduction. Ed. A.O. McKinnon and J.L. Voss. Lea and Feibiger, Philadelphia & London, p. 705.

Pickett, B.W. (1993). Seminal extenders and cooled semen. In: Equine Reproduction. Ed. A.O. McKinnon and J.L. Voss. Lea and Feibiger, Philadelphia & London, p. 746.

Pickett, B.W. and Amann, R.P. (1993). Cryopreservation of semen. In: Equine Reproduction. Ed. A.O. McKinnon and J.L. Voss. Lea and Feibiger, Philadelphia & London, p. 769

Pickett, B.W. and Back, D.G. (1973) Procedures for preparation, collection, evaluation and inseminating of stallion semen. Bull.Colo.St.Univ. Agric.Exp.Stn.Anim.Reprod.Lab.Gen.Ser. 935.

Pickett, B.W., Sullivan, J.J., Byers, M.S., Pace, M.M. and Rommenga, E.E. (1975). Effect of centrifugation and seminal plasma on motility and fertility of stallions and bull spermatozoa. Fert.Steril., 26, 167.

Pickett, B.W. and Voss, J.L. (1972) Reproductive management of stallions. Proc.18th Ann.Conv.Am.Ass.Equine Practnrs., 501.

Plewinska-Wierzobowska, D. and Bielanski, W. (1970) The methods of evaluation of the speed and sort of movement of spermatozoa. Medycyna Wet., 26, 237.

Roberts, S.J. (1971). Artificial Insemination in horses. Veterinary Obstetrics and Genital diseases (Theriogenology) 2nd. Ed., pp. 740–743. Ithaca. N.Y. Roberts.

Samper, J.C., Hellander, J.C. and Crabo, B.G. (1991). The relationship between the fertility of fresh and frozen stallion semen and semen quality. J.Reprod.Fert.Suppl., 44, 107.

Sano, M.E. (1949) Trichrome stain for tissue section culture or smear. Am.J.Clin.Path., 19, 898.

Swerczek, T.W. (1975) Immature germ cells in the semen of Thoroughbreds. J.Reprod.Fert.Suppl., 23, 135.

Tischer, M. (1979) Evaluation of deep frozen semen in stallions. J.Reprod. Fert.Suppl., 27,53.

Voss, J. L., Wallace, R.A., Squires, E.L., Pickett, B.W. and Shideler, R.K. (1979). Effects of Synchronisation and frequency of insemination on fertility. J.Reprod.Fert.Suppl., 27, 257.

CHAPTER 21

Embryo Transfer in the Mare

THE first reported successful equine embryo transfer was carried out surgically in Japan in 1974 (Oguri and Tsutsumi, 1974), though the success was only limited. Research over the last 15 years, including the development of non-surgical techniques, has considerably improved success rates but the commercial application of embryo transfer in horses has a long way to go before it catches up with the state of play in the cattle and sheep industries. One of the largest constraints on the development of embryo transfer in horses is the continued reluctance of breed societies to register foals conceived in this manner.

The main principle behind embryo transfer is the transfer of elite embryos from a genetically superior donor mare mated to a genetically superior stallion into a genetically inferior recipient mare who is perfectly capable of carrying a foal to term. This technique makes use of the fact that the genetic make-up of the mare carrying the foal has no effect whatever on the characteristics of that foal. All its genetic make-up and, therefore, its characteristics are determined by the mare that produced the ovum and the stallion whose sperm was used to fertilise it. The technique also makes use of the fact that for the first 20 days of life the embryo is free living within the uterus and has not formed an attachment. Therefore, moving it to another mare's uterus with only limited danger of damage is possible.

In order for embryo transfer to be successful the stage of the uterus into which the embryo is transferred must be synchronised with the uterus from which it has been collected, to ensure that the uterine secretions and development match the requirements of the embryo transferred. To achieve this the oestrous cycles of the donor and the recipient mares must be synchronised, and this is normally achieved artificially by using exogenous hormone therapy. Initial evidence suggested that the donor and recipient mares should ovulate within 24 hours of each other, but more recent research

422

suggests that better success rates are achieved if the recipient mare ovulates 24 hours after the donor mare. This presumably allows compensation for any developmental retardation that may occur in embryos which have undergone the stress of transfer.

Hormonal Treatment of Donor and Recipient

Methods of synchronisation are discussed in detail in Chapter 12. Both the donor and recipient are similarly synchronised, normally using prostaglandin F2α (PGF2α), which is commercially called Equimate (Palmer and Jousett, 1975). PGF2α may also be used in conjunction with hCG (Allen, Stewart, Trounson, Tischner and Bielanski, 1976). In an ideal transfer system a large number of embryos are collected at one collection from a donor. She is induced to produce many more embryos than she would as a result of her natural oestrous cycle, that is she is superovulated. In cattle and sheep superovulation is quite successful, though the results do tend to vary, especially in ewes. In the mare, however, success rates are much poorer and the failure to find a reliable superovulating agent is a major stalling point. Some success has been achieved in mares using equine pituitary extract, known commercially as Pitropin (Douglas, Nuti and Ginther, 1974; Lapin and Ginther, 1977; Douglas, 1979). Pregnant mare serum gonadotrophin (PMSG) as used in cattle and sheep has no effect on mares. Pitropin can be used with limited success, if injected as a course over seven days, but the ovulation rates achieved are not consistent either between different oestrous cycles in the same mare or between mares. Experiments are at present being conducted using gonadotrophin releasing hormone (GnRH) and look reasonably promising, though again this will involve a series of injections. Further along these lines some success has been achieved using human menopausal gonadotrophins (HMG) (Koene, Bader and Hoppen, 1990).

A general routine that can be used to synchronise and attempt to superovulate donor mares is as follows:

DAY 0	PGF2α (Equimate)
DAY 6	hCG
DAY 10	oestrus and ovulation in some mares
DAY 14	Equimate and Pitropin
DAY 15	Pitropin
DAY 16	Pitropin
DAY 17	Pitropin
DAY 18	Pitropin
DAY 19	Pitropin

DAY 20   Pitropin
DAY 21   hCG
DAY 25   oestrus
DAY 26   ovulation – MATING
DAY 31 ⎫
DAY 32 ⎭ embryo COLLECTION

Human chorionic gonadotrophin (hCG) is, as its name suggests, collected from women and is involved in the immunological aspects of pregnancy in much the same way as equine chorionic gonadotrophin (eCG) is in mares. HCG has properties like luteinising hormone (LH) and can be used as an added inducer to ovulation. Alternatives such as a single injection of Pitropin or multiple injections of GnRH have also been used with some success (Allen et al., 1976).

Recipients can be treated in a very similar manner to donors, except no attempt is made to superovulate them, and they are, of course, not mated. They should, however, be teased or rectally palpated to ensure that they have reacted to the synchronisation treatment and ovulated. Experiments have been done using ovariectomised mares as recipients. These mares alleviate the need for synchronisation, teasing, rectal palpation, etc but because of the lack of ovarian progesterone they have to receive artificial progesterone supplementation for the first 120 days of pregnancy. It is also not clear yet whether such treatment of the mares may have a long-term detrimental effect on the foals, especially on their development.

Embryo Recovery

The donor and recipient mares are started on their programme of synchronisation at the same time and are both rectally palpated to ensure that ovulation has occurred. Once ovulation has occurred the donor mare is either covered naturally or by artificial insemination. Embryos are normally recovered at 6–7 days after covering, by which time they are at the blastocyst stage (Figure 21.1). The age of embryos at collection is older than in other farm livestock, as equine conceptuses do not expand into elongated trophoblasts but remain as discrete structures. As a general rule the older the embryo, the more robust it is, but it is also more susceptible to any asynchrony in uterine secretions. Therefore, a happy medium between these two factors has to be reached, and in the mare the best results are achieved by collection and transfer at 6–7 days after fertilisation.

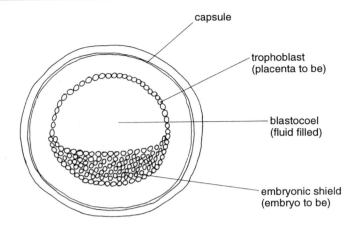

Figure 21.1   A blastocyst. This is the stage, 6–7 days post coitum, at which equine embryos are usually recovered.

Surgical Recovery

The initial method of recovery during the early work on equine embryo transfer was surgical, and this technique still has rates similar to the more recently developed non-surgical ones. It is carried out under general anaesthetic with the mare in a dorsal recumbency with her ventral side (abdomen) uppermost. The uterus is exteriorised through a ventral midline incision in the area between the mammary gland and the position of umbilical cord attachment. The uterine horn is canulated with a glass tube and ligated at the uterine body end and fluid is flushed from the oviduct down towards the uterine horn. As the fluid passes along the uterine horn it takes with it any embryos that are present and runs out via the glass canula to be collected in a warm collecting vessel (Allen and Rowson, 1975; Allen, Bielanskii, Cholewinski, Tischner and Zwolinski, 1977; Castleberry, Schneider and Griffin, 1980; Imel, Squires, Elsden and Shideler, 1981).

Phosphate buffered saline is the normal flushing medium with the addition of foetal calf serum as a source of protein; other sources of protein are being experimented with in order to reduce the cost. Penicillin may also be added to guard against infection. A total of up to 500 ml can be flushed through the uterus and collected. These flushings are then examined microscopically for embryos, a filter system being used to ease the search. Recovery rates in the order of 70–77% have been reported for surgical collection (Allen and Rowson, 1975).

Non-surgical Recovery

Non-surgical embryo recovery was first used in horses with any consistent success in Texas in 1979 (Douglas, 1979; Vogelsang, Sorensen, Potter, Burns and Draemer, 1979), though it had been attempted by several other researchers previously (Oguri and Tsutsumi, 1974; Allen and Rowson, 1975). It has become increasingly popular as it has the advantage of not requiring a general anaesthetic and hence carries a lower risk to the mare. Recovery rates are also reported to be as good as, if not better than, those for surgical recovery, though it does depend upon the stage of development of the embryos at collection. Recovery rates of up to 80% have been reported for the collection of 6- to 7-day-old embryos (Oguri and Tsutsumi, 1974), whereas rates for embryos younger than 5 days old are poor as this method will only collect embryos that are within the uterine horn. This technique, unlike surgical recovery, does not flush the mare's oviducts.

The techniques for non-surgical recovery are very similar to those used in cattle. The mare is restrained in stocks, having been prepared and washed as for natural covering. A three-way catheter (French or Rusch Foley catheter) is inserted through the cervix of the mare guided per rectum or by inserting a hand into the vagina and guiding the catheter through the cervix using the index finger (Figure 21.2).

The catheter is passed as high up into the uterine horn as possible without undue pressure. Once in position, the cuff of the catheter is inflated with 15–50 ml of air via the inlet tube occluding the base of the uterine horn and, therefore, preventing escape of the flushing medium through the uterus. Fluid is then flushed in through the entry catheter up into the top of the uterine horn, and returns, along with any embryos present, via an opening into the outlet tube for collection in a warm collecting vessel (Figure 21.3). Both horns may be flushed out simultaneously if the inflated cuff is situated within the body of the uterus (Imel, Squires and Elsden, 1980), or independently if the cuff is placed in each horn in turn (Douglas, 1979). Recovery of ova can be improved by palpation of the uterus, per rectum, at the same time as flushing.

The same flushing medium as for surgical collection can be used, and each donor is normally flushed twice with in the order of 0.5 litres of fluid per flush for ponies and up to 1 litre/flush for horses. All flushings are then passed through a filter system to ease the identification of the embryos in the very large volume of flushing medium. Ova are then evaluated to assess their viability and use for transfer. Ova at the appropriate stage for their age may be used. All unfertilised ova or those showing retarded growth

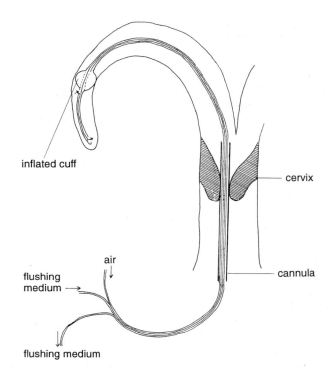

Figure 21.2   A diagrammatic representation of the Foley catheter illustrating the inlet and outlet tubes plus the air inlet for inflating the cuff.

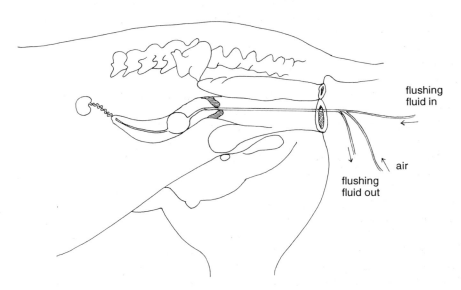

Figure 21.3   The Foley catheter in place ready for flushing and the collection of embryos.

should be discarded (Oguri and Tsutsumi, 1972; Allen, 1977).

One of the major advantages of non-surgical recovery is that, because of its non-invasive nature, it may be used repeatedly. If PGF2α is administered at the time of collection it is possible to collect another crop of ova from the same mare within 8–9 days (Griffin, Castleberry and Schneider, 1981).

Embryo Storage

Recovered embryos can be transferred immediately, the most common procedure with horses, or chilled or frozen for use at a later date. Equine embryos can be stored for up to 24 hrs at 42°C, that is slightly above body temperature prior to transfer, and this allows some limited transportation of embryos. Embryos have been successfully transported in ligated rabbit oviducts, providing they are transferred into recipients within 48 hours (Allen et al., 1977). Successful short-term storage is also possible in either physiological saline containing 2% gelatin or a 1:1 mix of mare's serum: Ringer's solution or modified Dulbecco's Phosphate Buffered Saline (PBS) (Whittingham, 1971; Oguri and Tsutsumi, 1972, 1974).

The only means of long-term storage is by cryopreservation, that is by freezing. This is as yet relatively unsuccessful in horses, although quite successful and commercially viable with sheep, cow and goat embryos. The first successful birth of a foal from a frozen embryo was not achieved until 1982 (Yamamoto, Oguri, Tsutsumi and Hachinohe, 1982). Success was limited with only one live foal from 14 ova. Czlonkowska, Boyle and Allen (1985) reported similar results. Better success rates have been achieved since, in the region of 20–40% (Slade, Takeda, Squires, Elsden and Seidel, 1985; Farinasso, De Faria, Mariante and Debem, 1989; Squires, Seidal and McKinnon, 1989). Success rates are still poor and variable in comparison to other species. It has been postulated that this is due to a variation in the size of embryos at freezing. Most of the embryos recovered from mares, especially using non-surgical techniques, are 6–7 days old and at this age approximately 500 μm in diameter. Experiments have shown that cryopreservation of equine embryos larger than 250 μm in diameter gives relatively poor results. This is confirmed by Skidmore, Boyle and Allen (1990), who achieved pregnancy rates from frozen embryos in the range of 40–55% when less than 225 μm in diameter. It seems then that the success rates for the transfer of equine morula and early blastocysts is quite acceptable, but the recovery rates for such young embryos from the mare by non-surgical means is low compared to that for the older and larger 6–7 day old blastocysts (Boyle, Sanderson, Skidmore and Allen, 1989)

In order to cryopreserve embryos a cryoprotectant, typically glycerol, has to be added and the embryos frozen in a series of steps. The temperature is initially dropped down to –6 or –7°C and then gradually dropped further to –33 to –35°C and stored in liquid nitrogen. They will survive stored like this indefinitely (Skidmore, Boyle and Allen, 1990). Prior to transfer they need to be thawed out by a gradual step-wise increase in temperature, with the possible addition of sucrose to aid rehydration and help prevent alterations in osmotic pressure.

Embryo Transfer

The transfer of embryos can, like collection, be done surgically or non-surgically. The ideal recipients are 5–10 years of age with a proven breeding record and no history of infection. They need be of no particular genetic merit as they will in no way affect the genetic make-up of the embryos transferred to them.

Surgical Transfer

Surgical transfer is not popular today. As with embryo collection, it requires a general anaesthetic. The mare is prepared as for surgical collection and a similar but smaller ventral midline incision is made. Just the uterine horn is exteriorised through this incision and a small hole is made at the top of the horn with a blunt needle. A Pasteur pipette containing the embryo is inserted through this hole and the embryo expelled into the lumen of the uterine horn. A similar, but less traumatic, procedure can be carried out through the flank of the mare using a local anaesthetic. A small incision is made in the mare's flank while she is restrained within a crush. The uterine horn is then exteriorised through this incision and the embryo inserted as described above.

Both these methods have the advantage that younger embryos, 2–4 days old, may be placed into the oviducts, as access into the top of the uterine horns and the fallopian tubes is possible. The reported success rates are very variable and are in the order of 50–70%, though they do tend to be higher than those in non-surgical transfer, especially with younger, less well-developed embryos (Allen and Rowson, 1975; Imel, Squires, Elsden and Shideler, 1981).

The site of embryo deposition in relation to the functional corpora lutea seems to be less crucial in the mare than in the ewe and the cow. This is presumably due to the naturally occurring high rate of transuterine migration of fertilised ova in the mare.

Non-surgical Transfer

The non-surgical technique is very similar to that used in cattle and also to that used in artificial insemination (Meira and Henry, 1991). The transfer gun is passed in through the mare's cervix and into the uterine body. With some difficulty it can be manoeuvred into either horn, allowing embryos to be placed into the horn ipsilateral to the ovary bearing the corpus luteum. Such replacement is reported by some to give slightly better success rates in ewes and cows but not conclusively so in mares. Once in place, the embryo and the associated fluids are expelled into the uterus by depressing the plunger at the end of the gun (Figure 21.4).

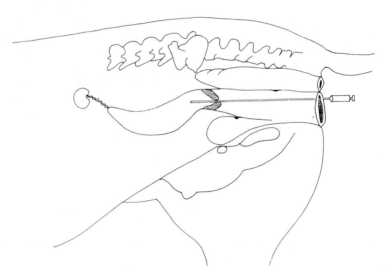

Figure 21.4   Non-surgical transfer of embryos into a recipient mare.

This non-surgical transfer is a relatively easy and quick procedure to carry out. However, it does increase the risk of introducing low-grade infections into the reproductive system, of damaging the embryo and of allowing the deposition of embryos into only the uterine body or the lower part of the uterine horns. Such techniques are of limited use with embryos less than 5 days old, which can only be successfully transferred surgically. For embryos of 5–8 days old pregnancy rates of 40–70% have been reported

(Allen and Rowson, 1975; Oguri and Tsutsumi, 1980; Castleberry, et al., 1980). The medium in which the embryos are transferred also seems to have an effect on pregnancy rates. Dulbecco's Phosphate Buffer Solution (PBS) gives better results as a transport medium than Medium 199 (Douglas, 1980).

Conclusion

Embryo transfer in horses has much development to undergo, especially in the area of cryopreservation, without which its successful commercial application is limited. At present, many breed societies refuse to accept the progeny of embryo transfer, and if they do they will only register one foal per mare per year. The expansion of embryo transfer within the equine industry is dependent upon several factors including a change in breed registration restrictions; the value of horses; the performance of embryo transfer foals; the cost of the procedures; refinement of techniques; and the attitude of the industry to its application. However, even within these restrictions embryo transfer in horses is a valuable experimental tool for research into the reproduction of the mare. Embryo transfer may find further application and commercial worth if a reliable superovulating agent can be found. The alternative to super ovulation, embryo bisection, resulting in two identical twins, is as yet relatively unsuccessful with success rates at present being only in the order of 10–20%.

References and Further Reading

Allen, W.R. (1977) Techniques and results in horses. In: Embryo Transfer in Farm Animals, Betteridge, K.J., Ed. Canada Agricultural Monograph, 16, pp 47–49.

Allen, W.R., Bielanski, W., Cholewinski, G., Tischner, M. and Zwolinski, J. (1977) Blood groups in horses born after double transplantation of embryos. Bull.Acad.Pol.Sci.Ser.Sci.Biol. 25, (11), 757.

Allen, W.R. and Rowson, L.E.A. (1975) Surgical and non-surgical egg transfer in horses. J.Reprod.Fert.Suppl., 23, 525.

Allen, W.R., Stewart, F., Trounson, A.O., Tischner, M. and Bielanski, W. (1976) Viability of horse embryos after storage and long distance transport in the rabbit. J.Reprod.Fert. 47, 387.

Boyle, M.S., Sanderson, M.W., Skidmore, J.A. and Allen, W.R. (1989) Use of serial progesterone measurements to assess cycle length, time of ovulation and timing of uterine flushes in order to recover equine morulae. Equine Vet.J.Suppl., 8, 10.

Castleberry, R.S., Schneider, H.J.,Jr. and Griffin, J.L. (1980) Recovery and transfer of equine embryos. Theriogenology, 13, 90.

Czlonkowska, M., Boyle M.S. and Allen W.R. (1985). Deep freezing of horse embryos. J.Reprod.Fert., 75, 2, 485.

Douglas, R.H. (1979) Review of induction of superovulation and embryo transfer in the equine. Theriogenology, 11, (1), 33.

Douglas, R.H. (1980) Pregnancy rates following non-surgical embryo transfer in the equine. J.Anim.Sci., 51, Suppl.1, 272.

Douglas, R.H., Nuti, L. and Ginther O.J. (1974) Induction of ovulation and multiple ovulation in seasonally an-ovulatory mares with equine pituitary fractions. Theriogenology, 2, (6), 133.

Embryo Transfer II (1989) Ed. W.R. Allen, D.F. Antczak and J.F. Wade. Proc of 2nd. International Symposium on Equine Embryo Transfer. Equine Veterinary Journal.

Farinasso, A., De Faria, C., Mariante, A.D. and DeBem, A.R. (1989) Embryo technology applied to the conservation of equids. Equine Vet.J.Suppl., 8, 84.

Griffin, J.L., Castleberry, R.S. and Schneider, H.S., Jr. (1981) Influence of day of recovery on collection rate in mature cycling mares. Theriogenology,15, 106.

Imel, K.J., Squires, E.L. and Elsden, R.P. (1980) Embryo recovery and effect of repeated uterine flushing of mares. Theriogenology, 13, 97.

Imel, K.J., Squires, E.L., Elsden R.P. and Shideler, R.K. (1981) Collection and transfer of equine embryos. J.Am.Vet.Med.Ass. 179, 987.

Koene, M.H., Boder, H. and Hoppen, H.O. (1990) Feasibility of using HNG as a superovulatory drug in a commercial embryo transfer programme. J.Reprod.Fert.Suppl., 44, 710.

Lapin, D.R and Ginther, O.J. (1977) Induction of ovulation and multiple ovulations in seasonally anovulatory and ovulatory mares with equine pituitary extract. J.Anim.Sci., 44, 834.

Meira, C. and Henry, M. (1991) Evaluation of two non-surgical equine embryo transfer methods. J.Reprod.Fert. Suppl., 44, 712.

Oguri, N. and Tsutsumi, Y. (1974) Non-surgical egg transfer in mares. J.Reprod.Fert., 41, 313.

Palmer, E. and Jousset, B. (1975) Synchronisation of oestrus in mares with a prostaglandin analogue and hCG. J.Reprod.Fert. Suppl., 23, 269.

Skidmore, J.A., Boyle, M.S. and Allen, W.R. (1990) A comparison of two different methods of freezing horse embryos. J.Reprod.Fert. Suppl., 44, 714.

Slade, N.P., Takeda, T., Squires, E.L., Elsden, R.P. and Seidel, G.E. Jr. (1985) A new procedure for the cryopreservation of equine embryos. Theriogenology, 24, 45.

Squires, E.L. (1993) Embryo transfer. In: Equine Reproduction. Ed. A.O. McKinnon and J.L. Voss. Lea and Feibiger, Philadelphia & London, p. 357.

Squires, E.L., Seidel, G.E. Jr. and McKinnon, A.O. (1989) Transfer of cryopreserved equine embryos to progestin treated ovarectomised mares. Equine Vet.J.Suppl., 8, 89.

Vogelsang, S.G., Sorensen, A.M.,Jr., Potter, G.D., Burns S.J. and Draemer,

D.C. (1979) Fertility of donor mares following non-surgical collection of embryos J.Reprod.Fert. Suppl., 27, 383.

Whittingham, D.G. (1971) Survival of mouse embryo after freezing and thawing. Nature Lond. 233, 125.

Yamamoto, Y., Oguri, N., Tsutsumi, Y. and Hachinohe, Y. (1982) Experiments in freezing and storage of equine embryos. J.Reprod.Fert. Suppl., 2, 399.

Conclusion

FUTURE research must target the gaps in our present understanding of equine reproduction so that a more complete picture can be built up. The more complete our knowledge, the more appropriate can be our management of the horse for both maximum reproductive efficiency and equine welfare. One of the major problems in equine research today is the lack of funding. There are only a few research centres in the world concentrating on equine reproduction, and funding for these is largely provided by funding councils, the private sector, charities and the industry itself. Government support for equine research in many countries is not forthcoming or, if it exists, is very limited in amount. The extent, and nature, of future research will be largely dependent upon the availability of funding.

The following areas present large gaps in our understanding of equine reproduction and warrant further investigation. The basics of reproductive control in both the mare and the stallion are largely understood, but the finer complexities remain to be clarified. Our understanding of the significance of hormones such as prolactin, melatonin, inhibin and activin is as yet limited. As far as physiology is concerned, gaps remain in our knowledge of fertilisation. The reasons for its failure and the mechanisms governing ova passage through the utero-tubular junction and fallopian tubes remain to be clarified. Along with fertilisation failure, embryo mortality, especially prior to implantation, is a prime cause of the relative inefficiency of reproduction in the mare and remains incompletely understood.

Testicular degeneration and dysfunction, along with abnormal sexual behaviour, place considerable limitations on the performance of many stallions. The causes, and hence treatment and prevention, are largely unknown. In addition, the significance of seminal plasma and details of the role of the accessory glands are not understood and require further investigation if the factors affecting sperm survival are to be understood and the refinement of semen extenders for use in AI is to be achieved. Along with the lack of knowledge of appropriate semen extenders, AI results are relatively poor in horses due to limited research into semen cryopreservation and refrigeration and the most appropriate freezing and thawing procedures. More successful preservation of semen will allow the expansion of semen exportation throughout the world, and will have a significant effect on the breeding possibilities within, and the economics of, the equine industry.

Embryo transfer is a further area warranting investigation. Our present knowledge is rather limited in comparison with the sophistication of embryo transfer in other farm livestock. At present, one of the major stumbling blocks is the lack of a consistently successful superovulating agent. In addition, further advances in *in vitro* fertilisation, along with the preservation of ova for use at a later date, would open up many other opportunities in the reproductive management of mares. Such developments would allow mares restricted by competition commitments or injury to produce a foal without the need to carry it herself and would allow valuable genetic material to be preserved for use at a future date.

Infection still remains one of the major causes of reproductive failure in the horse. Considerable work has been conducted into infections of the mare, but our knowledge of the infections of the stallion's reproductive tract is nowhere near as advanced. As the stallion is 50% responsible for the success or failure of reproduction, this is an area warranting much research.

Finally, lactation is a largely unexplored area in the horse, with much of the present information available being extrapolated from other species. Further knowledge on lactation will allow us to improve the management of the mare and foal, especially immediately post partum, a vital time in optimising the foal's survival and its ability to fulfil its potential in later life.

As more research into equine reproduction is carried out the question of the welfare of research animals must not be overlooked. In view of today's heightened awareness of animal rights, alternative research techniques will also have to be developed.

As our knowledge of equine reproduction becomes more complete, our management of the horse can be geared towards providing the ideal environment for our broodmares, stallions and youngstock and so achieve maximum efficiency in reproductive performance. This should not be at the expense of animal welfare, which will always form the basis of good management.

Full Names of Publications Cited in References

Acta Vet.Scand.	Acta Veterinaria Scandinavica
Am.J.Clin.Path.	American Journal of Clinical Pathology
Am.J.Vet.Res.	American Journal of Veterinary Research
Anat.Histol.Embryol.	Anatomy, Histology and Embryology
Anim.Reprod.Sci.	Animal Reproduction Science
Appl.Anim.Behav.Sci.	Applied Animal Behavioural Science
Aust.Vet.J.	Australian Veterinary Journal
Biol.Reprod.	Biology of Reproduction
Br.Med.Bull.	British Medical Bulletin
Br.Vet.J.	British Veterinary Journal
Bull.Acad.Pol.Sci.Ser.Sci.Biol.	Bulletin of the Polish Academy of Science. Biological Sciences Series
Bull.Colo.St.Univ.Agric.Exp.St.Reprod.Lab.Gen.Ser.	Bulletin of Colorado State University Agricultural Experimental Station. Reproduction Laboratory General Series.
Bull.Nat.Inst.Agric.Sci.	Bulletin of the National Institute of Agricultural Science
Bull.Wld.Hlth.Org.	Bulletin of the World Health Organisation
Can.J.Anim.Sci.	Canadian Journal of Animal Science
Compend.Cont.Ed.Pract.Vet.	Compendium on Continuing Education for Practising Veterinarians
Cornell Vet.	Cornell Veterinarian
Diss.Abstr.Int.B.Sci.Eng.	Dissertation Abstracts International B. Science, England
Dom.Anim.Endocri.	Domestic Animal Endocrinology
Equine Vet.J.	Equine Veterinary Journal
Exper.Physiol.	Experimental Physiology
Fert.Steril.	Fertility and Sterility
Int.Symp.Zootech.	International Symposium on Zootechnology
Ir.Vet.J.	Irish Veterinary Journal

436

J.Agric.Sci.Camb.	Journal of Agricultural Science, Cambridge
J.Am.Vet.Med.Ass.	Journal of the American Veterinary Association
J.Anim.Sci.	Journal of Animal Science
J.Cleavage	Journal of Cleavage
J.Clin.Microbiol.	Journal of Clinical Microbiology
J.Comp.Pathol.	Journal of Comparative Pathology
J.Endocri.	Journal of Endocrinology
J.Equine Med.Surg.	Journal of Equine Medical Surgery
J.Equine Vet.Sci.	Journal of Equine Veterinary Science
J.Physiol.Lond.	Journal of Physiology, London
J.Reprod.Fert.	Journal of Reproduction and Fertility
J.Reprod.Fert.Suppl.	Journal of Reproduction and Fertility, Supplement
J.S.Afr.Vet.Ass.	Journal of the South African Veterinary Association
Livestock Prod.Sci.	Livestock Production Science
Medycyna Wet.	Medycyna Weterynaryjna
Nature,Lond.	Nature, London
Neurosci.Behav.Rev.	Neuroscience and Behavioural Reviews
Physiol.Rev.	Physiology Review
Proc.Ann.Conf.Vet.College Vet.Med.Bio.Med.Sci.	
	Proceedings of the Annual Conference of Veterinary Colleges, Veterinary Medicine and Biological Medicine Science
Proc.Ann.Conv.Am.Ass.Equine Practnrs.	
	Proceedings of the Annual Convention of the American Association of Equine Practioners
Proc.Ann.Mtg.Vet.	Proceedings of the Annual Meeting of Veterinary Surgeons
Proc.Equine Pharm.Symposium	Proceedings of the Equine Pharmacological and Theraputics – Symposium
Proc.Int.Cong.Anim.Reprod. and AI.	
	Proceedings of the International Congress of Animal Reproduction and Artificial Insemination
Proc.Int.Symp.Equine Reprod.	Proceedings of the International Symposium on Equine Reproduction
Proc.Royal Soc.Med.	Proceedings of the Royal Society of Medicine
Proc.Soc.Theriogenology	Proceedings of the Society of Theriogenology
Proc.Soc.Exp.Biol.Med.	Proceedings of the Society for Experimental Biology and Medicine
Res.Vet.Sci.	Research in Veterinary Science
Vet.Clin.No.Am. (Large Anim. Pract.)	
	Veterinary Clinics of North America (Large Animal Practice)
Vet.Med.	Veterinary Medicine
Vet.Rec.	Veterinary Record

Index

Page numbers in bold refer to illustrations

Farming Press Books and Videos

Below is a sample of the wide range of agricultural and veterinary books and videos published by Farming Press. For more information or for a free illustrated catalogue listing all our publications please contact:

**Farming Press Books & Videos, Wharfedale Road
Ipswich IP1 4LG, United Kingdom
Telephone (0473) 241122   Fax (0473) 240501**

Horse and Pony Ailments   Eddie Straiton

300 photographs and concise text give clear guidance to the recognition and treatment of horse ailments.

Harnessed to the Plough   Roger and Cheryl Clark with Paul Heiney
(VHS video)

Roger and Cheryl Clark demonstrate a year of contemporary horse-drawn cultivations and harvesting on their Suffolk farm. Additional commentary by Paul Heiney.

The Horse in Husbandry   Jonathan Brown

Photographs of horses working on farms from 1890 to 1950, with an account of how they were managed.

The Hired Lad   Ian Thomson

A young man's first work on a Scottish farm when horses were yielding to tractor and bothy life was rough and ready.

Farming Through the Ages   Robert Trow-Smith

From earliest times to World War II, an account of Britain's farming history built around a remarkable collection of pictures.

Farming Press Books & Videos is part of the Morgan-Grampian Farming Press Group which publishes a range of farming magazines: *Arable Farming, Dairy Farmer, Farming News, Pig Farming, What's New in Farming.* For a specimen copy of any of these please contact the address above.